ALLE ZEIT WACH 1842

Chemical Waste
Handling and Treatment

Managing Editor: K. R. Müller

Editors: J. Bromley, J. T. Farquhar, P. T. Gidley, S. James, D. Martinetz, A. Robin, N. B. Schomaker, R.T. Stephens, D. B. Walters

With 64 Figures and 56 Tables

Springer-Verlag
Berlin Heidelberg New York Tokyo

ISBN-13: 978-3-642-69627-5 e-ISBN-13: 978-3-642-69625-1
DOI: 10.1007/978-3-642-69625-1

Library of Congress Cataloging-in-Publication Data.
Main entry under title: Chemical waste.
Includes index.
1. Chemical industry—Waste disposal—Handbooks, manuals, etc. 2. Sewage—Purification—Handbooks, manuals, etc. I. Müller, K. R. (Karl Robert), 1929–. II. Bromley, J. (John).
TD899.C5C48 1985 363.7'28 85-18365

Softcover reprint of the hardcover 1st edition 1986

Typesetting and Offsetprinting: Th. Müntzer, GDR; Bookbinding: Lüderitz & Bauer, Berlin
2152/3020-543210

Contributors

Dr. John Bromley	Harwell Environmental Safety Group Bldg. 7.12, Harwell Lab. Oxfordshire, OX11 ORA/England
Dr. M. Chambon	Rhône-Poulenc S.A. Délégation à l'Environnement Tour Crédit Lyonnais 129, rue Servient, P.B. 3085 F-69398 Lyon Cedex 03/France
Dr. H. K. Hatayama	Hazardous Materials Laboratory California Department of Health 2151 Berkeley Way Berkeley, CA 94704/USA
Dr. D. Martinetz	Forschungsstelle für chemische Toxikologie der Akademie der Wissenschaften der DDR DDR-7050 Leipzig Permoserstraße 15/GDR
Dr. K. R. Müller	BASF Aktiengesellschaft DUR/E — K 210 D-6700 Ludwigshafen/FRG
Prof. A. Navarro	Institut National des Sciences Appliquees 20 Avenue Albert Einstein F-69621 Villeurbanne Cedex/France
Dr. M. Palmark	Chemcontrol A/S Dagmarhus DK-1553 Copenhagen V/Denmark
Dr. H. T. Propfe	Heinr. Propfe GmbH Postfach 107 D-6800 Mannheim 24/FRG
Dr. A. Robin	107, avenue de Combault F-94420 Le Plessis-Trevise/France
H. Schirmer	Carl-Zeiss-Str. 24 D-6500 Mainz 42/FRG
Prof. Dr. O. Tabasaran	Institut für Siedlungswasserbau Universität Stuttgart Bandtele D-7000 Stuttgart 80/FRG
Dr. E. Thomanetz	Institut für Siedlungswasserbau Universität Stuttgart Bandtele D-7000 Stuttgart 80/FRG

Preface

During the past few years the worlds has reverberated of names like Seveso, Love Canal, Lekkerkerk, Times Beach, just to name the most publicized ones. All these names are connected with hazardous or toxic waste, waste from business and industry, especially the chemical industry. The list is endless because there are, all over the world, many thousands of "points noirs": not yet discovered or identified old lagoons and landfills, polluted rivers, estuaries, and harbors needing remedial action, which undoubtedly will reveal more unpleasant secrets of the chemical industry's past. It is not an exaggerated statement that chemists of the past have paid too much attention to the *com*position of new products while neglecting the *dis*position of byproducts, i.e., chemical waste.

Admittedly, during the last decade this attitude has changed dramatically. Although we cannot yet properly speak of a new science of periontology (the theory of residues), we seem to be headed towards substantiated rules, analyses, disposal protocols, definitions and remedial practices in handling the problems of chemical waste. Especially during the last two years comprehensive treatises of the whole complex subject as well as monographs dealing with assorted aspects of waste technology have appeared.

Notwithstanding this fact, this is the first work generated by an *international* group of renowned and recognized scientists. They present, in the form of handbook, recently acquired and ascertained knowledge and demonstrated techniques. It is the aim of this handbook to be helpful in solving future problems and to present enlightening discussions, thus helping to move chemical waste problems out of the tabloid headlines and practice into the scientific literature.

Apart from the professional knowledge they present, the contributors of this book reflect the attitude of their respective countries and give indications of how to improve and correct the situation. We did not wish to write a tedious compendium or to compile lexicographically all established techniques for waste treatment, as this would not suit the scope of this book. Instead, preference is given to the presentation of either the application of old methods to new problems or of new methods to old problems. (The references are listed in the original language of publication.)
The topics of "Waste Disposal at Sea" (including incineration at sea) and "Transport of Non-Nuclear Toxic and Dangerous Wastes" will be treated in a separate volume to be published by Springer-Verlag in the near future.

This contribution is intended to stimulate further work in the area and provide scientifically accurate and environmentally sound approaches to the subject of waste handling and treatment.

Ludwigshafen, September 1985 Dr. K. R. Müller

Erratum

Table of Contents

Table of Contents

A. Introduction

Waste Legislations in 1985

The legal aspects of industrial wastes is in a worldwide state of transition. The UNEP and OECD are now attempting to work out internationally compulsory directives, primarily in view of the most urgent problems: definitions of hazardous wastes and their transfrontier transport to appropriate treatment facilities.

The Council of Ministers of the European Community has passed two directives: one in 1978 defining toxic and dangerous wastes and in 1984 another regulating the transfrontier transport of such wastes. To date both directives have not been fully transposed into respective national laws.

When scrutinizing the national legislation in many states, we become aware of some incongruities prevailing in many places; this is mostly due to the exaggerated public concer leading to "immature" legislation. See for instance the current waste legislation in the Federal Republic of Germany: The general waste law (Abfallbeseitigungsgesetz) passed in 1972 has been amended twice alone in 1985 along with several decrees and ordinances — on federal or country level — all of which overlap and contradict each other.

Switzerland and France have shown that legislation can be reached in a more acceptable way; in 1985, both countries issued a practicable legislation which cleared up formerly existing obscurities and which could serve as an example elsewhere. In Great Britain cautious legislative adaptations to the changing conditions are made at a pace inherent to anglosaxon jurisdiction. In the USA the results especially of the federal Superfund-Regulation and of the private Clean Sites-Inc.-Initiative are followed with great interest. Considerable *attention is focused on the outcomes of the Environmental Insurance Liability (EIL) which will be compulsory starting in the beginning of 1986.*

These few selected examples may indicate to which extent waste legislations are presently in a dynamic state of transition all around the world. The moment has not yet come to give a valid thorough survey. It can be hoped that regulations will soon be found to ensure that industrial wastes can be transported to places where best treatment is guaranteed, for the sake of the public and the environment.

K. R. Müller

Legislation for Waste Treatment in the United Kingdom

James T. Farquhar

Up until the year 1972, there was no single national Act of the UK Parliament which was specifically aimed at the control of those who disposed of waste. This did not however mean that prior to 1972, the disposal of waste was free from all controls. A number of local Acts were in existence under which Parliament had given powers to certain local authorities to control aspects of waste disposal in their areas. Also, under the traditional law of England and Wales it was always possible in certain cases to take legal action against anyone where nuisance or negligence could be established as a result of waste disposal operations. Nuisance was traditionally defined as the unlawful interference with a person's use or enjoyment of land or of some right in connection with it. The possible forms of interference are innumberable, but some of the most usual are, noise, smell and the pollution of air or water.

Up until 1945, it would have been difficult to take action against anyone depositing waste on land unless trespass could be shown or unless damage or nuisance to the interests of a third party could be established. In specific individual cases however, an application could be made to a Court to prohibit a waste disposal operation if it could be established that the operation was likely to result in nuisance.

A section of the Water Act of 1945 stated: "If any person is guilty of an act or negligent whereby any spring, well or adit, the water from which is used or is likely to be used for human consumption, is polluted or is likely to be polluted, he shall be guilty of an offence against this act". The importance of this provision was that it was possible to take action by showing that, although pollution had not occurred and therefore no damage had been done, pollution was likely to occur at sometime in the future. Under the Rivers (Prevention of Pollution) Acts of 1951 and 1961, the powers of the authorities to control the disposal of waste under circumstances where they felt that pollution of a watercouse might arise, were further strengthened.

In 1948, Parliament passed the first national Town and Country Planning Act under which a wide range of development activities came to require the formal consent of the local authority. Further additions and refinements to the requirement of this legislation occurred until 1971 when the Town and Country Planning Act of that year was passed by Parliament in an attempt to consolidate the previous enactments. In the 1971 Act, development was defined as the carrying out of building, engineering, mining or other operations in, on, over or under land, or the making of any material change in the use of any buildings or other land. The status of landfill waste disposal sites were further clarified in this 1971 Act for the avoidance of doubt by the provision that "Deposit of refuse or waste on land involves a material change in the use thereof, notwithstanding that the land is comprised in a site already used for that purpose, if either the superficial area of the deposit is thereby extended, or the height of the deposit is thereby extended and exceeds the level of the land adjoining the site".

The powers of planning authorities to refuse planning applications or to grant consent subjects to specific conditions are very extensive, but are always subject to appeal to a Central Government Minister. The main limitations on the use of the Town and Country Planning legislation as a means of control of landfilling operations has been firstly, that decisions fall to be made by committees with wide overall control of development and not necessarily with any specific knowledge of the finer aspects of waste disposal site management and secondly, that once consent is given and provided the subsequent operation of the site is conducted within the terms of the consent, it is usually impossible for the planning authority to enforce changes in the method of working.

In the late 1960s and early 1970s public indignation was aroused by the activities of some unscrupulous waste disposal operators who were disposing of waste materials improperly. Cases were reported in the press and radio where drums containing poisonous waste chemicals had been dumped on roadsides and even in public parks. On of the excuses offered for such actions was that adequate facilities were not available for the disposal of hazardous wastes in many areas. Public anger arose from the realisation that even if such operators were detected, they were only subject to comparatively minor penalties unless actual damage to life or property could be established.

As early as 1964, the UK Government had undoubtedly been aware of the lack of information available about toxic waste handling and the lack of legal constraints on its disposal. A committee under the Chairmanship of a senior Civil Servant, Dr. Andrew Key, had been set up "To consider present methods of disposal of solid and semi-solid wastes from the chemical and allied industries, to examine suggestions for improvement, and to advise what, if any, changes are desirable in current practice, in the facilities available for disposal and in control arrangements, in order to ensure that such wastes are disposed of safely and without risk to water supplies and rivers". The Key Committee did not issue its report until 1970 and during the six years of its work, it discovered a most unsatisfactory state of affairs. Indeed, the true position has since been shown to be a great deal worse than the Key Committee reported and data which has become available since indicates that the Committee seriously underestimated the mass of toxic wastes being disposed of by various means in the UK. The thirtyeight recommendations made by the Committee were based on its findings that wastes in general were not being disposed of in an acceptable manner, that too little scientific study was being carried out into the subject, that the legal constraints on these disposing of waste were inadequate and that waste disposal operations should be "Specifically authorised by somebody with adequate, technical and local knowledge and financial independence of what is decided". Thus the Key Committee in 1970, made the first recommendation for the formal licencing of waste disposal facilities in UK.

Following the publication of the Key report, the Government expressed the intention of bringing in such a licencing system. Unfortunately, the public outcry resulting from adverse publicity on toxic waste disposal blew up with such intensity that the Government could not wait to work out the detailed procedures for a full, and it was hoped, long term system for imposing the recommended controls.

As a result, a Bill was introduced into Parliament and was rushed through all stages during the single month of March 1972. This emerged as the Deposit of Poisonous Waste Act 1972 and although it was a short Act, its effects were far reaching. It laid down that anyone who deposited waste on land, where the waste was "poisonous, noxious or polluting and its presence on land was liable to give rise to an environmental hazard", could be charged and if found guilty could be fined and given a sentance of up to five years imprisonment. No actual damage to life or property had to be established. Under the 1972 Act an environmental hazard was defined as covering risk of death, injury or impairment of health whether to persons or to animals or a threat to the quality of surface or underground waters.

In addition, the Deposit of Poisonous Waste Act 1972, required that anyone removing a waste material from a site with a view to having it disposed of elsewhere, must notify his intentions in advance to the local authority and the water authority in the area. Likewise, anyone depositing waste on land was required to notify the same authorities after the disposal had taken place. There is no doubt that the primary objective of these notifications was to provide meaningful statistics on waste disposal operations rather than to allow the authorities to detect any illegal disposals. The Deposit of Poisonous Waste Act 1972 covered all waste materials irrespective of their properties but it allowed for regulations to be introduced to exempt certain wastes where their properties did not justify notification. Shortly after the Act came into operation, a list defining wastes which would be exempt from the notification requirements was issued. All wastes not mentioned on the exempt list automatically required to be notified.

Shortly after the Deposit of Poisonous Waste Act 1972 came into force, the Government published its proposals for a more comprehensive system of controls to take into account many of the recommendations of the Key Committee. Although these proposals were included as part of a comprehensive environment Act, the Control of Pollution Act passed by Parliament in 1974, the passing of the Act did not in itself, bring its requirements into force as each of its sections was subject to a commencement order from the relevant Minister before it took effect. Several sections in the Act were not specific in their requirements and depended in their provisions on power given to the Minister to bring in detailed regulations on specific matters.

Partly because of the necessity to reduce public expenditure and partly because of controversy over the details of specific regulations to be brought under the 1974 Act, many of its provisions including some on waste disposal lay dormant for several years and indeed a minority of sections of the Act are not yet implemented (January 1984). Because of this the provisions of the temporary Deposit of Poisonous Waste Act 1972 remained in force until 1981 which was far longer than had been originally envisaged.

Controls on waste disposal to land in the UK are now dependent on two principal instruments. Firstly, the Planning Act of 1971 the provisions of which have not been changed in so far as they effect waste disposal and secondly, the Control of Pollution Act 1974, the first 30 sections of which deal exclusively with matters connected with the collection, handling and disposal of domestic, commercial and industrial waste. These are referred to collectively as controlled waste.

The principle offence of disposing of waste improperly is contained in section 3 (1) of the Control of Pollution Act 1974 which states that:

"Except in prescribed cases, a person shall not

(a) Deposit controlled waste on any land or cause or knowingly permit controlled waste to be deposited on any land; or

(b) Use any plant or equipment, or cause or knowingly permit any plant or equipment to be used, for the purpose of disposing of controlled waste or of dealing in a prescribed manner with controlled waste, unless the land on which the wasteis deposited or, as the case may be, which forms the site of the plant or equipment is occupied by the holder of a licence issued in pursuance of section 5 of this Act (in this part of this Act referred to as a "disposal licence"), which authorises the deposit or use in question and the deposit or use is in accordance with the conditions, if any, specified in the licence.

Both the 1974 Act and regulatons made under it exempt a limited range of waste disposal operations from control. For example, it is not an offence to deposit household waste within the cortilage of a private dwelling by or with the permission of the occupier of the dwelling. The maintenance of the traditional compost heap for recycling garden refuse, is not prohibited by the Act. Apart from this limited range of exemptions, the statute would appear to be comprehensive and anyone depositing waste on land other than the Waste Disposal Authority itself must be in possession of a site licence. This licence may set out conditions which must be complied with in the operation of the site.

Although there was no statutory limit on the Authority to impose conditions, a statement issued by the Central Government Minister has advised that conditions on a licence should relate to the operating conditions to be imposed on the site. This was an attempt to ensure that the planning requirements and site licencing requirements did not involve a duplication of controls over the same issues.

On receiving an application for a waste disposal site licence, the Waste Disposal Authority is required under the law to consult with a number of other official bodies including the Regional Water Authority of the area. The areas under the control of Water Authorities are not the same as those of the waste disposal authorities (the County Councils) and are usually much larger. England is covered by nine Regional Water Authorities and Wales by one such authority. The law is clear that no waste disposal site licence can be issued by a Waste Disposal Authority against the wishes of the relevant Water Authority and in the event of a disagreement, either party may appeal for a decision to the Central Government Minister. It is stated in the 1974 Act that "It shall be the duty of the authority not to reject the application (for a licence) unless the authority is satisfied that its rejection is necessary for the purpose of preventing pollution of water or danger to public health". It is notable that the authority may not reject the application on the sole grounds that the operation might cause a nuisance. The fact that the disposal operation might cause a nuisance either directly or indirectly could however be the basis of a rejection by the Planning Authority.

Another limit to the powers of waste disposal authorities arises from the entitlement of any holder of a waste disposal site licence to cancel his licence at any time by giving notice to the authority that it is no longer required. Following the cancellation of a licence, the authority has no remaining powers in connection with the terms of the former licence and this means that it is not possible for a waste disposal site licence to include enforceable conditions for the reinstatement of the site after waste disposal ceases. Here again, no such limitation applies to the planning consents and it is through conditions applied by Planning Authorities that site owners can be required to reinstate or monitor their sites even after disposal of waste has ended and after the waste disposal licence has been cancelled.

It is clear from the wording of the 1974 Act, that the intention was to require the licencing of sites carrying plants used for the purpose of treating wastes even when no landfilling was taking place. It was not even stated that plants would only require licencing when the material treated was destined to be disposed of by landfilling elsewhere. When and under what conditions, a plant can be held to be a waste disposal or waste treatment plant and under what circumstances a waste disposal site licence would be required for its operation is so unclear in the legislation that no consistent code of enforcement is being applied in UK.

The general position in UK at the present time is that anyone depositing domestic, commercial or industrial waste on a landfill site must be satisfied that the site is the subject of a current waste disposal site licence and that any conditions laid down in the licence are being respected. No waste disposal site licence can be issued unless a planning consent for use of the site as a landfill site has been obtained. Any specific conditions in the planning consent must also be honoured.

The only further obligation which may arise in connection with the disposal follows if the composition of the waste is such as to bring it under the additional provisions laid down in Government regulations issued under special powers given under Section 17 of the 1974 Act. This section gives power to the Central Government Minister to impose special requirements in cases where "Waste is or may be so dangerous or difficult to dispose of that special provision in pursuance of this subsection is required for the disposal of waste of that kind by disposal authorities or other persons . . .". The basis of the requirements in the regulations enforced in UK under Section 17 stem from the provisions of the original Deposit of Poisonous Waste Act.

As already stated, that Act had provision for the sending of notifications to the local authorities and also directly to the Water Authorities about waste disposal movements. The revised notification procedures brought in under Section 17 of the 1974 Act, exclude any notification to the Water Authorities, and have brought about a radical change in the means of defining those wastes requiring notification. The 1972 Act had by itself required notification for all controlled waste without exception unless these were exempted by regulation. Under the 1974 Act, the wastes which *are* subject to notification are defined in terms of their properties. Wastes not possessing the properties in the definition are not subject to the notification requirements. The definition therefore changed from the use of an exclusive list to an inclusive list. This undoubtedly reduced to some extent the classes of waste requiring notification. This was a deliberate change on account of the fact that too many wastes were being notified under the old system thus giving rise to so much paperwork that the data therein could not be adequately handled. In addition, former justification for notifying a wider range of wastes under the 1972 Act on the grounds that this was the only means by which the Authorities could, as of right, find out what wastes were being disposed of in their area, no longer applied because of the introduction of site licencing under the 1974 Act. The fact that no site licence could be issued without the knowledge of the Water Authority also justified the exclusion of these Authorities from the notification procedures.

The regulations brought under Section 17 of the 1974 Act and which are currently under review, seek to do three things. Firstly, they define mainly in terms of waste properties, those wastes which will be regarded as special. Secondly, they set out a system of paperwork to be used in association with the transportation of special wastes. Thirdly, they require the keeping of registers and site records so that the location where special wastes have been deposited are known to future developers.

To be a special waste within the meaning of the Section 17 regulations, the first requirement is that either the waste must be of a class included on a list of 31 substances (see Appendix A) or it must be a medicinal product which, under the appropriate UK law, can only be sold to the public under a prescription signed by an appropriate practitioner. In the latter case, the law states "the substance is a medicinal product" thus suggesting that a well mixed waste which merely contains a medicinal product would not, for that reason alone, be considered special.

If the waste has a composition as defined in the list of 31 substances, this does not in itself establish the waste as special and further investigation is necessary. In addition to being on the list, a waste is considered special if it is inflammable and has a flashpoint, as determined by a defined procedure, of 21 °C or less. Alternatively, a material on the list is considered special if best estimates indicate, that either 5 cubic centimeters of the waste would cause death or serious damage to tissue if ingested by a 20 kilogram child or if best estimates indicate that exposure to the waste for 15 minutes or less would be likely to cause serious damage to human tissue by inhalation, skin contact or eye contact.

If the producer of a waste or anyone importing such a waste into the United Kingdom having determined its composition and properties, concludes that the waste is special, the notification requirements must be applied. A standard form of consignment note (shown in Appendix B at reduced size), must be used. The producer or importer of a consignment of waste which is considered to be special, must complete sections A and B of up to six copies of the notes as far as the information is known. Normally, one copy must be despatched to the office of the Waste Disposal Authority for the area in which the waste is to be disposed of. This will usually be the authority which has issued the site licence for the disposal facility. This pre-notification must be in the hands of the authority, not more than one month nor less than three clear days before the despatch of the waste from the producer's or importer's site.

The transporter of the waste must, prior to removal from the producer's or importer's premises, complete parts C and D of the consignment note and he will return one of these to the producer or importer who must retain it in a register for at least two years. In cases where the waste producer or importer has arranged for disposal in the area controlled by a Waste Disposal Authority other than his own, he must also at this stage furnish a further copy of the consignment note to his own area Authority. When the waste is being disposed of in the same authority area in which is has been produced or into which it has been imported, the Waste Disposal Authority in that area will already have received the pre-notification and so the post despatch notification is not required.

On arrival at the disposal site, the transporter must furnish the disposer with the remaining copies of the consignment note and the disposer must then complete part E. One copy of this fully completed note will be returned to the transporter who must retain it in a register for at least two years. The disposer must now return one completed copy to the Waste Disposal Authority in whose area the waste originated and in addition, he must retain one copy himself in his own register. In the case of the disposer, his register of consignment notes must be retained indefinately and all accumulated consignment notes must be returned to the Waste Disposal Authority in the event that the site licence is surrendered.

Where a waste material is transported by means of a static pipeline or where it is transported entirely within the curtilage of the site where it is produced, the notification provisions and the use of consignment notes is not required under the regulation. It is possible to agree certain exemptions with the authorities under which notification of individual loads of waste need not be made if they form part of a series or regular consignments.

The site record requirements of the Section 17 Regulations require that when a deposit of waste is made on land, the location where the deposit has taken place shall be recorded on a site plan. References on the plan must relate to the consignment notes in the disposer's register where these have been used in connection with the transport of the waste to the disposer's site.

The disposal of waste materials by dumping at sea from ships and aircraft is subject to the Dumping at Sea Act 1974. Under this Act, by which UK gives effect to the requirements of the Oslo and London Conventions, licences for disposal of waste by dumping at sea are issued at Central Government level, by the Ministry of Agriculture, Fisheries and Food in England and Wales and the Department of Agriculture and Fisheries for Scotland in Scotland. The dumping at sea of waste which is not special waste is not a matter for Waste Disposal Authority. If the waste concerned is special however the documentation required for the transport of special waste must be applied unless the waste is loaded on to a ship or aircraft without being transported on land out of the site where it was produced.

All operations involving the production, handling, transporting and disposal of waste materials in UK are covered by the general requirements of the Health & Safety at Work Act 1974. The general requirement under this Act is that "it shall be duty of every employer to ensure, so far as is reasonably practicable, the health, safety and welfare at work of all his employees".

A further feature of waste disposal legislation and policies in UK centres on a series of Waste Management Papers which have been issued over the period of 10 years by the Environment Ministry. At the present time, 25 such papers have been issued each dealing with a specific aspect of waste disposal and many giving detailed advice and comments on the means which are recommended for the disposal of specific classes of wastes. Although nothing in the Waste Management Papers carries the force of law, they obviously have a major influence on UK thinking on waste disposal and they are closely studied by both industry and controlling authorities.

Appendix A

Listed Substances

Acids and alkalis

Antimony and antimony compounds

Arsenic compounds
Asbestos (all chemical forms)
Barium compounds
Beryllium and beryllium compounds
Biocides and phytopharmaceutical substances
Boron compounds
Cadmium and cadmium compounds
Copper compounds
Heterocyclic organic compounds containing oxygen, nitrogen or sulphur
Hexavalent chromium compounds
Hydrocarbons and their oxygen, nitrogen and sulphur compounds
Inorganic cyanides
Inorganic halogen-containing compounds
Inorganic sulphur-containing compounds
Laboratory chemicals
Lead compounds
Mercury compounds
Nickel and nickel compounds
Organic halogen compounds, excluding inert polymeric materials
Peroxides, chlorates, perchlorates and azides
Pharmaceutical and veterinary compounds
Phosphorus and its compounds
Selenium and selenium compounds
Silver compounds
Tarry materials from refining and tar residues from distilling
Tellurium and tellurium compounds
Thallium and thallium compounds
Vanadium compounds
Zinc compounds

Appendix B

Department of the Environment/Welsh Office/
Scottish Development Department

Serial No. ……………

CONSIGNMENT NOTE FOR THE CARRIAGE
& DISPOSAL OF HAZARDOUS WASTES

Producer's Certificate

A

(1) The material described in B is to be collected from ....

....

and (2) taken to ....

....

Signed	Name ....
On behalf of	Position ....
Address and telephone	Date ....
....	Estimated date of collection ...

Description of the Waste **B**	(1) General description and physical nature of waste (2) Relevant chemical and biological components and maximum concentrations (3) Quantity of waste and size, type and number of containers (4) Process(es) from which waste originated
Carrier's Collection Certificate **C**	I certify that I collected the consignment of waste and that the information given in A(1) & (2) and B(1) & (3) is correct, subject to any amendment listed in this space: I collected this consignment on .... at .... hours Signed .... Name .... Vehicle Registration No ... On behalf of .... Address and telephone .... Date ....
Producer's Collection Certificate **D**	I certify that the information given in B & C is correct and that the carrier was advised of appropriate precautionary measures. Signed .... Name .... Telephone ..... Date ..
Disposer's Certificate **E**	I certify that Waste Disposal Licence No. ..... issued by .... Council, authorises the treatment/disposal at this facility of the waste described in B (and as amended where necessary at C). Name and address of facility .... This waste was delivered in vehicle (Reg. No.) .... at .....: ..... hours on (date) .... and the carrier gave his name as .... on behalf of .... Proper instructions were given that the waste should be taken to .... Signed ...... Name ...... Position .... Date .... on behalf of ....
For use by Producer/ Carrier/ Disposer	

Editor: J. T. Farguhar
Received September, 1984

Legislation for Waste Treatment in the United States of America

Howard K. Hatayama

1 Introduction

In the United States, the two primary laws which relate to the *handling* of chemical wastes or hazardous wastes are the Resource Conservation and Recovery Act of 1976 [1] and less specifically, the Comprehensive Environmental Response, Compensation, and Liability Act of 1980 [2]. The first is known as RCRA and has been amended twice, in 1978 and in 1980. Although RCRA also addresses solid waste disposal and recycling, its primary impact was the creation of the U.S. hazardous waste management program. That section of the law which relates specifically to hazardous waste management is known as Subtitle C which includes Sections 3001 to 3013. The second law is known as CERCLA or "Superfund" and was enacted to provide for liability, compensation, clean-up and emergency response for hazardous substances released into the environment and the clean-up of inactive hazardous waste disposal sites. The following discussion will focus on RCRA and its pertinent regulations because it relates specifically to the subject matter of this volume, *Chemical Waste Handling*.

2 Identification and Listing of Hazardous Wastes

Section 3001 of the law directs the Administrator of the U.S. Environmental Protection Agency to develop and promulgate criteria for identifying hazardous wastes and for listing hazardous wastes. Such criteria takes into account the characteristics of toxicity persistence and degradability in nature, potential for bioaccumulation, flammability, corrosiveness and other hazardous characteristics. These criteria can be revised from time to time as appropriate. They are identified in Part 261, 40 Code of Federal Regulations [3].

2.1 Criteria for Identifying Hazardous Wastes

A waste that is hazardous may: "(1) Cause or significantly contribute to an increase in mortality or an increase in serious irreversible, or incapacitating reversible illness or (2) Pose a substantial present or potential hazard to human health or the environment when it is improperly treated, stored, transported, disposed of or otherwise managed".

A waste is listed as hazardous if it exhibits the characteristics of: (1) Ignitability, (2) Corrosivity, (3) Reactivity, or (4) E.P. toxicity. Specific criteria for these characteristics are given in Subpart C of Part 261, however, it is noteworthy to discuss in more detail the characteristics of EP toxicity.

EP toxicity is determined by a prescribed extraction test [4]. If the extract of a representative sample of the waste contains certain contaminants in excess of maximum concentration limits, the waste exhibits the characteristics of EP toxicity. These contaminants and limits are: (see Table 1).

A waste is also listed as hazardous if it has been found to be fatal to humans in low doses, or in the absence of human toxicity data has been shown by studies to have an oral rat LD_{50} < 50 mg/kg or is capable of causing or contributing to serious irreversible or incapacitating reversible illness. If a waste contains any of approximately 350 hazardous constituents listed in Appendix VIII of Part 261, it is also listed as hazardous.

Table 1. EP Toxicity — Maximum Concentration Limits

Constituent	Concentration (mg/l)	Constituent	Concentration (mg/l)
Arsenic	5.0	Silver	5.0
Barium	100.0	Endrin	0.02
Cadmium	1.0	Lindane	0.4
Lead	5.0	Methoxychlor	10.0
Mercury	0.2	Toxaphene	0.5
Selenium	1.0	2,4-D	10.0
Chromium	5.0	2,4,5-TP Silvex	10.0

2.2 Listing of Hazardous Wastes

The above criteria for listing hazardous wastes has resulted in a list of wastes from non-specific sources (Section 261.31), from specific sources (Section 261.32) and a list of commercial chemical products (Section 261.33) when discarded as off-specification products. Container residues, and spill residues would be considered hazardous wastes. For purposes of recordkeeping, all listed waste types and waste types that meet the hazard characteristics of ignitability, corrosivity, reactivity, and EP toxicity are assigned hazardous waste numbers.

3 RCRA and Generators of Hazardous Wastes

Section 3002 of RCRA provides for the development of standards for generators of hazardous wastes relating to recordkeeping practices, labeling practices for containers of hazardous wastes, use of appropriate containers, furnishing of information on the chemical composition of wastes, use of a manifest system, and submission of reports on waste quantities generated and disposed. If generators of hazardous wastes also store wastes for greater than ninety days, or treat or dispose of wastes on-site, they are also subject to Sections 3004 of RCRA. These standards will be discussed in a later section.

3.1 Standards Applicable to Generators

The standards applicable to generators are contained in Part 262, 40 Code of Federal Regulations [3]. The key aspect of this regulation is that it requires generators to initiate the process of tracking hazardous wastes from "cradle to grave". The chain custody document or manifest is prepared by the generator who identifies itself, the transporter, the designated and alternate treatment, storage or disposal facility, the waste type, and the total quantity of each waste type. In certifying the above information by signing, the generator is ultimately responsible for any harm or damage that may be caused by the waste at any time in the future. The manifest consists of at least enough copies to provide one copy for the generator, one for each transporter, one for the treatment, storage or disposal facility, and one to be returned to the generator by the facility. In practice, two other copies are necessary. The generator sends one and the facility sends one to the responsible agency. The last set allows for reconciliation and verification of the "cradle to grave" concept.

4 RCRA and Transporters of Hazardous Wastes

Section 3004 of RCRA directs the U.S. Environmental Protection Agency to promulgate standards for transporters of hazardous wastes related to: (1) recordkeeping of points of pick-up and delivery, (2) transportation of only properly labeled wastes, (3) compliance with the manifest system, and (4) transportation of wastes only to facilities designated by the manifest and which hold valid permits issued by the responsible agencies.

4.1 Standards Applicable of Transporters

The pertinent regulations are presented in Part 263, 40 Code of Federal Regulations [3] and focus on the recordkeeping which is necessary to insure the documentation of the proper movement of hazardous wastes from generator to transporter to disposer. A transporter may not accept hazardous waste from a generator unless it is accompanied by a manifest signed by the generator. A transporter must ensure that the manifest accompanies the hazardous waste and must deliver the entire quantity of waste to the designated or alternate recipient. In delivering the waste, the transporter must retain a copy of the manifest after it has been signed by the recipient and give the remaining copies to the recipient.

5 Standards Applicable to Treatment, Storage and Disposal of Hazardous Waste

Section 3004 of RCRA directs the U.S. Environmental Protection Agency to develop and promulgate requirements applicable to owners and operators of hazardous waste treatment, storage, and disposal facilities. These regulations are the most significant and extensive set of standards promulgated under RCRA and exist as Part 264, 40

Code of Federal Regulations [3]. In general, these regulations fall into three major categories [4]: (1) Procedural requirements such as recordkeeping, financial responsibility, and contingency plans, (2) Permitting requirements, and (3) Compliance with substantive controls that pertain to construction and operation of the facility.

5.1 General Facility Standards

General facility standards apply to all treatment, storage, and disposal facilities. They relate to facility identification, required notices, waste analysis, personnel training, general requirements for ignitable, reactive, or incompatible wastes, and location standards. Location standards relate to seismic and flood considerations where a new facility cannot be located within 61 meters of an active fault, and a facility located in a 100 year floodplain must be designed, constructed and operated to prevent washout by a 100 year flood.

5.2 Contingency Plan for Accidents and Emergencies

All facilities are also required to have an accident prevention program and a detailed contingency plan for sudden or non-sudden releases of hazardous waste constituents. The plan is required to: (1) Describe actions of all facility personnel, (2) Contain a spill prevention control plan, (3) Describe agreements made with local public safety officials, (4) List means, addresses and phone numbers of all persons qualified to act as an emergency coordinator, (5) List of emergency equipment, and (6) Include an evacuation plan for facility personnel.

5.3 Recordkeeping Requirements

The recordkeeping requirements relate primarily to the proper use of a manifest, keeping an operating record of the facility and periodic reporting to responsible agencies. It is the facility owner's responsibility to reconcile any significant discrepancies between the manifest and the actual waste composition or volume disposed. If such discrepancies are not resolved within 15 days, the facility owner is required to report this to the responsible agency.

5.4 Groundwater Monitoring

A very important section of these Part 264 regulations relate to groundwater monitoring requirements for treatment, storage and disposal facilities. These requirements apply to facilities which treat, store or dispose of hazardous waste on land (surface impoundments, piles, land treatment units and landfills). In general, there are two types of groundwater monitoring programs required: (1) Compliance monitoring and (2) Detection monitoring.

A compliance monitoring program is required when hazardous constituents from a regulated unit (i.e. surface impoundment) are detected in designated wells around the unit. Such a compliance program involves monitoring for specific compounds

Table 2. Maximum Concentration of Constituents for Groundwater Protection (3)

Constituent	Maximum Concentration (mg/l)	Constituent	Maximum Concentration (mg/l)
Arsenic	0.05	Silver	0.05
Barium	1.0	Endrin	0.0002
Cadmium	0.01	Lindane	0.004
Chromium	0.05	Methoxychlor	0.1
Lead	0.05	Toxaphene	0.005
Mercury	0.002	2,4-D	0.1
Selenium	0.01	2,4,5-TP Silvex	0.01

related to the regulated unit to determine if established concentration limits are exceeded. For some hazardous constituents, these limits are specified by the following (see Table 2).

For other hazardous constituents, the concentration limit is the background concentration in the groundwater at the time the permit is issued for the regulated unit. An alternate limit can be specified based on the environmental toxicology and human health effects of the constituent, local hydrogeology, quality of the groundwater, and use patterns in the area. In the event that these limits are exceeded, the owner or operator is required to take corrective action to bring the regulated unit into compliance with the groundwater protection standards.

All facilities that store, treat or dispose of hazardous waste on land are required at a minimum to establish a detection monitoring program. Rather than monitor for specific hazardous constituents that are related to the regulated unit, owners and operators are allowed to monitor for such indicator parameters as specific conductance, total organic carbon, total organic halogen, or any other reliable indicator of the presence of hazardous constituents. If these indicator parameters are found to significantly exceed background values, the owner is required to immediately monitor for specific hazardous constituents and establish a compliance monitoring program. The minimum sampling period for a detection monitoring program is semi-annual while the period for the compliance monitoring program is quarterly.

Part 264 also specifies some general groundwater monitoring requirements that relate to: (1) Well placement and construction, (2) Sampling and analysis procedures, (3) Determination of background water quality, and (4) Determination of exceedences.

5.5 Closure and Post Closure Requirements

Closure requirements apply to all facilities while post-closure requirements apply to all hazardous waste *disposal* facilities and to surface impoundments and waste piles where waste residues are encapsulated in place upon closure. These facilities must be closed in a manner that: (1) Minimize the need for further maintenance, and (2) Controls, minimizes or eliminates threats to human health and the environment from the escape of hazardous constituents.

The owner of a facility must submit a closure plan with the permit application. The plan must at minimum describe how and when the facility will be closed or partially closed. The plan must also estimate the expected operating life of the facility, its maximum inventory of wastes, the maximum extent of activity during its operating life and the time necessary for final closure. Procedures for disposal or decontamination of equipment must also be included. The plan can be amended at any time during the active life of the facility.

If maintenance and monitoring is required after closure, this must continue for at least thirty years. This period may be shortened if that period is sufficient to protect human health and the environment. It can also be lengthened if continued protection is necessary beyond the 30 year period.

Use of property on which post-closure care is necessary is restricted such that any component of the waste containment system or the monitoring network must not be disturbed. These systems can be disturbed only if the potential hazard is not increased or if it is necessary to reduce a threat to human health and the environment. The owner is also required to notify the local land use agency and place a notice on the deed of the property indicating the existence of hazardous wastes and the restrictions on uses.

5.6 Financial Requirements

All facilities which are subject to closure requirements must prepare a cost estimate for closure based on the point in the facility's operating life when costs for closure would be at a maximum. This estimate must be adjusted annually for inflation and whenever a change in the closure plan increases the cost of closure.

All facilities which are subject to post-closure must prepare a written estimate of the annual and total costs for maintenance and monitoring. This estimate must also be adjusted annually for inflation and whenever changes in post-closure requirements increase costs.

In order to ensure that there are sufficient funds for implementing closure and post-closure requirements, all subject facilities must provide financial assurance. They can choose from the following options: (1) Closure or post-closure trust fund, (2) Surety bond guaranteeing payment to a trust fund for performance of the closure or post-closure requirements, (3) Closure or post-closure letter of credit, (4) Closure or post-closure insurance, (5) Financial test and corporate guarantee for closure and post-closure. A combination of the above for a given facility or application of one of the above for more than one facility owned by the same entity is allowed. Financial assurance for closure and post-closure can also be combined.

All facilities are also required to demonstrate financial responsibility for injuries and property damage to third parties which may result from sudden and non-sudden accidents arising from operations at the facility. Coverage for sudden occurrences must include at least 1 million dollars U.S. per occurrence and 2 million dollars U.S. per year. Non-sudden coverage must include at least 3 million dollars U.S. per occurrence and 6 million dollars U.S. per year. These liabilities may be covered by insurance or by passing a financial test, or a combination.

5.7 Standards Applicable to Specific Treatment, Storage and Disposal Techniques

The remainder of Part 264 describe standards applicable to specific hazardous waste treatment, storage and disposal techniques. These techniques include: (1) Containers,

(2) Tanks, (3) Surface impoundments, (4) Waste piles, (5) Land treatment, (6) Landfills, and (7) Incineration.

In general, these standards apply to such aspects as: (1) Design, (2) Operating requirements, (3) Monitoring and inspection, (4) Recordkeeping, (5) Special requirements for ignitable wastes, (6) Special requirements for incompatible wastes, (7) Closure and post-closure care, and (8) Other technique specific requirements.

6 Summary

The requirements discussed in the above sections constitute the most important set of laws and regulations related to the handling of hazardous waste in the United States. They describe in significant detail the "cradle to grave" concept of tracking hazardous waste, the definition of a hazardous waste and standards applicable to all treaters, storers, and disposers of hazardous waste. They seek to ensure proper handling of wastes, adequate monitoring, adequate closure and post-closure care, and adequate financial assurance.

Others aspects of the Resource Conservation and Recovery Act and its related regulations are also important, but not as critical to understanding the hazardous management scheme in the United States as those discussed above. These other sections include: (1) Permit requirements, (2) Authorized state hazardous waste management programs, (3) Inspections, (4) Enforcement, (5) Retention of state authority, (6) Effective dates, (7) Authorization and assistance to states, (8) Restrictions on recycled oil, (9) Hazardous waste site inventory and, (10) Monitoring, analysis and testing.

References

1. Resource Conservation and Recovery Act of 1976. 42 U.S. Code 6901 et. seq.
2. Comprehensive Environmental Response, Compensation and Liability Act of 1980. 42 U.S. Code 9601 et. seq.
3. 40 Code of Federal Regulations, Subchapter 1, Parts 261, 262, 263, and 264.
4. Test Methods for Evaluating Solid Waste: Physical Chemical Methods. U.S. Environmental Protection Agency, Office of Water and Waste Management: Washington, D.C., 1980, SW-846.

Legislation for Waste Treatment in the German Democratic Republic

D. Martinetz

In the German Democratic Republic (GDR) the handling of dangerous goods — solids, liquids, and gases — is regulated by various laws and national standards, as e.g.:

— 1. Implementing Regulation of the 5. Implementing Order of the National Culture Legislation (Landeskulturgesetz) — Prevention of Air Pollution — Limitation and Control of Immissions and Emissions (Air Pollutants) — issued June 28, 1979
— Water Legislation (Wassergesetz) and 1. Implementing Regulation — issued July 2, 1982
— Legislative Direction on general conditions for the connection of realty to, and for the introduction of waste water into public sewerage systems — Abwassereinleitungsbedingungen — issued July 20, 1978
— 6. Implementing Order of the National Culture Legislation (Landeskulturgesetz) — Nonhazardous Disposal of Nonreusable End Products — issued Sept. 1, 1983

Special attention is given to toxic waste products, e.g.:

— Legislation on the Transportation/Handling of Toxins — Toxin Law (Giftgesetz) — issued April 7, 1977 and Implementing Regulations
— Department Standard: Utilization and nonhazardous deposition of waste products and community wastes. Overground depositing of toxic waste products and other harmful substances — issued May 8, 1981; TGL 37597

In addition, national standards specifically regulate the handling and disposal of specific wastes produced within the GDR, as e.g.:

— GDR Standard: Handling of Polyurethane Raw Materials; nonhazardous disposal of contaminants and raw material wastes — issued Feb. 4, 1980; TGL 34222

In general, legal regulations are based on the principle of reusage of waste materials. The reusage of waste materials as secondary raw materials is the responsibility of the producer; the producer also is responsible for reusing waste materials resulting from the social and individual comsumption of his goods. Necessary research and technical developments for such processing must be supplied by the producer. If reusage is by no means feasible, then this must be proven. The necessary research and technical developments for the nonhazardous disposal of the material in question — if not already supplied — must also be delivered by the producer.

Companies producing toxic waste materials must report — in compliance with the GDR Toxin Law (Giftgesetz) — to the council of the district. Furthermore, companies are required to submit proposals for appropriate means of waste disposal if proof is given that the materials in question are not reusable.

Nonreusable toxins — in compliance with the Toxin Law (Giftgesetz) — must be disposed of in preventing possible hazard to the life and health of humans, domesticated animals, and economic plants and without risk of other forms of damage to or negative effects on the environment.

Nonreusable toxic waste products are considered to be toxins as well as solid, viscous, or liquid wastes and residues which cannot be utilized for another purpose and which contain components categorized as highly toxic (toxin category 1) or toxic (toxin category 2).

Harmful substances are considered to be nonreusable solid, viscous, or liquid wastes and residues which damage or negatively affect natural resources, especially underground water and surface waters.

Nonhazardous disposal of toxic waste products and other harmful substances is considered by the legislation as the transformation into nontoxic or nondamaging substances or as the depositing of these substances by excluding any possibility of hazard to humans, domestic animals, and economic plants as well as preventing any economic damage or negative environmental effects. Nonhazardous disposal, necessary research and technical developments, required investigations on disposable substances, and documentation of actions are the responsibility of the producer. The mode of disposal is decided by the council of the district upon submission of proposals by the producer. The council may prescribe other means of disposal or locations of disposal according to territorial conditions. Decisions by the council must be justified.

Legislation for Waste Treatment in France

A. Robin

French legislation on waste control is principally based on the law of 15 July 1975 [1] relating to waste disposal and recovery of materials. This text is supplemented by the law of 19 July 1976 [2] relating to the classification of installations for environmental protection and by special provisions concerning certain categories of waste or certain procedures for waste disposal.

The law of 15 July 1975 [1] defines the responsibility of the company according to which waste is discharged, how waste is disposed of and its possible harmful effects. The company is responsible for safe disposal of the waste produced.

The law applies to waste in general, although it also aims at defining a special category of waste which is liable to have harmful effects. Companies discharging or storing such waste must comply with the following requirements:

- The government body in charge must be notified of the nature and quantities of waste (article 8).
- Proof must be provided of its final destination and the process by which it has been disposed of.

This latter point has not yet been completely settled, because, although article 9 provides for official approval of plants authorised to treat toxic and dangerous waste, no implementing decree has defined or listed the waste to which this provision applies, or the conditions which plants must meet before they can be approved.

The law provides for penal sanctions, including fines and imprisonment (section VIII) in cases of failure to comply with the regulations that it lays down. In certain cases, in addition to the sentence, the offender may be ordered to restore damaged areas. In other cases the court may order plants to be closed down.

It is of interest to note that this law on waste disposal also contains provisions on the recovery of materials (section V) and set up the ANRED (Agence Nationale pour la Récupération et l'Elimination des Déchets — National Agency for Waste Disposal and Recovery).

The ANRED is an official body responsible for promoting the building of waste disposal and recovery facilities, helping the development of new technologies and organising campaigns to inform and educate the public. Its means are based on technical assistance, financial aid and a documentation centre.

In addition, the law of 19 July 1976 [2] on installations that are classified for environmental protection applies to the actual source of waste. It lays down regulations for the construction and operation of industrial plants, and stipulates that they are subject to declaration or authorisation, depending on the seriousness of the environmental dangers or disadvantages that their operation may involve (article 2). This text throws new light on the law of 19 December 1917 on plants classified as dangerous, unhealthy, noisy or noxious.

These two laws are supplemented by regulations on certain categories of waste or certain disposal techniques.

As regards categories of waste, the following may be mentioned:

For mineral or synthetic oils, a special tax [4] is charged and paid over to the ANRED. It is used to finance information campaigns, assistance and collection of waste oil. Regulations are laid down for the collection of this tax [5] and for the incineration of waste oil [6].

For polychlorobiphenyls (PCB) a decree [7] lays down regulations for conditions of use and disposal, which must avoid any risk of dispersion in the environment.

Regarding disposal techniques:

Very restrictive texts cover the conditions of incineration at sea [8] and dumping of waste [9, 10], limiting both these practices.

Technical recommendations of 22 January 1980 [11] on tipping lay down criteria for assessing and classifying tips in three categories. They forbid the tipping of certain types of waste (which are listed), which might be very harmful or toxic when leached by rainwater.

Finally, to assume the control of transfrontier shipments, a decree of 5.7.83 [13] lays down a system of preliminary declaration which allows a systematic information of the Government Service.

Because of the delay in the publication of the regulations on tipping (January 1980), French regulations gave priority for some time to the building of incineration centres. Subsequently, under the combined effects of energy recovery, economies of materials and technical improvements, the activity of these centres was greatly disturbed and readjustments were necessary.

At all events, the texts clearly show the main objectives of waste control policy in France:

- To reduce the amount of waste.
- To increase recovery.
- To treat and dispose of waste which cannot be recycled in ways which do not damage the environment.

The laws and regulations are both preventive and restrictive, allowing action upstream and downstream of waste production:

- Upstream, at the level of the products, as manufacturing methods determine the conditions of their disposal as waste ([1], article 6),
- Downstream from waste producers, by organising inspection of disposal plants, particularly strict for toxic and dangerous waste, for which the joint responsibility of the producer and the disposer is clearly mentioned ([1], article 11).

As has already been said, the decrees implementing article 9 of the law [1] have not yet been issued, and "toxic and dangerous" waste is therefore not strictly defined.

Projects in this direction in France have come up against the same difficulties as on the European level (the implementation of EEC Directive 78/319 of 20 March 1978). Apart from some special cases, such as arsenic-containing sludge, cyanide solutions, PCB and soluble salts of heavy metals, it has proved very difficult to define toxic and dangerous waste on the basis of the toxic elements that it contains. The behaviour of waste in contact with other waste, water and bacteria in the soil involves numerous physico-chemical phenomena which this approach does not take into account.

To fill this gap, the French authorities recently published a nomenclature for waste [12].

In this double-entry nomenclature, each type of waste is designated by combinating two items of information — the category to which it belongs and the activity which produced it.

Deliberately limited to about a hundred categories of waste and the same number of originating activities, the nomenclature can be broken down in more detail according to need. It has in fact proved necessary, for waste from the chemical industry, to add information on the chemical nature of the most characteristic element of potential toxicity in the waste, in the list of waste-generating activities.

For the moment this nomenclature is being used experimentally in France. Through the common language which it provides, however, the nomenclature may contribute to progress in the delicate problem of monitoring the transport of waste and of better information for the authorities involved, which is currently a matter of concern for all the European countries.

References

1. Law N° 75–033 of 15 July 1975 relating to waste disposal and material recovery.
2. Law N° 75–663 of 19 July 1976 relating to installations that are classified for environmental protection.
3. Decree N° 77–974 of 19 August 1977 relating to information required for toxic waste.
4. Decree N° 79–517 of 30 June 1979 establishing a special tax to benefit the French Government's agency for waste recovery and disposal.
5. Decree N° 79–981 of 21 November 1979 on used oil recovery legislation.
6. Decree of 21 May 1981 relating to equipment and exploitation of thermal units fired by used oil.
7. Decree of 8 July 1975 relating to the conditions of use of polychlorobiphenyls.
8. Law N° 76–600 of 7 July 1976 relating to prevention and control of sea pollution by incineration.
9. Law N° 76–599 of 7 July 1976 relating to the prevention and control of sea pollution due to dumping and accidental discharge of pollutants by ships and aircraft.
10. Decree N° 82 842 of 29 September 1982 implementing law N° 76–599 of 7 July 1976.
11. Technical recommendation of 22 January 1980 on the control of industrial waste discharge.
12. 1983 nomenclature concerning waste classification published by the French State Secretariat for Environment and Quality of Life.
13. Import of hazardous waste. Decree of 5 July 1983. J.O. 2 august 83.

B. The Nature of Chemical Wastes

Types of Chemical Wastes

Mogens Palmark, Chemocontrol AIS

Contents

Type of Chemical Wastes

The paper deals with regulatory definitions and classification of hazardous waste types with emphasis on the situation in the U.S.A. and in the European Common Market.

These classification systems are regulatory based and it is discussed whether they are sufficient as practical tools within the daily operation of treatment systems and for planning purposes.

Examples of various chemical waste types which are based instead on classification according to treatability are presented.

1 Introduction

Before considering the handling or disposal of the different types of chemical wastes, it is necessary to define what is meant by a chemical waste in general terms. This is not an easy task because of the difficulty in drawing a distinct line between materials which have no conceivable monetary value and other by-products of chemical processes which might be utilised for some useful purpose.

When a company solves its own waste problem by treatment on site, are the wastes then interim products or "real waste"? Some products which are modified in an irreversible way, for example by incineration, would normally always be wastes but to categorize contaminated solvents as waste might sometimes seem more doubtful, as they could be recovered.

Are "waste products" to be regarded as waste only when they leave the company's gate, for example for off site disposal?

Many so-called hazardous waste types can also be described as potential raw material for another enterprise. It therefore has a value and should it now be regarded as a waste?

Hazardous waste should perhaps be defined as hazardous products which are no longer used for the original intended purpose, or chemical products which can no longer be used at all and which are therefore discarded.

Defining of waste can also be a question about concentration. If a compound, despite all difficulties mentioned above, has been defined as a hazardous waste — to what extent could it be "diluted" with non-hazardous products and still be regarded as "hazardous"?

In the years from 1970 environmental legislations concerning hazardous wastes were passed in several countries, many of these using statutory definition based on harmful effect as criteria.

Examples of this, compiled as part of the work of a Nato committee [2] are:[1]

France — categories of waste may be defined by decree and the enterprises that produce, import, transport or dispose of wastes which belong to these categories and which are in a state such that they cause, or at the time of their disposal may cause, a nuisance such as ... injurious effects on the soil, plants or animals, to degrade the scenery or the countryside, to pollute the air or water, to create a noise or odor, or, ... (are) harmful to human health or the environment ... (Art. 8 and 2; Law No. 75–633; July 16, 1975).

Federal Republic of Germany — Special wastes are such wastes from commercial or trade companies which, due to their nature, composition or quantities, are especially

[1] see also the Introduction of this work

hazardous to human health, air or water, or which are explosive, flammable or may cause diseases. Their disposal must be subject to additional requirements according to the Act. (Federal Act on the Disposal of Waste, 1972, as amended, 1976).
Netherlands — Chemical wastes are: (1) wastes consisting wholly or partly of chemicals indicated by General Administrative Order and (2) wastes produced by chemical processes designated by General Administrative Order. (Chemical Waste Act, 1977).
United Kingdom — Waste "of a kind which is poisonous, noxious or polluting and whose presence on the land is liable to give rise to an environmental hazard". (Deposit of Poisonous Waste Act, 1972); special wastes are those which "may be . . . dangerous or difficult to dispose of" (Control of Pollution Act, 1974).
United States — Hazardous waste means a solid waste or combination of solid wastes, which because of its quantity, concentration or physical, chemical or infectious characteristics may (A) cause, or significantly contribute to an increase in mortality or an increase in serious irreversible or incapacitating reversible illness; or (B) pose a substantial present or potential hazard to human health or the environment when improperly treated, stored, transported or disposed of, or otherwise managed". *Solid waste* includes "any garbage, refuse, sludge from a waste treatment plant, water supply treatment plant, or air pollution control facility and other discarded material, including solid, liquid, semisolid, or contained gaseous material . . ." (Resource Conservation and Recovery Act of 1976).

Based on these statutory definitions which in some cases have been revised, the responsible agencies within each country have later worked out a regulatory context. In the following section some examples on this are further outlined.

2 Definition and Grouping of Chemical Wastes

2.1 Definitions According to USA's Environmental Protection Agency (EPA)

The EPA has published regulations for hazardous waste management as required by the Resource Conservations and Recovery Act of 1976 [3].
According to the regulations Hazardous Wastes are defined as solid wastes which:
1. exhibit one or more of the characteristics:
   - ignitability,
   - corrosivity,
   - reactivity,
   - toxicity;
2. is listed as a hazardous waste (table sub-part D)
3. is a mixture of solid waste and which contains one or more of the above-mentioned listed hazardous wastes.

Providing the listed wastes are not excluded according to a specific § and in the regulation (§ 261.4).
As hazardous wastes are included in the term solid wastes, the definition of the latter is a prerequisite as well for the understanding. RCRA (and EPA) have the following definition:

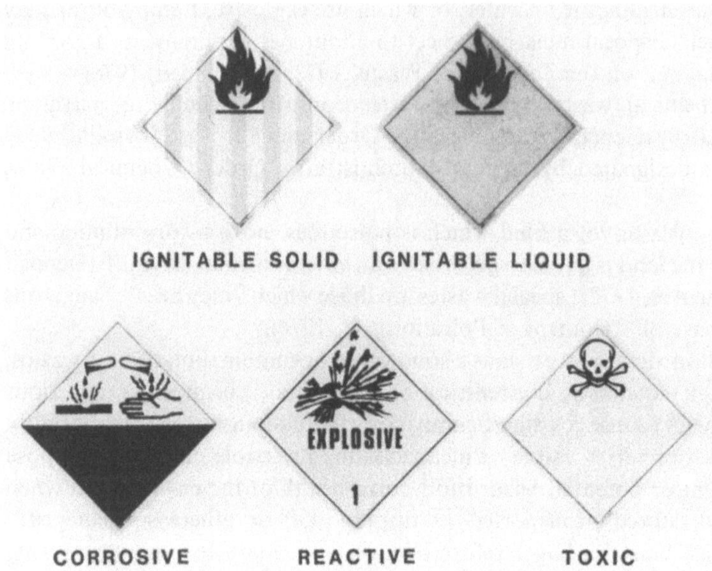

Fig. 2.1. Symbols for characterization of hazardous materials

Solid Waste

Any garbage, refuse, sludge from a wastewater treatment plant, water supply treatment plant or air pollution control facility and other discarded material, including solid, liquid, semisolid, or contained gaseous material resulting from industrial, commercial or mining and agricultural operations, and from community activities, but which does not include solid or dissolved materials in domestic sewage, or solid or dissolved materials in irrigation return flows, or industrial discharges which are point sources subject to permits under section 402 of the Federal Water Pollution Control Act, or source, special nuclear, or byproduct material as defined by the Atomic Energy of Act of 1954. It has to be observed that "solid waste" also includes semi-solid, and even liquid constituents!
The characteristics are defined as follows:

Ignitability

(a) A solid waste exhibits the characteristics of ignitability if a representative sample of the waste has any of the following properties:
    (1) It is a liquid, other than an aqueous solution containing less than 24 per cent alcohol by volume, and has a flash point less than 60 °C (140 °F), as determined by a Pensky-Martens Closed Cup Tester, using the test method specified in ASTM Standard D-93-79, or a Setaflash Closed Cup Tester, using the test method specified in ASTM standard D-3278-78, or as determined by an equivalent test method approved by the Administrator under the procedures set forth in §§ 260.20 and 260.21.
    (2) It is not a liquid and is capable, under standard temperature and pressure, of causing fire through friction, absorption of moisture or spontaneous

chemical changes and, when ignited, burns so vigorously and persistently that is creates a hazard.
(3) It is an ignitable compressed gas as defined in 49 CFR 173.300 and as determined by the test methods described in that regulation or equivalent test methods approved by the Administrator under §§ 260.20 and 260.21.
(4) It is an oxidizer as defined in 49 CFR 173.151.

(b) A solid waste that exhibits the characteristics of ignitability, but is not listed as a hazardous waste in Sub-part D, has the EPA Hazardous Waste Number of D001.

Corrosivity

§ 261.22 Characteristics of corrosivity

(a) A solid waste exhibits the characteristics of corrosivity if a representative sample of the waste has either of the following properties:
(1) It is aqueous and has a pH less than or equal to 2 or greater than or equal to 12.5, as determined by a pH meter using either the test method specified in the "Test Methods for the Evaluation of Solid Waste, Physical/Chemical Methods" (also described in "Methods for Analysis of Water and Wastes" EPA 600/4-79-020, March 1979), or an equivalent test method approved by the Administrator under the procedures set forth in §§ 260.20 and 260.21.
(2) It is a liquid and corrodes steel (SAE 1020) at a rate greater than 6.35 mm (0.250 inch) per year at a test temperature of 55 °C (130 °F) as determined by the test method specified in NACE (National Association of Corrosion Engineers) Standard TM-01-69[3] as standardized in "Test Methods for the Evaluation of Solid Waste, Physical/Chemical Methods", or an equivalent test method approved by the Administrator under the procedures set forth in §§ 260.20 and 260.21.
(b) A solid waste that exhibits the characteristics of corrosivity, but is not listed as a hazardous waste in Sub-part D, has the EPA Hazardous Waste Number of D002.

Reactivity

(a) A solid waste exhibits the characteristics of reactivity if a representative sample of the waste has any of the following properties:
(1) It is normally unstable and readily undergoes violent change without detonating.
(2) It reacts violently with water.
(3) It forms potentially explosive mixtures with water.
(4) When mixed with water, it generates toxic gases, vapors or fumes in a quantity sufficient to present a danger to human health or the environment.
(5) It is a cyanide or sulfide bearing waste which, when exposed to pH conditions between 2 and 12.5, can generate toxic gases, vapors or fumes in a quantity sufficient to present a danger to human health or the environment.

(6) It is capable of detonation or explosive reaction if it is subjected to a strong initiating source or if heated under confinement.
(7) It is readily capable of detonation or explosive decomposition or reaction at standard temperature and pressure.
(8) It is a forbidden explosive as defined in 49 CFR 173.51, or a Class A explosive as defined in 49 CFR 173.53 or a Class B explosive as defined in 49 CFR 173.88.

(b) A solid waste that exhibits the characteristics of reactivity, but is not listed as a hazardous waste in Subpart D, has the EPA Hazardous Waste Number of D003.

Toxicity

(a) A solid waste exhibits the characteristics of EP[2] toxicity if, using the test methods described in Appendix II or equivalent methods approved by the Administrator under the procedures set forth in §§ 260.20 and 260.21, the extract from a representative sample of the waste contains any of the contaminants listed in Table 2.1 at a concentration equal to or greater than the respective value given in that Table. Where the waste contains less than 0.5 per cent filterable solids,

Table 2.1. Maximum concentration of contaminants for characteristic of EP toxicity

EPA hazardous waste number	Contaminant	Maximum concentration (milligrams per litre)
D004	Arsenic	5.0
D005	Barium	100.0
D006	Cadmium	1.0
D007	Chromium	5.0
D008	Lead	5.0
D009	Mercury	0.2
D010	Selenium	1.0
D011	Silver	5.0
D012	Endrin (1,2,3,4,10,10-hexachloro-1,7-epoxy-1,4,4a,5,6,7,8,8a-octahydrol, 4-endo, endo-5, 8-dimethano naphthalene	0.02
D013	Lindane (1,2,3,4,5,6-hexachlorocyclohexane, gamma isomer)	0.4
D014	Methoxychlor (1,1,1-Trichloro-2,2-bis (p-methoxyphenyl)ethane)	10.0
D015	Toxaphene ($C_{10}H_{10}Cl_5$) Technical chlorinated camphene, 67–69 percent chlorine)	0.5
D016	2,4-D(2,4-Dichlorophenoxyacetic acid)	10.0
D017	2,4,5-TP Silvex (2.4.5-Trichlorophenoxypropionic acid)	1.0

[2] EP = Extraction procedures

the waste itself, after filtering, is considered to be the extract for the purposes of this section.

(b) A solid waste that exhibits the characteristics of EP toxicity, but is not listed as a hazardous waste in Subpart D, has the EPA Hazardous Waste Number specified in Table 1.1, which corresponds to the toxic contaminant causing it to be hazardous.

2.1.1 Hazardous Waste Lists, U.S.A.

Hazardous wastes are wastes which are listed specifically by EPA in three tables, examples of which are given (Table 2.2) below.

The waste listed is:

- Hazardous waste from non-specific sources.
- Hazardous waste from specific sources.
- Discarded commercial chemical products, off-specification species, containers and spill residues thereof.

The basis for listing the wastes is that the solid waste meet one of the criteria:

- They exhibit any of the characteristics mentioned above
- They have been found to be fatal to human beings in small doses or it has been shown that they have low lethal doses (rats, rabbits) as to cause serious, irreversible or incapacitating reversible illness.
  Waste listed according to these criteria is designated Acute Hazardous Wastes.
- It contains certain hazardous constituents unless it is concluded, based on specifical evaluation, that no hazard is posed. Waste listed according to these criteria is designated Toxic Wastes.

2.2 Definition According to the Regulations from the European Common Market

In a directive from EEC on Toxic and Hazardous Wastes [4], the following definition is presented:

"Toxic and dangerous wastes: Waste containing, or which has been identified with the compounds or substances mentioned in the appendix to the directive, or substances in a condition or which are present in such quantities or concentrations that they pose a risk to health or environment."

Wastes in turn are classified as being:

"Every compound or every object, which the owner rids of or is obliged to get rid of according to national regulations in force".

The appendix includes the following items:

Hazardous and Toxic Material

1. Arsenic; arsenic compounds
2. Mercury; mercury compounds
3. Cadium; cadium compounds
4. Thallium; thallium compounds
5. Beryllium; beryllium compounds
6. Chrome 6 compounds

Table 2.2 Example of Hazardous Waste Lists [EPA, USA [3]]

§ 261.31 Hazardous waste from nonspecific sources.

Industry and EPA hazardous waste No.	Hazardous waste	Hazard code
Generic:		
F001	The spent halogenated solvents used in degreasing, tetrachloroethylene, trichloroethylene, methylene chloride, 1,1,1-trichloroethane, carbon tetrachloride, and the chlorinated fluorocarbons; and sludges from the recovery of these solvents in degreasing operations.	(T)
F002	The spent halogenated solvents, tetrachloroethylene, methylene chloride, trichloroethylene, 1,1,1-trichloroethane, chlorobenzene, 1,1,2-trichloro-1,2,2-trifluoroethane, o-dichlorobenzene, trichlorofluoromethane and the still bottoms from the recovery of these solvents.	(T)
F003	The spent non-halogenated solvents, xylene, acetone, ethyl acetate, ethyl benzene, ethyl ether, n-butyl alcohol, cyclohexanone, and the still bottoms from the recovery of these solvents.	(I)
F004	The spent non-halogenated solvents, cresols and cresylic acid, nitrobenzene, and the still bottoms from the recovery of these solvents	(T)
F005	The spent non-halogenated solvents, methanol, toluene, methyl ethyl ketone, methyl isobutyl ketone, carbon disulfide, isobutanol, pyridine and the still bottoms from the recovery of these solvents.	(I, T)
F006	Wastewater treatment sludges from electroplating operations	(T)
F007	Spent plating bath solutions from electroplating operations	(R, T)
F008	Plating bath sludges from the bottom of plating baths from electroplating operations	(R, T)
F009	Spent stripping and cleaning bath solutions from electroplating operations	(R, T)
F010	Quenching bath sludge from oil baths from metal heat treating operations	(R, T)
F011	Spent solutions from salt bath pot cleaning from metal heat treating operations	(R, T)
F012	Quenching wastewater treatment sludges from metal heat treating operations	(T)
F013	Flotation tailings from selective flotation from mineral metals recovery operations	(T)
F014	Cyanidation wastewater treatment tailing pond sediment from mineral metals recovery operations	(T)
F015	Spent cyanide bath solutions from mineral metals recovery operations	(R, T)
F016	Dewatered air pollution control scrubber sludges from coke ovens and blast furnaces	(T)

§ 261.32 Hazardous waste from specific sources.

Industry and EPA hazardous waste No.	Hazardous waste	Hazard code
Wood Preservation: K001	Bottom sediment sludge from the treatment of wastewaters from wood preserving processes that use creosote and/or pentachlorophenol	(T)
Inorganic Pigments:		
K002	Wastewater treatment sludge from the production of chrome yellow and orange pigments	(T)
K003	Wastewater treatment sludge from the production of molybdate orange pigments	(T)
K004	Wastewater treatment sludge from the production of zinc yellow pigments	(T)
K005	Wastewater treatment sludge from the production of chrome green pigments	(T)
K006	Wastewater treatment sludge from the production of chrome oxide green pigments (anhydrous and hydrated)	(T)
K007	Wastewater treatment sludge from the production of iron blue pigments	(T)
K008	Oven residue from the production of chrome oxide green pigments	(T)
Organic Chemicals:		
K009	Distillation bottoms from the production of acetaldehyde from ethylene	(T)
K010	Distillation side cuts from the production of acetaldehyde from ethylene	(T)
K011	Bottom stream from the wastewater stripper in the production of acrylonitrile	(R, T)
K012	Still bottoms from the final purification of acrylonitrile in the production of acrylonitrile	(T)
K013	Bottom stream from the acetonitrile column in the production of acrylonitrile	(R, T)
K014	Bottoms from the acetronitrile purification column in the production of acrylonitrile	(T)
K015	Still bottoms from the distillation of benzyl chloride	(T)
K016	Heavy ends or distillation residues from the production of carbon tetrachloride	(T)
K017	Heavy ends (still bottoms) from the purification column in the production of epichlorohydrin	(T)
K018	Heavy ends from fractionation in ethyl chloride production	(T)
K019	Heavy ends from the distillation of ethylene dichloride in ethylene dichloride production	(T)
K020	Heavy ends from the distillation of vinyl chloride in vinyl chloride monomer production	(T)
K021	Aqueous spent antimony catalyst waste from fluoromethanes production	(T)
K022	Distillation bottom tars from the production of phenol/acetone from cumene	(T)
K023	Distillation light ends from the production of phthalic anhydride from naphthalene	(T)
K024	Distillation bottoms from the production of phthalic anhydride from naphthalene	(T)
K025	Distillation bottoms from the production of nitrobenzene by the nitration of benzene	(T)
K026	Stripping still tails from the production of methyl ethyl pyridines	(T)
K027	Centrifuge residue from toluene diisocyanate production	(R, T)
K028	Spent catalyst from the hydrochlorinator reactor in the production of 1,1,1-trichloroethane	(T)
K029	Waste from the product stream stripper in the production of 1,1,1-trichloroethane	(T)
K030	Column bottoms or heavy ends from the combined production of trichloroethylene and perchloroethylene	(T)
Pesticides:		
K031	By-products salts generated in the production of MSMA and cacodylic acid	(T)
K032	Wastewater treatment sludge from the production of chlordane	(T)
K033	Wastewater and scrub water from the chlorination of cyclopentadiene in the production of chlordane	(T)
K034	Filter solids from the filtration of hexachlorocyclopentadiene in the production of chlordane	(T)
K035	Wastewater treatment sludges generated in the production of creosote	(T)
K036	Still bottoms from toluene reclamation distillation in the production of disulfoton	(T)
K037	Wastewater treatment sludges from the production of disulfoton	(T)
K038	Wastewater from the washing and stripping of phorate production	(T)
K039	Filter cake from the filtration of diethylphosphorodithoric acid in the production of phorate	(T)
K040	Wastewater treatment sludge from the production of phorate	(T)
K041	Wastewater treatment sludge from the production of toxaphene	(T)
K042	Heavy ends or distillation residues from the distillation of tetrachlorobenzene in the production of 2,4,5-T	(T)
K043	2,6-Dichlorophenol waste from the production of 2,4-D	(T)
Explosives:		
K044	Wastewater treatment sludges from the manufacturing and processing of explosives	(R)
K045	Spent carbon from the treatment of wastewater containing explosives	(R)
K046	Wastewater treatment sludges from the manufacturing, formulation and loading of lead-based initiating compounds	(T)
K047	Pink/red water from TNT operations	(R)
Petroleum Refining:		
K048	Dissolved air flotation (DAF) float from the petroleum refining industry	(T)
K049	Slop oil emulsion solids from the petroleum refining industry	(T)
K050	Heat exchanger bundle cleaning sludge from the petroleum refining industry	(T)
K051	API separator sludge from the petroleum refining industry	(T)
K052	Tank bottoms (leaded) from the petroleum refining industry	(T)
Leather Tanning Finishing:		
K053	Chrome (blue) trimmings generated by the following subcategories of the leather tanning and finishing industry: hair pulp/chrome tan/retan/wet finish; hair save/chrome tan/retan/wet finish; retan/wet finish; no beamhouse; through-the-blue; and shearling.	(T)
K054	Chrome (blue) shavings generated by the following subcategories of the leather tanning and finishing industry: hair pulp/chrome tan/retan/wet finish; hair save/chrome tan/retan/wet finish; retan/wet finish; no beamhouse; through-the-blue; and shearling.	(T)

Table 2.2. (cont.)

Industry and EPA hazardous waste No.	Hazardous waste	Hazard code
K055	Buffing dust generated by the following subcategories of the leather tanning and finishing industry: hair pulp/chrome tan/retan/wet finish; hair save/chrome tan/retan/wet finish; retan/wet finish; no beamhouse; and through-the-blue.	(T)
K056	Sewer screenings generated by the following subcategories of the leather tanning and finishing industry: hair pulp/chrome tan/retan/wet finish; hair save/chrome tan/retan/wet finish; retan/wet finish; no beamhouse; through-the-blue; and shearling.	(T)
K057	Wastewater treatment sludges generated by the following subcategories of the leather tanning and finishing industry: hair pulp/chrome tan/retan/wet finish; hair save/chrome tan/retan/wet finish; retan/wet finish; no beamhouse; through-the-blue and shearling.	(T)
K058	Wastewater treatment sludges generated by the following subcategories of the leather tanning and finishing industry: hair pulp/chrome tan/retan/wet finish; hair save/chrome tan/retan/wet finish; and through-the-blue.	(R, T)
K059	Wastewater treatment sludges generated by the following subcategory of the leather tanning and finishing industry: hair save/non-chrome tan/retan/wet finish.	(R)
Iron and Steel:		
K060	Ammonia still lime sludge from coking operations	(T)
K061	Emission control dust/sludge from the electric furnace production of steel	(T)
K062	Spent pickle liquor from steel finishing operations	(C, T)
K063	Sludge from lime treatment of spent pickle liquor from steel finishing operations	(T)
Primary Copper: K064	Acid plant blowdown slurry/sludge resulting from the thickening of blowdown slurry from primary copper production	(T)
Primary Lead: K065	Surface impoundment solids contained in and dredged from surface impoundments at primary lead smelting facilities	(T)
Primary Zinc:		
K066	Sludge from treatment of process wastewater and/or acid plant blowdown from primary zinc production	(T)
K067	Electrolytic anode slimes/sludges from primary zinc production	(T)
K068	Cadmium plant leach residue (iron oxide) from primary zinc production	(T)
Secondary Lead: K069	Emission control dust/sludge from secondary lead smelting	(T)

§ 261.33 Discarded Commercial Chemical Products, Off-Specification Species, Containers, and Spill Residues Thereof.

Hazardous waste No.	Substance [1]
	1080 see P058
	1081 see P057
	(Acetato)phenylmercury see P092
	Acetone cyanohydrin see P069
P001	3-(alpha-Acetonylbenzyl)-4-hydroxycoumarin and salts
P002	1-Acetyl-2-thiourea
P003	Acrolein
	Agarin see P007
	Agrosan GN 5 see P092
	Aldicarb see P069
	Aldifen see P048
P004	Aldrin
	Algimycin see P092
P005	Allyl alcohol
P006	Aluminum phosphide (R)
	ALVIT see P037
	Aminoethylene see P054
P007	5-(Aminomethyl)-3-isoxazolol
P008	4-Aminopyridine
	Ammonium metavanadate see P119
P009	Ammonium picrate (R)
	ANTIMUCIN WDR see P092
	ANTURAT see P073
	AQUATHOL see P088
	ARETIT see P020
P010	Arsenic acid
P011	Arsenic pentoxide
P012	Arsenic trioxide
	Athrombin see P001
	AVITROL see P008
	Aziridene see P054
	AZOFOS see P061
	Azophos see P061
	BANTU see P072
P013	Barium cyanide
	BASENITE see P020
	BCME see P016
P014	Benzenethiol
	Benzoepin see P050
P015	Beryllium dust
P016	Bis(chloromethyl) ether
	BLADAN-M see P071
P017	Bromoacetone
P018	Brucine
P019	2-Butanone peroxide
	BUFEN see P092
	Butaphene see P020
P020	2-sec-Butyl-4,6-dinitrophenol
P021	Calcium cyanide
	CALDON see P020
P022	Carbon disulfide
	CERESAN see P092
	CERESAN UNIVERSAL see P092
	CHEMOX GENERAL see P020
	CHEMOX P.E. see P020
	CHEM-TOL see P090
P023	Chloroacetaldehyde
P024	p-Chloroaniline
P025	1-(p-Chlorobenzoyl)-5-methoxy-2-methylindole-3-acetic acid
P026	1-(o-Chlorophenyl)thiourea
P027	3-Chloropropionitrile
P028	alpha-Chlorotoluene
P029	Copper cyanide
	CRETOX see P108
	Coumadin see P001
	Coumafen see P001
P030	Cyanides
P031	Cyanogen
P032	Cyanogen bromide
P033	Cyanogen chloride
	Cyclodan see P050
P034	2-Cyclohexyl-4,6-dinitrophenol
	D-CON see P001
	DETHMOR see P001
	DETHNEL see P001
	DFP see P043
P035	2,4-Dichlorophenoxyacetic acid (2,4-D)
P036	Dichlorophenylarsine
	Dicyanogen see P031
P037	Dieldrin
	DIELDREX see P037
P038	Diethylarsine
P039	0,0-Diethyl-S-(2-(ethylthio)ethyl)ester of phosphorothioic acid
P040	0,0-Diethyl-0-(2-pyrazinyl)phosphorothioate
P041	0,0-Diethyl phosphoric acid, 0-p-nitrophenyl ester
P042	3,4-Dihydroxy-alpha-(methylamino)-methyl benzyl alcohol
P043	Di-isopropylfluorophosphate
	DIMETATE see P044
	1,4:5,8-Dimethanonaphthalene, 1,2,3,4,10,10-hexachloro-1,4,4a,5,8,8a-hexahydro endo, endo see P060
P044	Dimethoate
P045	3,3-Dimethyl-1-(methylthio)-2-butanone-O-[(methylamino)carbonyl] oxime
P046	alpha,alpha-Dimethylphenethylamine
	Dinitrocyclohexylphenol see P034
P047	4,6-Dinitro-o-cresol and salts
P048	2,4-Dinitrophenol
	DINOSEB see P020
	DINOSEBE see P020
	Disulfoton see P039
P049	2,4-Dithiobiuret
	DNBP see P020
	DOLCO MOUSE CEREAL see P108
	DOW GENERAL see P020
	DOW GENERAL WEED KILLER see P020
	DOW SELECTIVE WEED KILLER see P020
	DOWICIDE G see P090
	DYANACIDE see P092
	EASTERN STATES DUOCIDE see P001
	ELGETOL see P020
P050	Endosulfan
P051	Endrin
	Epinephrine see P042
P052	Ethylcyanide
P053	Ethylenediamine
P054	Ethyleneimine
	FASCO FASCRAT POWDER see P001
	FEMMA see P091
P055	Ferric cyanide
P056	Fluorine
P057	2-Fluoroacetamide
P058	Fluoroacetic acid, sodium salt
	FOLODOL-80 see P071
	FOLODOL M see P071
	FOSFERNO M 50 see P071
	FRATOL see P058
	Fulminate of mercury see P065
	FUNGITOX OR see P092
	FUSSOF see P057
	GALLOTOX see P092
	GEARPHOS see P071
	GERUTOX see P020
P059	Heptachlor
P060	1,2,3,4,10,10-Hexachloro-1,4,4a,5,8,8a-hexahydro-1,4:5,8-endo, endo-dimethanonaphthalene
	1,4,5,6,7,7-Hexachloro-cyclic-5-norbornene-2,3-dimethanol sulfite see P050
P061	Hexachloropropene
P062	Hexaethyl tetraphosphate
	HOSTAQUICK see P092
	HOSTAQUIK see P092
	Hydrazomethane see P068
P063	Hydrocyanic acid
	ILLOXOL see P037
	INDOCI see P025
	Indomethacin see P025
	INSECTOPHENE see P050
	Isodrin see P060
P064	Isocyanic acid, methyl ester
	KILOSEB see P020
	KOP-THIODAN see P050
	KWIK-KIL see P108
	KWIKSAN see P092
	KUMADER see P001
	KYPFARIN see P001
	LEYTOSAN see P092
	LIQUIPHENE see P092
	MALIK see P050
	MAREVAN see P001
	MAR-FRIN see P001
	MARTIN'D MAR-FRIN see P001
	MAVERAN see P001
	MEGATOX see P005
P065	Mercury fulminate
	MERSOLITE see P092
	METACID 50 see P071
	METAFOS see P071
	METAPHOR see P071
	METAPHOS see P071
	METASOL 30 see P092
P066	Methomyl
P067	2-Methylaziridine
	METHYL-E 605 see P071
P068	Methyl hydrazine
	Methyl isocyanate see P064
P069	2-Methyllactonitrile
P070	2-Methyl-2-(methylthio)propionaldehyde-o-(methylcarbonyl) oxime
	METHYL NIRON see P042
P071	Methyl parathion
	METRON see P071
	MOLE DEATH see P108
	MOUSE-NOTS see P108
	MOUSE-RID see P108
	MOUSE-TOX see P108
	MUSCIMOL see P007

Table 2.2. (cont.)

Hazardous Waste No.	Substance[1]
P072	1-Naphthyl-2-thiourea
P073	Nickel carbonyl
P074	Nickel cyanide
P075	Nicotine and salts
P076	Nitric oxide
P077	p-Nitroaniline
P078	Nitrogen dioxide
P079	Nitrogen peroxide
P080	Nitrogen tetroxide
P081	Nitroglycerine (R)
P082	N-Nitrosodimethylamine
P083	N-Nitrosodiphenylamine
P084	N-Nitrosomethylvinylamine
	NYLMERATE see P092
	OCTALOX see P037
P085	Octamethylpyrophosphoramide
	OCTAN see P092
P086	Oleyl alcohol condensed with 2 moles ethylene oxide
	OMPA see P085
	OMPACIDE see P085
	OMPAX see P085
P087	Osmium tetroxide
P088	7-Oxabicyclo[2.2.1]heptane-2,3 dicarboxylic acid
	PANIVARFIN see P001
	PANORAM D-31 see P037
	PANTHERINE see P007
	PANWARFIN see P001
P089	Parathion
	PCP see P090
	PENNCAP-M see P071
	PENOXYL CARBON N see P048
P090	Pentachlorophenol
	Pentachlorophenate see P090
	PENTA-KILL see P090
	PENTASOL see P090
	PENWAR see P090
	PERMICIDE see P090
	PERMAGUARD see P090
	PERMATOX see P090
	PERMITE see P090
	PERTOX see P090
	PESTOX III see P085
	PHENMAD see P092
	PHENOTAN see P020
P091	Phenyl dichloroarsine
	Phenyl mercaptan see P014
P092	Phenylmercury acetate
P093	N-Phenylthiourea
	PHILIPS 1861 see P008
	PHIX see P092
P094	Phorate
P095	Phosgene
P096	Phosphine
P097	Phosphorothioic acid, 0,0-dimethyl ester, 0-ester with N,N-dimethyl benzene sulfonamide
	Phosphorothioic acid 0,0-dimethyl-0-(p-nitrophenyl) ester see P071
	PIED PIPER MOUSE SEED see P108
P098	Potassium cyanide
P099	Potassium silver cyanide
	PREMERGE see P020
P100	1,2-Propanediol
	Propargyl alcohol see P102
P101	Propionitrile
P102	2-Propyn-1-o1
	PROTHROMADIN See P001
	QUICKSAM see P092
	QUINTOX see P037
	RAT AND MICE BAIT see P001
	RAT-A-WAY see P001
	RAT-B-GON see P001
	RAT-O-CIDE #2 see P001
	RAT-GUARD see P001
	RAT-KILL see P001
	RAT-MIX see P001
	RATS-NO-MORE see P001
	RAT-OLA see P001
	RATOREX see P001
	RATTUNAL see P001
	RAT-TROL see P001
	RO-DETH see P001
	RO-DEX see P108
	ROSEX see P001
	ROUGH & READY MOUSE MIX see P001
	SANASEED see P108
	SANTOBRITE see P090
	SANTOPHEN see P090
	SANTOPHEN 20 see P090
	SCHRADAN see P085
P103	Selenourea
P104	Silver Cyanide
	SMITE see P105
	SPARIC see P020
	SPOR-KIL see P092
	SPRAY-TROL BRAND RODEN-TROL see P001
	SPURGE see P020
P105	Sodium azide
	Sodium coumadin see P001
P106	Sodium cyanide
	Sodium fluoroacetate see P056
	SODIUM WARFARIN see P001
	SOLFARIN see P001
	SOLFOBLACK BB see P048
	SOLFOBLACK SB see P048
P107	Strontium sulfide
P108	Strychnine and salts
	SUBTEX see P020
	SYSTAM see P085
	TAG FUNGICIDE see P092
	TEKWAISA see P071
	TEMIC see P070
	TEMIK see P070
	TERM-I-TROL see P090
P109	Tetraethyldithiopyrophosphate
P110	Tetraethyl lead
P111	Tetraethylpyrophosphate
P112	Tetranitromethane
	Tetraphosphoric acid, hexaethyl ester see P062
	TETROSULFUR BLACK PB see P048
	TETROSULPHUR PBR see P048
P113	Thallic oxide
	Thallium peroxide see P113
P114	Thallium selenite
P115	Thallium (I) sulfate
	THIFOR see P092
	THIMUL see P092
	THIODAN see P050
	THIOFOR see P050
	THIOMUL see P050
	THIONEX see P050
	THIOPHENIT see P071
P116	Thiosemicarbazide
	Thiosulfan tionel see P050
P117	Thiuram
	THOMPSON'S WOOD FIX see P090
	TIOVEL see P050
P118	Trichloromethanethiol
	TWIN LIGHT RAT AWAY see P001
	USAF RH-8 see P069
	USAF EK-4890 see P002
P119	Vanadic acid, ammonium salt
P120	Vanadium pentoxide
	VOFATOX see P071
	WANADU see P120
	WARCOUMIN see P001
	WARFARIN SODIUM see P001
	WARFICIDE see P001
	WOFOTOX see P072
	YANOCK see P057
	YASOKNOCK see P058
	ZIARNIK see P092
P121	Zinc cyanide
P122	Zinc phosphide (R,T)
	ZOOCOUMARIN see P001
	AAF see U005
U001	Acetaldehyde
U002	Acetone (I)
U003	Acetonitrile (I,T)
U004	Acetophenone
U005	2-Acetylaminoflourene
U006	Acetyl chloride (C,T)
U007	Acrylamide
	Acetylene tetrachloride see U209
	Acetylene trichloride see U228
U008	Acrylic acid (I)
U009	Acrylonitrile
	AEROTHENE TT see U226
	3-Amino-5-(p-acetamidophenyl)-1H-1,2,4-triazole, hydrate see U011
U010	6-Amino-1,1a,2,8,8a,8b-hexahydro-8-(hydroxymethyl)8-methoxy-5-methylcarbamate azirino(2',3':3,4) pyrrolo(1,2-a) indole-4, 7-dione (ester)
U011	Amitrole
U012	Aniline (I)
U013	Asbestos
U014	Auramine
U015	Azaserine
U016	Benz[c]acridine
U017	Benzal chloride
U018	Benz[a]anthracene
U019	Benzene
U020	Benzenesulfonyl chloride (C,R)
U021	Benzidine
	1,2-Benzisothiazolin-3-one, 1,1-dioxide see U202
	Benzo[a]anthracene see U018
U022	Benzo[a]pyrene
U023	Benzotrichloride (C,R,T)
U024	Bis(2-chloroethoxy)methane
U025	Bis(2-chloroethyl) ether
U026	N,N-Bis(2-chloroethyl)-2-naphthylamine
U027	Bis(2-chloroisopropyl) ether
U028	Bis(2-ethylhexyl) phthalate
U029	Bromomethane
U030	4-Bromophenyl phenyl ether
U031	n-Butyl alcohol (I)
U032	Calcium chromate
	Carbolic acid see U188
	Carbon tetrachloride see U211
U033	Carbonyl fluoride
U034	Chloral
U035	Chlorambucil
U036	Chlordane
U037	Chlorobenzene
U038	Chlorobenzilate
U039	p-Chloro-m-cresol
U040	Chlorodibromomethane
U041	1-Chloro-2,3-epoxypropane
	CHLOROETHENE NU see U226
U042	Chloroethyl vinyl ether
U043	Chloroethene
U044	Chloroform (I,T)
U045	Chloromethane (I,T)
U046	Chloromethyl methyl ether
U047	2-Chloronaphthalene
U048	2-Chlorophenol
U049	4-Chloro-o-toluidine hydrochloride
U050	Chrysene
	C.I. 23060 see U073
U051	Cresote
U052	Cresols
U053	Crotonaldehyde
U054	Cresylic acid
U055	Cumene
	Cyanomethane see U003
U056	Cyclohexane (I)
U057	Cyclohexanone (I)
U058	Cyclophosphamide
U059	Daunomycin
U060	DDD
U061	DDT
U062	Diallate
U063	Dibenz[a,h]anthracene
	Dibenzo[a,h]anthracene see U063
U064	Dibenzo[a,i]pyrene
U065	Dibromochloromethane
U066	1,2-Dibromo-3-chloropropane
U067	1,2-Dibromoethane
U068	Dibromomethane
U069	Di-n-butyl phthalate
U070	1,2-Dichlorobenzene
U071	1,3-Dichlorobenzene
U072	1,4-Dichlorobenzene
U073	3,3'-Dichlorobenzidine
U074	1,4-Dichloro-2-butene
	3,3'-Dichloro-4,4'-diaminobiphenyl see U073
U075	Dichlorodifluoromethane
U076	1,1-Dichloroethane
U077	1,2-Dichloroethane
U078	1,1-Dichloroethylene
U079	1,2-trans-dichloroethylene
U080	Dichloromethane
	Dichloromethylbenzene see U017
U081	2,4-Dichlorophenol
U082	2,6-Dichlorophenol
U083	1,2-Dichloropropane
U084	1,3-Dichloropropene
U085	Diepoxybutane (I,T)
U086	1,2-Diethylhydrazine
U087	0,0-Diethyl-S-methyl ester of phosphorodithioic acid
U088	Diethyl phthalate
U089	Diethylstilbestrol
U090	Dihydrosafrole
U091	3,3'-Dimethoxybenzidine
U092	Dimethylamine (I)
U093	p-Dimethylaminoazobenzene
U094	7,12-Dimethylbenz[a]anthracene
U095	3,3'-Dimethylbenzidine
U096	alpha,alpha-Dimethylbenzylhydroperoxide (R)
U097	Dimethylcarbamoyl chloride
U098	1,1-Dimethylhydrazine
U099	1,2-Dimethylhydrazine
U100	Dimethylnitrosoamine
U101	2,4-Dimethylphenol
U102	Dimethyl phthalate
U103	Dimethyl sulfate
U104	2,4-Dinitrophenol
U105	2,4-Dinitrotoluene
U106	2,6-Dinitrotoluene
U107	Di-n-octyl phthalate

Table 2.2. (cont.)

Hazardous Waste No.	Substance[1]
U108	1,4-Dioxane
U109	1,2-Diphenylhydrazine
U110	Dipropylamine (I)
U111	Di-n-propylnitrosamine
	EBDC see U114
	1,4-Epoxybutane see U213
U112	Ethyl acetate (I)
U113	Ethyl acrylate (I)
U114	Ethylenebisdithiocarbamate
U115	Ethylene oxide (I,T)
U116	Ethylene thiourea
U117	Ethyl ether (I,T)
U118	Ethylmethacrylate
U119	Ethyl methanesulfonate
	Ethylnitrile see U003
	Firemaster T23P see U235
U120	Fluoranthene
U121	Fluorotrichloromethane
U122	Formaldehyde
U123	Formic acid (C,T)
U124	Furan (I)
U125	Furfural (I)
U126	Glycidylaldehyde
U127	Hexachlorobenzene
U128	Hexachlorobutadiene
U129	Hexachlorocyclohexane
U130	Hexachlorocyclopentadiene
U131	Hexachloroethane
U132	Hexachlorophene
U133	Hydrazine (R,T)
U134	Hydrofluoric acid (C,T)
U135	Hydrogen sulfide
	Hydroxybenzene see U188
U136	Hydroxydimethyl arsine oxide
	4,4'-(Imidocarbonyl)bis(N,N-dimethyl)aniline see U014
U137	Indeno(1,2,3-cd)pyrene
U138	Iodomethane
U139	Iron Dextran
U140	Isobutyl alcohol
U141	Isosafrole
U142	Kepone
U143	Lasiocarpine
U144	Lead acetate
U145	Lead phosphate
U146	Lead subacetate
U147	Maleic anhydride
U148	Maleic hydrazide
U149	Malononitrile
	MEK Peroxide see U160
U150	Melphalan
U151	Mercury
U152	Methacrylonitrile
U153	Methanethiol
U154	Methanol
U155	Methapyrilene
	Methyl alcohol see U154
U156	Methyl chlorocarbonate
	Methyl chloroform see U226
U157	3-Methylcholanthrene
	Methyl chloroformate see U156
U158	4,4'-Methylene-bis-(2-chloroaniline)
U159	Methyl ethyl ketone (MEK) (I,T)
U160	Methyl ethyl ketone peroxide (R)
	Methyl iodide see U138
U161	Methyl isobutyl ketore
U162	Methyl methacrylate (R,T)
U163	N-Methyl-N'-nitro-N-nitrosoguanidine
U164	Methylthiouracil
	Mitomycin C see U010
U165	Naphthalene
U166	1,4-Naphthoquinone
U167	1-Naphthylamine
U168	2-Naphthylamine
U169	Nitrobenzene (I,T)
	Nitrobenzol see U169
U170	4-Nitrophenol
U171	2-Nitropropane (I)
U172	N-Nitrosodi-n-butylamine
U173	N-Nitrosodiethanolamine
U174	N-Nitrosodiethylamine
U175	N-Nitrosodi-n-propylamine
U176	N-Nitroso-n-ethylurea
U177	N-Nitroso-n-methylurea
U178	N-Nitroso-n-methylurethane
U179	N-Nitrosopiperidine
U180	N-Nitrosopyrrolidine
U181	5-Nitro-o-toluidine
U182	Paraldehyde
	PCNB see U185
U183	Pentachlorobenzene
U184	Pentachloroethane
U185	Pentachloronitrobenzene
U186	1,3-Pentadiene (I)
	Perc see U210
	Perchlorethylene see U210
U187	Phenacetin
U188	Phenol
U189	Phosphorous sulfide (R)
U190	Phthalic anhydride
U191	2-Picoline
U192	Pronamide
U193	1,3-Propane sultone
U194	n-Propylamine (I)
U196	Pyridine
U197	Quinones
U200	Reserpine
U201	Resorcinol
U202	Saccharin
U203	Safrole
U204	Selenious acid
U205	Selenium sulfide (R,T)
	Silvex see U233
U206	Streptozotocin
	2,4,5-T see U232
U207	1,2,4,5-Tetrachlorobenzene
U208	1,1,1,2-Tetrachloroethane
U209	1,1,2,2-Tetrachloroethane
U210	Tetrachloroethene
	Tetrachloroethylene see U210
U211	Tetrachloromethane
U212	2,3,4,6-Tetrachlorophenol
U213	Tetrahydrofuran (I)
U214	Thallium (I) acetate
U215	Thallium (I) carbonate
U216	Thallium (I) chloride
U217	Thallium (I) nitrate
U218	Thioacetamide
U219	Thiourea
U220	Toluene
U221	Toluenediamine
U222	o-Toluidine hydrochloride
U223	Toluene diisocyanate
U224	Toxaphene
	2,4,5-TP see U233
U225	Tribromomethane
U226	1,1,1-Trichloroethane
U227	1,1,2-Trichloroethane
U228	Trichloroethene
	Trichloroethylene see U228
U229	Trichlorofluoromethane
U230	2,4,5-Trichlorophenol
U231	2,4,6-Trichlorophenol
U232	2,4,5-Trichlorophenoxyacetic acid
U233	2,4,5-Trichlorophenoxypropionic acid alpha, alpha, alpha- Trichlorotoluene see U023
	TRI-CLENE see U228
U234	Trinitrobenzene (R,T)
U235	Tris(2,3-dibromopropyl) phosphate
U236	Trypan blue
U237	Uracil mustard
U238	Urethane
	Vinyl chloride see U043
	Vinylidene chloride see U078
U239	Xylene

[1] The Agency included those trade names of which it was aware; an omission of a trade name does not imply that it is not hazardous. The material is hazardous if it is listed under its generic name.

Basis for Listing Hazardous Wastes

EPA hazardous waste No.	Hazardous constituents for which listed
F001	tetrachloroethylene, methylene chloride trichlorоethylene, 1,1,1-trichloroethane chlorinated fluorocarbons, carbon tetrachloride
F002	tetrachloroethylene, methylene chloride, trichloroethylene, 1,1,1-trichloroethane, chlorobenzene, 1,1,2-trichloro-1,2,2-trifluoroethane, o-dichlorobenzene, trichlorofluoromethane
F003	N.A.
F004	cresols and cresylic acid, nitrobenzene
F005	methanol, toluene, methyl ethyl ketone, methyl isobutyl ketone, carbon disulfide, isobutanol, pyridine
F006	cadmium, chromium, nickel, cyanide (complexed)
F007	cyanide (salts)
F008	cyanide (salts)
F009	cyanide (salts)
F010	cyanide (salts)
F011	cyanide (salts)
F012	cyanide (complexed)
F013	cyanide (complexed)
F014	cyanide (complexed)
F015	cyanide (salts)
F016	cyanide (complexed)
K001	benzene, benz(a)anthracene, benzo(a)pyrene, chrysene, 4-nitrophenol, toluene, naphthalene phenol, 2-chlorophenol, 2,4-dimethyl phenol, 2,4,6-trichlorophenol, pentachlorophenol, 4,6-dinitro-o-cresol, tetrachlorophenol
K002	chromium, lead
K003	chromium, lead
K004	chromium
K005	chromium, lead
K006	chromium
K007	cyanide (complexed), chromium
K008	chromium
K009	chloroform, formaldehyde, methylene chloride, methyl chloride, paraldehyde, formic acid
K010	chloroform, formaldehyde, methylene chloride, methyl chloride, paraldehyde, formic acid, chloroacetaldehyde
K011	acrylonitrile, acetonitrile, hydrocyanic acid
K012	acrylonitrile, acetonitrile, acrolein, acrylamide
K013	hydrocyanic acid, acrylonitrile, acetonitrile
K014	acetonitrile, acrylamide
K015	benzyl chloride, chlorobenzene, toluene, benzotrichloride
K016	hexachlorobenzene, hexachlorobutadiene, carbon tetrachloride, hexachloroethane, perchloroethylene
K017	epichlorohydrin, chloroethers [bis(chloromethyl) ether and bis (2-chloroethyl) ethers], trichloropropane, dichloropropanols
K018	1,2-dichloroethane, trichloroethylene, hexachlorobutadiene, hexachlorobenzene
K019	ethylene dichloride, 1,1,1-trichloroethane, 1,1,2-trichloroethane, tetrachloroethanes (1,1,2,2-tetrachloroethane and 1,1,1,2-tetrachloroethane), trichloroethylene, tetrachloroethylene, carbon tetrachloride, chloroform, vinyl chloride, vinylidene chloride
K020	ethylene dichloride, 1,1,1-trichloroethane, 1,1,2-trichloroethane, tetrachloroethanes (1,1,2,2-tetrachloroethane and 1,1,1,2-tetrachloroethane), trichloroethylene, tetrachloroethylene, carbon tetrachloride, chloroform, vinyl chloride, vinylidene chloride
K021	antimony, carbon tetrachloride, chloroform
K022	phenol, tars (polycyclic aromatic hydrocarbons)
K023	phthalic anahydride, maleic anhydride
K024	phthalic anhydride, polynuclear tar-like materials, naphthoquinone
K025	meta-dinitrobenzene, 2,4-dinitrotoluene
K026	paraldehyde, pyridines, 2-picoline
K027	toulene diisocyanate, toluene-2,4-diamine, tars (benzidimidazapone)
K028	1,1,1-trichloroethane, vinyl chloride
K029	1,2.-dichloroethane, 1,1,1-trichloroethane, vinyl chloride, vinlyidene chloride, chloroform
K030	hexachlorobenzene, hexachlorobutadiene, hexachloroethane, 1,1,1,2-tetrachloroethane, 1,1,2,2-tetrachloroethane, ethylene dichloride
K031	arsenic
K032	hexachlorocyclopentadiene

Table 2.2. (cont.)

EPA hazardous waste No.	Hazardous constituents for which listed
K033	hexachlorocyclopentadiene
K034	hexachlorocyclopentadiene
K035	cresote, benz(a)anthracene, benz(b)fluoroanthene, benzo(a)pyrene
K036	toulene, phosphorodithioic and phosphorothioic acid esters
K037	toulene, phosphorodithioic and phosphorothioic acid esters
K038	phorate, formaldehyde, phosphorodithioic and phosphorothioic acid esters
K039	phosphorodithioic and phosphorothioic acid esters
K040	phorate, formaldehyde, phosphorodithioic and phosphorothioic acid esters
K041	toxaphene
K042	hexachlorobenzene; ortho-dichlorobenzene
K043	2,4-dichlorophenol, 2,6-dichlorophenol, 2,4,6-trichlorophenol
K044	N.A.
K045	N.A.
K046	lead
K047	N.A.
K048	chromium, lead
K049	chromium, lead
K050	chromium
K051	chromium, lead
K052	lead
K053	chromium
K054	chromium
K055	chromium, lead
K056	chromium, lead
K057	chromium, lead
K058	chromium, lead
K059	N.A.
K060	cyanide, naphthalene, phenolic compounds, arsenic
K061	chromium, lead, cadmium
K062	chromium, lead
K063	chromium, lead
K064	lead, cadmium
K065	lead, cadmium
K066	lead, cadmium
K067	lead, cadmium
K068	lead, cadmium
K069	chromium, lead, cadmium

N.A.—Waste is hazardous because it meets either the ignitability, corrosivity or reactivity characteristic.

Lead and compounds, N.O.S.[1]
Lead acetate
Lead phosphate
Lead subacetate
Maleic anhydride
Malononitrile
Melphalan
Mercury and compounds, N.O.S.
Methapyrilene
Methomyl
2-Methylaziridine
3-Methylcholanthrene
4,4'-Methylene-bis-(2-chloroaniline)
Methyl ethyl ketone (MEK)
Methyl hydrazine
2-Methyllactonitrile
Methyl methacrylate
Methyl methanesulfonate
2-Methyl-2-(methylthio)propionaldehyde-o-(methylcarbonyl) oxime
N-Methyl-N'-nitro-N-nitrosoguanidine
Methyl parathion
Methylthiouracil
Mustard gas
Naphthalene
1,4-Naphthoquinone
1-Naphthylamine
2-Naphthylamine
1-Naphthyl-2-thiourea
Nickel and compounds, N.O.S.
Nickel carbonyl
Nickel cyanide
Nicotine and salts
Nitric oxide
p-Nitroaniline
Nitrobenzene
Nitrogen dioxide
Nitrogen mustard and hydrochloride salt
Nitrogen mustard N-oxide and hydrochloride salt
Nitrogen peroxide
Nitrogen tetroxide
Nitroglycerine
4-Nitrophenol
4-Nitroquinoline-1-oxide
Nitrosamine, N.O.S.
N-Nitrosodi-N-butylamine
N-Nitrosodiethanolamine
N-Nitrosodiethylamine
N-Nitrosodimethylamine
N-Nitrosodiphenylamine
N-Nitrosodi-N-propylamine
N-Nitroso-N-ethylurea
N-Nitrosomethylethylamine
N-Nitroso-N-methylurea
N-Nitroso-N-methylurethane
N-Nitrosomethylvinylamine
N-Nitrosomorpholine
N-Nitrosonornicotine
N-Nitrosopiperidine
N-Nitrosopyrrolidine
N-Nitrososarcosine
5-Nitro-o-toluidine
Octamethylpyrophosphoramide
Oleyl alcohol condensed with 2 moles ethylene oxide
Osmium tetroxide
7-Oxabicyclo[2.2.1]heptane-2,3-dicarboxylic acid
Parathion
Pentachlorobenzene
Pentachloroethane
Pentachloronitrobenzene (PCNB)
Pentacholorophenol
Phenacetin
Phenol
Phenyl dichloroarsine
Phenylmercury acetate
N-Phenylthiourea
Phosgene
Phosphine
Phosphorothioic acid, O,O-dimethyl ester, O-ester with N,N-dimethyl benzene sulfonamide
Phthalic acid esters,
Phthalic anhydride
Polychlorinated biphenyl, N.O.S.
Potassium cyanide
Potassium silver cyanide
Pronamide
1,2-Propanediol
1,3-Propane sultone
Propionitrile
Propylthiouracil
2-Propyn-1-ol
Pryidine
Reserpine
Saccharin
Safrole
Selenious acid
Selenium and compounds, N.O.S.
Selenium sulfide
Selenourea
Silver and compounds, N.O.S.
Silver cyanide
Sodium cyanide
Streptozotocin
Strontium sulfide
Strychnine and salts
1,2,4,5-Tetrachlorobenzene
2,3,7,8-Tetrachlorodibenzo-p-dioxin (TCDD)
Tetrachloroethane, N.O.S.
1,1,1,2-Tetrachloroethane
1,1,2,2-Tetrachloroethane
Tetrachloroethene (Tetrachloroethylene)
Tetrachloromethane
2,3,4,6-Tetrachlorophenol
Tetraethyldithiopyrophosphate
Tetraethyl lead
Tetraethylpyrophosphate
Thallium and compounds, N.O.S.
Thallic oxide
Thallium (I) acetate
Thallium (I) carbonate
Thallium (I) chloride
Thallium (I) nitrate
Thallium selenite
Thallium (I) sulfate
Thioacetamide
Thiosemicarbazide
Thiourea
Thiuram
Toluene
Toluene diamine
o-Toluidine hydrochloride
Tolylene diisocyanate
Toxaphene
Tribromomethane
1,2,4-Trichlorobenzene
1,1,1-Trichloroethane
1,1,2-Trichloroethane
Trichloroethene (Trichloroethylene)
Trichloromethanethiol
2,4,5-Trichlorophenol
2,4,6-Trichlorophenol
2,4,5-Trichlorophenoxyacetic acid (2,4,5-T)
2,4,5-Trichlorophenoxypropionic acid (2,4,5-TP) (Silvex)
Trichloropropane, N.O.S.
1,2,3-Trichloropropane
0,0,0-Triethyl phosphorothioate
Trinitrobenzene
Tris(1-azridinyl)phosphine sulfide
Tris(2,3-dibromopropyl) phosphate
Trypan blue
Uracil mustard
Urethane
Vanadic acid, ammonium salt
Vanadium pentoxide (dust)
Vinyl chloride
Vinylidene chloride
Zinc cyanide
Zinc phosphide

[1] The abbreviation N.O.S. signifies those members of the general class "not otherwise specified" by name in this listing.

7. Lead; lead compounds
8. Antimony; antimony compounds
9. Phenols; phenol compounds
10. Cyanides, organic and inorganic
11. Isocyanates
12. Organic-halogens compounds, excluding inert polymeric materials and other substances referred to in this list or covered by other Directives concerning the disposal of toxic or dangerous waste
13. Chlorinated solvents
14. Organic solvents
15. Biocides and phyto-pharmaceutical substances
16. Tarry materials from refining and tar residues from distilling
17. Pharmaceutical compounds
18. Peroxides, chlorates, perchlorates and azides
19. Ethers
20. Chemical laboratory materials, not identifiable and/or new, whose effects on the environment are not known
21. Asbestos (dust and fibres)
22. Selenium; selenium compounds
23. Tellurium; tellurium compounds
24. Aromatic polycyclic compounds (with carcinogenic effects)
25. Metals carbonyls
26. Soluble copper compounds
27. Acids and/or basic substances used in surface treatment and finishing of metals

It is noticed that no instructions and guidelines are given to judge whether the chemical containing compounds are present in a hazardous concentration or modification.

The working out of procedures for testing those materials are an obligation for the individual Member States.

Many of the Member States follow the EEC directive rather closely. (For example the United Kingdom and France). Others, especially those with a more extensive treatment system, have a more detailed and specified waste categorization system (see Section 2.3.2 and 2.3.3).

2.3 Grouping According to Types of Generators

The above mentioned (Section 2.1 and 2.2) examples on classification systems might be important as guidelines for defining whether an actual waste product should be regarded as hazardous wastes or not.

However, the systems are not quite so useful as practical planning tools for designing of a hazardous waste management system or for the control and operation of such systems and treatment facilities.

Moreover, it might be difficult for many waste generators, having no special chemical knowledge, to evaluate whether their waste products should be regarded as hazardous wastes.

A system based more directly on generator or process types might overcome this problem.

Examples of this way of categorizing hazardous wastes are found in California, West Germany and in Denmark.

2.3.1 Categorization in California (Example)

The US-EPA classification systems are followed by many of the individual states in the U.S.A. However, other classifications have also been proposed.

For example California has investigated the possibility of using as a supplement the "Standard Industrial Classification" (SIC) which groups industrie and business by particular activities.

Table 2.3. Summary of waste generation according to investigation in California (two months: September 1979 + May 1980)

SIC-Code	Activity	Tonnes Waste	% of total
00	Generators with UCD augmented codes	6,087	2.69
07	Agricultural Services	625	0.28
13	Oil & Gas extraction (1311 & 138)	44,194	19.52
20	Food & kindred products	815	0.36
24	Lumber & wood products	165	0.07
26	Paper & allied products	1,707	0.75
27	Printing & publishing	304	0.13
28	Chemicals & allied products	42,808	18.90
29	Petroleum & coal products	63,507	28.04
30	Rubber & misc. plastics products	191	0.08
31	Leather tanning & finishing (3111)	1,949	0.86
32	Stone, clay & glass products	3,625	1.60
33	Primary metal industries	4,467	1.97
34	Fabricated metal products	10,711	4.73
35	Machinery, except electrical	3,060	1.35
36	Electric & electronic equipment	9,250	4.08
37	Transportation equipment	7,932	3.50
38	Instruments & related products	1,290	0.57
40	Railroad transportation	578	0.26
42	Trucking & sanitary services[a]	2,049	0.90
44	Water transportation	263	0.12
45	Air transportation	622	0.27
46	Pipe lines, except natural gas	625	0.28
49	Electric, gas & sanitary services	11,545	5.10
51	Wholesale trade-nondurable goods	362	0.16
73	Business services	826	0.36
75	Auto repair, services & garages	307	0.14
96	Dept. of Agriculture	327	0.14
**	Other minor 0100–1999 SIC groups	200	0.09
**	Other minor 2000–3999 SIC groups	149	0.07
**	Other minor 4000–9700 SIC Groups	945	0.42
**	Generators with unidentifiable SIC	4,966	2.19
	Total =	226,451	100.00

a Two month total/Sept. 1979 + May 1980).

Table 2.3 shows the result of a two months period investigation carried out by the University of California, Davis, of waste generation by SIC-code companies [5].

2.3.2 Classification of Hazardous Wastes in West Germany

According to a "Verordnung" from 1977 [6], 34 categories of hazardous wastes are defined and for each of them the generator type is presented also.

Naturally, several waste categories may be generated by the same generator type as shown in Table 2.5, which is a translation of the appendix to the "Verordnung".

It is observed that the wastes are characterized by type (sludge, emulsion, solvents, etc.), as well as indications of the "active" component. Hospital wastes are included.

2.3.3 Classification of Hazardous Wastes in Denmark

In Denmark the classification of hazardous wastes is found in the "Notification on Chemical Wastes" of July 3rd 1980 [7]. In the appendix to this notification 50 categories of hazardous wastes are listed (Table 2.5) and for each of the waste categories so-called "waste cards" have been worked out presenting more details including the origin of the waste (process).

In October 1980 a guide book for generators and controlling authorities was worked out by the Danish EPA. This guide book was revised in November 1983 [8].

Tables in this book inform about the waste categories which might be expected from various process types. 272 industry types are now included.

A shortened form of the waste category list is presented below (Table 2.6).

2.4 Classification of Wastes According to Common Treatment Practice/Treatability

Characterization of hazardous (chemical) wastes might often be very complex if one uses the individual chemical constituent as a basis.

A step to simplify the way of categorizing waste is to use the treatment technology which can be applied, see Table 2.7 [2].

When planning management systems for hazardous wastes, it is necessary to divide the wastes into groups for which waste management methods are practicable.

The following simplified division can be useful:

- Wastes which can be recovered/recycled.
- Wastes which need (pre-)treatment before disposal.
- Wastes which can be disposed of without (pre-)treatment.

It is obvious that some waste categories can belong to more than one of the above divisions. In the following the different groups are described briefly.

2.4.1 Wastes Which can be Recovered/Recycled

Recovery and recycling of wastes on site is considered here as a part of the manufacturing process. Only commercial (off-site) recovery/recycling is regarded as waste treatment alternatives.

Table 2.4. Classification of Hazardous Waste in West Germany

Waste Type	Characteristics	Waste Code No.	Origin
Liming sludge		14401	Leather production
Tanning sludge		14402	Tanneries
Kiln excavations/ from metal process	Containing arsenic, lead, cadmium, cyanide, mercury	31108	Non-ferrous metal production Blast-furnace works
Light metal slag	Aluminium containing	31205	Light metal producing industry
	Magnesium containing	31206	Light metal finishing industry (mills, foundries, melting works).
Salt slag	Aluminium containing	31211	Light metal melting works and parts of such
Used filter and absorption mass (diatomite, activated earth, activated coal)	Containing halogenous organic solvents or mixtures of solvents (waste code No. 55201 to 55213, 55220).		Chemical industries
Asbestos dust		31437	Asbestos mining
Beryllium containing dust		35318	Beryllium production Beryllium processing industry
Electroplating sludge	Cyanide	51101	Electroplating factory and parts of such (surface finishing, surface treatment)
	Chrome +6	51102	
	Cadmium	51106	
Arsenic lime		51513	Non-ferrous metal production
Hardening salts	Cyanide	51533	Hardening salts production and parts of such
	Nitrate Nitrite	51534	
Acids acid mixtures pickling (acids)		52102	Chemical industry, surface metal treatment factories, and parts of such Surface metal treatment Surface metal finishing factories and parts of such (Surface metal finishing)
Lye Lye mixtures Pickling (basic)		52402	Same as above
Concentrate, baths	Sulfur containing	52711	Metal and synthetic Surface treatment factories and parts of such (Metal and synthetic surface finishing factories and parts of such)
	Chrome +6 containing	52712	
	Cyanide containing	52713	
	Metal salts containing	52716	
Semi-concentrate	Chrome +6	52717	Same as above
	Cyanide	52718	
	Metal salts	52719	

Table 2.4. (cont.)

Waste Type	Characteristics	Waste Code No.	Origin
Waste from herbicide and pesticides (insecticides, herbicides, fungicides, rodenticide, acaricide, nematicide, molluskizide)		53104	Chemical industry production of vegetation protection and pest control agents
Waste from the production of pharmaceutical products		53502	Industrial production
Synthetic cooling agent and lubricants		54401	Metal processing industry Metal processing industry
Drilling and grinding oil emulsion, emulsion mixtures	Mineral oil containing	54801	Waste oil refinery Lubricating oil refinery
Fullers earth	Mineral oil containing	54802	Same as above
Sludge from mineral oil refinery		54803	Mineral oil refinery
Sludge from the treatment of crude oil and coal processing	Phenol containing	54903	Petro-chemical industry
	Mercaptan containing	54904	Cokeries
	Cyanide containing	54923	Gas works
Halogenous organic solvents	Containing:		
	Ethyl chloride	55201	
	Chlorobenzene	55202	
	Chloroform	55203	
	Dichlorophenol	55204	
	Cooling agent	55205	Chemical industry
	Methyl chloride	55206	Production of dyes, varnish, varnish color, coating agents
	Monochloracetic phenol	55207	Gas works Production of cooling agents
	Paraffin (chlorinated)	55208	Cokeries Plastic production industry
	Perchloro ethylene	55209	Plastic finishing factories
	PVC-softener	52210	Production of solvents Metal works industry
	Carbon tetra chloride (tetra)	52211	Surface treatment industry Petrochemical industry
	Trichlorethane Trichlorethylene (tri)	55212	Pharmaceutical industry
Halogenous organic solvents	Containing: Halogenous organic solvents (waste code No. 55201 to 55213)	55220	
Halogen free organic solvents	Containing:		
	Acetone	55301	
	Ethyl acetate	55302	
	Ethylene glycol	55303	
	Ethyl glycol	55304	

Table 2.4. (cont.)

Waste Type	Characteristics	Waste Code No.	Origin
	Ethyl phenol	55305	
	Benzene	55306	
	Butyl acetate	55307	
	Cyclohexanone	55308	
	Decahydronaphthalen	55309	
	Diethyl ether	55310	
	Dimethyl formamide	55311	Chemical industry Production of dye & varnish, varnish color, coating agents
	Dimethyl sulfide	55312	
	Dimethyl sulfoxide	55313	Gas works Cooling agent production
	Dioxane	55314	Cokeries
	Methyl alcohol	55315	Plastic industry Plastic finishing
	Methyl acetate	55316	Solvent production
	Methyl ethyl ketone	55317	Metal finishing industry Surface treatment industry
	Methyl isobutyl ketone	55318	Petrochemical industry
	Methyl phenol	55319	Pharmaceutical industry
	Pyridine	55320	
	Carbon disulphide	55321	
	Tetrahydrofurane	55322	
	Tetrahydronaphthaline	55323	
	Terpentine	55324	
	Toluene	55325	
	Benzene (cleansing)	55326	
	Xylene	55327	
Halogen free solvents	Containing: Halogen free organic solvents (waste code No. 55301 vis 55327)	55370	
Solvents containing sludge	Containing: Halogen containing organic solvents or other solvent mixtures (waste code No. 55201 vis. 55213, 55220)	55401	Chemical industry Production of dyes & varnish, varnish color, coating agents, Plastic manuf. industry Plastic finishing industry Distillation of solvents
	Containing: Halogen free organic solvents or solvent mixtures (waste code No. 55301 vis . 55327, 55370)	55402	Refining of solvents Metal manuf. industry Metal processing industry Surface treatment industry
Varnish sludge Paint sludge	Containing: Halogenous or halogen free organic solvents	55503	

Table 2.4. (cont.)

Waste Type	Characteristics	Waste Code No.	Origin
	or solvent mixtures (waste code No. 55201 vis. 55213, 55220 or 55327, 55370) and/or heavy metal		
Paints		5507	Industrial production of paints and coating agents
Coating agents	Containing: Same as above	55508	Industrial production of paints and coating agents
Cautchouch solvents		55704	Rubber producing industry Rubber processing industry
Catalysts	Heavy metal containing	59507	Chemical industry Mineral oil processing
Polychlorinated Biphenyles and terphenyles		59901	Industrial production and processing
Hospital wastes	Containing: Infectious matter which can transfer diseases	97101	Hospitals and clinics with one or more of the following sections: Bloodbank Surgical section Dialysis section Obstetric section Gynaecological section Infectious section Microbiological section Pathological section Virological section

Table 2.5. List of chemical wastes, from the Nodification, Danish EPA 1980 [7]

Animal and Vegetable Fats
Soap and detergent wastes.
Organic Halogen Containing Compounds
Solvents, containing halogen.
Solvents, containing halogen, and mixed with flameable solvents.
Solvents, halogen free, mixed with halogen and/or sulphur containing compounds.
PCB — including PCT wastes.
Liquid residues from organic syntheses, containing toxic compounds and organic halogen and/or sulphur.
Liquid residues from organic syntheses containing organic halogen and/or sulphur.
PVC containing sludge from plastic covering.
Solid residues from recovery of halogen containing solvents.
Solid residues from organic syntheses containing toxics and halogen and/or sulphur.
Solid residues from organic syntheses containing halogen and/or sulphur.
Organic Halogen Free Compounds
Organic solvents containing aromatic solvents.
Organic solvents containing aromatic solvents or compounds containing halogen or sulphur.
Paint residues containing solvents.
Paint residues not containing solvents.

Table 2.5. (cont.)

Tar and rust protecting oils.
Alcohol/water mixtures.
Distillation residues from mixtures of acetone and unhardened polyester.
Metal organic compounds.
Liquid organic residues from distillation, not containing halogen or sulphur.
Formaldehyde solutions.
Phenol- and formaldehyde emulsions.
Di-isocyanates.
Anti-freeze liquids.
Latex sludge.
Acid sludges.
Waste from adhesives.
Solid residues from organic syntheses.
Grinding dust.
Beetroot mush.
Inorganic Compounds
Acid solutions containing chromium.
Acid solutions containing nitric acids.
Acid solutions containing fluoric acids and/or salts of fluoric acid.
Acid solutions containing for example hydrochloric acid, sulphuric acid, or phosphoric acid.
Photographic developing baths.
Photographic baths containing chromium.
Fixation baths.
Alkaline solutions without cyanides.
Alkaline solutions, containing cyanides.
Alkaline solutions, which might contain metals.
Metalhydroxide — or metaloxide sludge.
Sludge produced from fume washings and fume filter dust.
Paint wastes.
Sludge from wood preservation processes.
Hardening salts.
Mercury wastes.
Other Waste Types
Rags contaminated with solvents.
Wastes from production and trade with pesticides.
Medical wastes.
Laboratory wastes.
Phenol containing glass-wool and mineral-wool.

Some of the most important categories of wastes are:

- Liquid organic chemicals and solvents (recovery process for example: distillation).
- Heavy metals containing wastes (recovery processes are for example: reduction, precipition, electrolysis).

In principle, a lot of chemicals from hazardous wastes are technically recoverable, but the price competition from the product fabricated normally is prohibitive in most areas at present [9].

It is often technically possible to recover marketable chemical from hazardous wastes but consideration of the economics of such processes usually show them to be unattractive.

Table 2.6. Listing of industries generating hazardous wastes [8]

Production and processing, including surface treatment of iron, steel and metals.
Glass production and processing, mineral wool industries
Plants for refining of mineral oil
Gas production
Production of asphalt, roof felt and tar
Processing or destruction of waste oil and other wastes from refinery products
Petrochemical industry
Pharmaceutical industry
Plants for production of pesticides
Detergent producing industry
Paint, lacquer and wood protection production
Photographical industry and development companies
Hazardous waste treatment plants
Plastic processing
Glue producing industry
Wood protection enterprises
Printing and graphical companies
Wood cutting, production of plywood or fiberboards, and furniture production
Textile production
Washing, dry cleaning
Treatment, processing and distribution of grain and animal foodstuffs
Sugar production
Food processing and animal food production, tanneries, leather production and shoe production
Power plants, heat plants
Means of transport, transport facilities
Hospitals
Various service companies (for example tank cleaning)

2.4.2 Wastes Which Need (Pre-)Treatment Before Disposal/Wastes Which can be Disposed of Without (Pre-)Treatment

The available commercial operating all-round treatment facilities are frequently based on:

- Thermal treatment
  - Rotary kiln incinerators
  - Liquid injection incinerators
- Chemical treatment
  - Neutralization
  - Detoxification
  - Precipitation
  - Ion exchange
- Physical treatment (separation)
  - Filtration
  - Flocculation
  - Sedimentation
  - Centrifugation

Table 2.7. Application matrix of chemical, physical, and biological treatment process to wastes (see ref. 2)

Process	Relative Cost Factor	Waste: Petrol Refin	Petrochemicals	Pharmaceutical	Munitions	Pant & Ink Ind.	Painting Waste	PVC Production	Leather Tan.	Food Processing	Metal Fin. Liquid	M. F. Sludges	Pickle Liquor	Metallurg. Slag	Waste Oil	Chlor-Alkali	Chemical Groups	Carbonyls	Fluoride	Organic Acids	Inorganic Acids	Mercury	Cadmium	PNA's	Amines	Aromatics	Aliphatics	Ethers	Phthalates	Phenolics	Chlor. HC's	Cyanide	Sulfides	Pesticides	PCB's	Specific Chemicals: PCP	2378-Dioxin
carbon absorption	Low	1	3	1	1	3	3	3			1					1						2	3	2	2	2	2	2	2	1	1		3	1	1	2	2
oil sep techniques	Low	1	2					2							1																						
wet air oxidation	High		3	2	3			3		3															3	3				3	3			3			
neutralization	Low				1						1	1			1				2	1												1					
chemical fixation/encaps.	Med	3					2				1	1		2							2		2							3	3		2				
filtration	Low	1		1		3	2			1		1			1	1																		2			
ox-reduction	Med			3							1					1				2										1	2	1	1	3			
chem-precip./settling	Low	1		1		1	2		2		1	2	1		1	1							1			1		1	1			1	1				
evaporation	Low	2	3		1			3			1	1	2										3	2	2	1	2										
acid treatment	Low											3		1	1																						
distillation	Med	1	2	1		2		2							2					2		2		1	1	1	1	1		1	1		2				
biotreatment	Low	1		1		1			1	1						3				1					1	1	1	1	1	1			1	2			
calcination	High	2					1				3	3																				3					
flotation	Low					1					3							3	2																		
hydrolysis	Low																																	3			
liq-liq extraction	Low to High	1												2	3					2				3		3	3	2	3	2	1				3	2	2
ozonation	Med			2							2							3						3						2	3	2	2	3			
chemical dechlorination	High																														3			3	3		3
UV dechlorination	High																														3			2			2
catalysis	High																	2												3	3	3		3			
electrodialysis	Low									2	3																										
reverse osmosis	Med					3					2												2			3	2	3	2	3			2	2			
dissolution	Low to Med									2		3																									
ion exchange	Med	2			2						1		2			1					2		2							3	3	2	3	3	2		
microwave	High					3																									3			3			
chlorinalysis	High																														2			2			
electrolysis	High									3	3	3			2								3														
ultrafiltration	Med					2	2								2																						
liquid ion exchange	Low to High									3	3	3	3										3									3					
resin absorption	Low																							2	2	2	2	2	2	3	3						

Use Codes: 1. Full Scale Common Use, 2. Moderate Application, 3. Research Stage

— Disposal
  Direct disposal on landfill
  Disposal on landfill after pretreatment
  Discharge of waste water
  Discharge to the atmosphere

A practical way of classifying hazardous wastes, especially suited for planning purposes is obtained when combining the treatment possibilities with various parameters such as:

- Chemical properties
  - inorganic chemicals
    - acids
    - alkalies
    - cyanide content
    - dichromate content
    - heavy metal content
  - organic chemicals
    - halogen and/or sulfur content
- Physical properties
  - liquid or sludge
    - pumpable/non-pumpable
  - solid
  - heat value
  - flash point
- Physiological properties
  - toxic
- Environmental protection
  - characteristics of waste after processing (properties of end products).

Classification of wastes used by commercial operating integrated facilities is often based on above mentioned principles (combination of treatment possibilities and chemical/physical characteristics).

Table 2.8 presents items of the categorization of wastes used by Gesellschaft zur Beseitigung von Sondermüll in Bayern (GSB), West Germany. Table 2.9 presents the categorization of wastes used by Kommunekemi, Denmark.

2.5 Waste Classification Useful for Planning of Hazardous Waste Management Systems

As mentioned in the foregoing section, hazardous waste treatment is carried out in a rather limited range of commercialized processes. For planning and design purposes regarding hazardous waste management a practical waste characterization oriented towards the treatment process is required rather than a very detailed waste classification which is normally relevant.

Some of the needed planning parameters are often not catered for in many waste assessments if it is not possible to obtain information regarding such data. Practical experience is therefore required.

Table 2.8. Examples of waste categorization at GSB, Germany

Type	Examples
Wastes which can be disposed of	Slag and ashes Sludge from metal finishing industry Sludge from painting industry, without solvents
Wastes for incineration	Solvents up to 2% halogen and sulphur Organic liquid pumpable wastes up to 2% halogen and sulphur Organic solid wastes up to 2% halogen and sulphur Compounds with halogen, sulphur or phosphor from 2 to 10% Compounds with halogen, sulphur or phosphor more than 10% Pesticides, fungicides, insecticides
Organic loaded waste water and sludge, oil containing, can be separated, not containing solvents, pH 5–9, not containing heavy metals, cyanide, nitrite, chromate, phenol, salts, etc.	Sludges with various solid contents, (up to 60%), oil emulsions, etc.
Inorganic contaminated waste water and sludge (solids up to 60%), pumpable, acids, not containing organic wastes, heavy cyanide, nitrite, chromate, phenol, salts, etc.	Acids, bases, inorganic sludges
Waste water and sludges containing heavy metals, cyanide, nitrite, chromate, phenol, salts, etc.	
Solvents which can be regenerated	
Other compounds	

Table 2.9. Waste categorization at Kommunekemi, Denmark

Group	Type	Examples	Treatment Process (simplified)
A	Mineral Oil Waste	Pumpable wastes containing mineral oil, for example: lubricating oil, hydraulic oil, heat transmission oil, drilling oil, cutting oil, fuel oil, synthetic oil, oil from intercepting traps	Recovery & incineration

Table 2.9. (cont.)

Group	Type	Examples	Treatment Process (simplified)
B	Halogenous Solvent Waste	Pumpable wastes containing halogenated solvents such as trichlor-ethylene, perchlor-ethylene, tetrachlor-ethylene, chloroform, chlorotene, genklene, freon Pumpable halogen or sulphur containing organic chemical wastes	Incineration
C	Solvent Waste	Pumpable wastes containing nonhalogenous solvents, such as gasoline, turpentine, solvent, xylene, ethyl alcohol, propyl alcohol, thinners, octane, MIBK, MEK, ethylacetate, butylacetate	Incineration
H	Organic Chemical Wastes, Halogen and Sulphur Free	Used paint, paint sludge, distillation residues, organic chemical bi-products, tar, deep frying oil, organic acids and their salts, glue waste, used developer, alkaline, cyanide free washing baths, bitumen, grease, solid fuel oil, soap wastes	Incineration
T	Pesticide Containing Wastes	Insecticides, fungicides, weed killers, rat poison Seed grain containing mercury	Special sorting + Incineration or special disposal
X	Inorganic Chemical Wastes	Used pickling acid, electroplating baths, metal hydroxide sludge, wastes from regeneration of ion exchange, contaminated sulfuric acid, hydrochloric acid, nitric acid, hydrofluoric acid, block metal acid, soda lye, ammonia water, alkaline cyanide containing degreasing baths, hardening salts, salts, caustic soda	Neutralisation, detoxification and precipitation
Z	Other Wastes	Non-pumpable, halogen or sulfur containing organic-chemical wastes	Incineration
		Isocyanate (MDI and TDI)-containing wastes	Pre-treatment + incineration
		Oil polluted soil	Incineration
		Pharmaceutical wastes	Incineration
		Wastes from laboratories, in small containers	Incineration after sorting
		Spray cans	Incineration
		Chemical wastes from households	Incineration after sorting
		Used mercury batteries	Special disposal
		Used drums and used small containers packed in drums, sacks or the like	Incineration

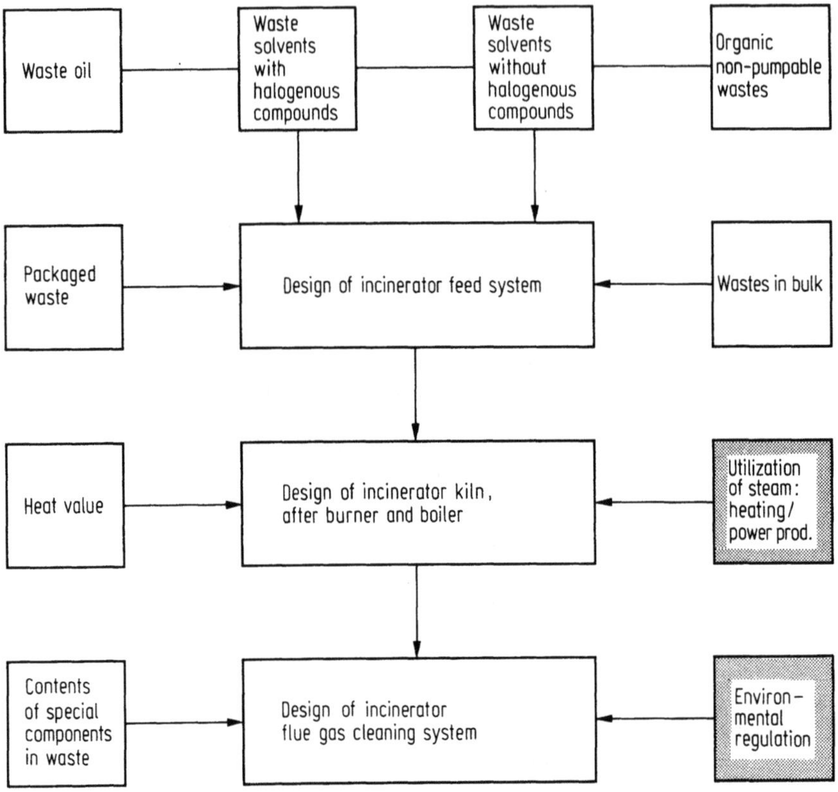

Fig. 2.2. Planning a hazardous waste incinerator. Simplified example of waste categorization applicable. ("Gray" boxes indicate necessary other planning parameters.)

Fig. 2.2 presents some of the main waste data necessary for planning an incinerator train and for calculating the performance parameters.

Data for the heat value of the waste categories are necessary for estimating the kiln and after burner capacity, or the proportion of wastes which can be loaded into the kiln and after burner. Also, heat value data forms the basis for estimating the possible needed supporting fuel.

Information about the content of halogen, sulphur and heavy metals are used for designing the flue gas cleaning system.

The parameters needed for designing an inorganic treatment facility for plating waste are presented in Fig. 2.3. The design is related to a plant based upon detoxification of cyanide and chromate, followed by precipitation of metals as hydroxides.

The parameters indicated on the figure is therefore a prerequisite for dimensioning the holding tanks reaction vessels, filters, centrifuges, pumps etc.

The waste streams scenario at the Danish centralized hazardous waste plant, Kommunekemi, is presented in Fig. 2.4 (rounded values) [10].

The width of the waste streams symbolize the waste quantity in percent (W/W) of the incoming waste amount.

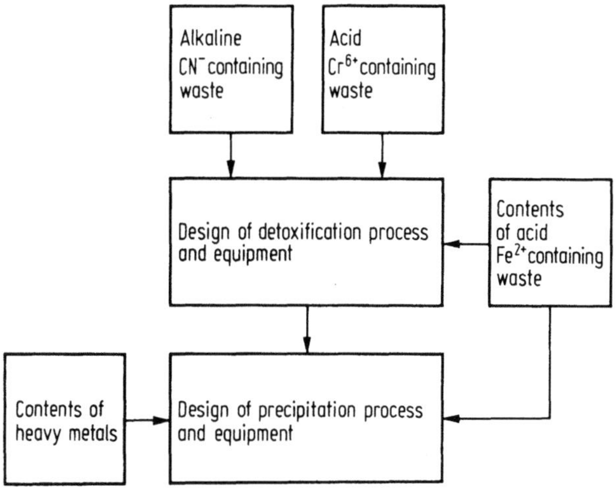

Fig. 2.3. Planning an inorganic waste treatment plant. Simplified example of waste categorization applicable

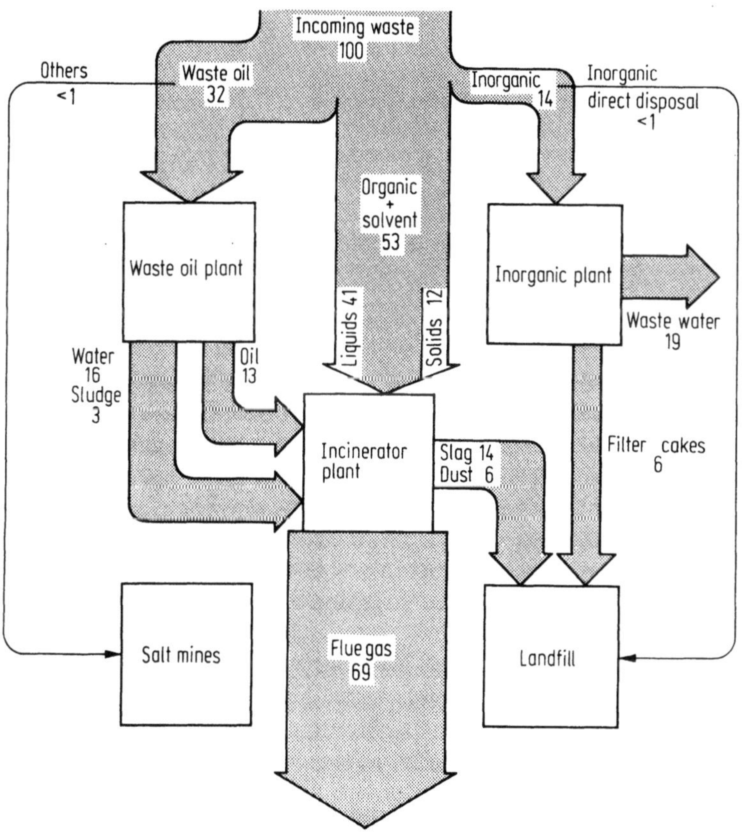

Fig. 2.4. Waste streams at Kommunekemi. The Danish Hazardous Waste Plant. Quantities in per cent of incoming waste amount

Observe that the largest position of the waste is destinated for incineration, including waste processed in the waste oil plant.

For further illustration the waste quantities received at Kommunekemi in 1985 are presented in Table 2.10.

Table 2.10. Waste received at Kommunekemi, Denmark, 1983

Waste Group	Waste Type	Quantity (metric tons)
A	Mineral oil	21555
B	Halogenous Solvents	1296
C	Solvents	4147
H	Organic, Halogen and Sulphur Free	25392
T	Pesticides	247
X	Inorganic	7276
Z	Other	5439
	Total	65352

3 Examples of Chemical Waste Types

3.1 Recoverable Wastes

The main hazardous waste types which can be treated in recovery plants are either contaminated organic solvents or metal containing waste water. Also, waste oil is often recovered.

Contaminated solvents

Raw materials for recovery of solvents are often contaminated halogen containing solvents (trichlorethane, trichlorethylene, perchlor-ethylene) from degreasing baths and cleansing solvents from color and paint production.

The composition of wastes from paint production might be:

	%	b.p. °C
Toluene, xylene, higher aromatics		
Esters, alcohols, alcohols and water	80	70–200
Pigments and binders	20	

By stripping and distillation it is possible to recover 70–80% of the solvent mixture which could then be used in the paint production.

By products from the process is water with water-soluble solvents and sludge containing organics, water and ash.

Metal Containing Wastes

Metal from metal containing waste water with not too many different metals can sometimes be economically recovered by physical-chemical methods. As an example waste water from the electroplating industry can be mentioned. Metals can be recovered by means of ion exchange or by formation of complexes followed by solvent extraction.

The photographic industry gives rise to waste waters containing silver and traditionally, this is recovered as the free element by a combination of electrolysis and ion exchange or one of the said methods alone.

Research on separation of metals from complex mixtures of wastes is carried out at several companies or research centers.

Waste Oil

Waste oil containing water and sludge and oil originating from various oil types can be recovered by using sedimentation, stripping of solvents and "polishing" by filtering, centrifugation or other separation techniques.

Transformer Oil containing PCB.

Transformer oil with PCB is no longer allowed in most countries of the world.

Special processes have been developed for reclaiming such products by destroying PCB.

The oil without PCB can be used as transformer oil again when additives have been added.

In the USA mobile units for reclaiming of PCB containing transformer oil are now in operation.

3.2 Burnable Waste Types

Practically all hazardous wastes except inorganic hazardous wastes and explosives can be regarded as combustible. However, the destruction and removal efficiency (DRE) for principal organic hazardous constituents (POHC's) is a function of the temperature and the residence time.

The DRE for an incinerator is calculated from the equation:

$$\mathrm{DRE} = \frac{1\text{-}W_{out}}{W_{in}} \times 100 \text{ per cent}$$

Where: W_{in} = the mass feed rate of 1 POHC in the waste stream going into the incinerator

W_{out} = the mass emission rate of the same POHC in the exhaust prior to release to the atmosphere.

The DRE for a POHC is a function of the combustion temperature and the residence time.

The following equation expresses the time (t) required for 99,99% destruction of a specific compound:

$$\ln t = \ln \frac{9.21}{A} + \frac{E}{RT}$$

Where: A = Arrhenius pre-exponent frequency factor (S^{-1})
E = Energy af activation (J/kg-mole)
R = Universal gas constant (8314 J/kg-mole-°K)
T = Absolute temperature (°K).

If the A and E values are known the required retention time can be calculated for a selected temperature (and visa versa).

The retention time-temperature relationship for some organic compounds mentioned in Section 2.1.2 is shown in Fig. 3.1.

Some organic compounds are shown in Table 3.1 in order of destructability [11].

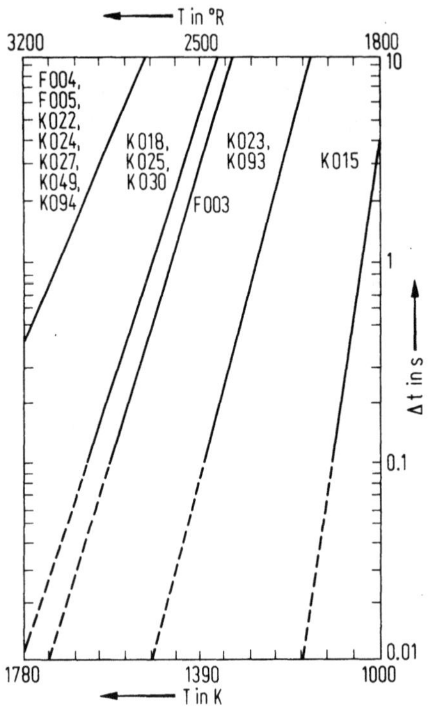

Fig. 3.1. Plots of time-temperature requirements for 99.99 per cent destruction of selected hazardous waste categories. (US, EPA, April 1983)

3.2.1 Toxic Wastes, Which can be Incinerated

For many organic toxic wastes such as pesticides the incineration process if often the most practical destruction alternative.

If the temperature and residence time has been selected correctly the destruction will be almost complete.

Compounds containing toxic metals (for example mercury) pose a problem however. Such waste might be treated by special thermal methods pyrolysis or distillation. If mercury containing wastes are incinerated special flue gas cleaning systems should be used.

Table 3.1. Order of increasing destructibility based on temperature required for 99.99 percent destruction at 1-second residence time [11]

Order	Compound	Temperature[a]	
		°C	(°F)
1	Methyl chloride	844	(1550)
2	Methane	836	(1535)
3	Phenol	807	(1484)
4	Methylene chloride	797	(1466)
5	Pyridine	791	(1455)
6	Chlorobenzene	768	(1413)
7	Dichlorobenzene	765	(1408)
8	Hexachlorobenzene	763	(1405)
9	Ethane	761	(1401)
10	Vinyl chloride	746	(1374)
11	Ethyl chloride	737	(1358)
12	Benzene	737	(1358)
13	Cresol	722	(1331)
14	Ethylene	721	(1329)
15	Toluene	720	(1327)
16	Nitrobenzene	720	(1327)
17	Hexachlorobutadiene	719	(1325)
18	Trichlorobenzene	716	(1320)
19	Vinylidene chloride	712	(1313)
20	Acetophenol	711	(1310)
21	Propane	708	(1305)
22	1,2,2-trichloro-1,1,2-trifluoroethane	692	(1277)
23	Dichloropropane	686	(1266)
24	Trichloroethane	683	(1260)
25	Phthalic anhydride	672	(1240)
26	Isobutanol	666	(1229)
27	Trichloroethylene	662	(1222)
28	Naphthalene	661	(1221)
29	Methyl ethyl ketone	657	(1213)
30	Dichloroethane	656	(1212)
31	Epichlorohydrin	618	(1143)
32	Maleic anhydride	607	(1123)
33	Bis(2-chloroethyl)ether	602	(1115)
34	Methyl isobutyl ketone	599	(1109)
35	Trichloropropane	571	(1060)
36	Benzotrichloride	528	(982)
37	Carbon disulfide	400	(751)

a Temperatur required for 99.99 per cent destruction at a residence time of 1 second.

3.3 Toxic Wastes Which can be Detoxified

3.3.1 Inorganic Toxic Wastes

Among the wastes which can be detoxified by chemical methods, the cyanide or chromate containing wastes originating from metal plating industry should be mentioned.

The cyanide containing wastes are often detoxified by hypochlorite oxidization.

$$CN^- + ClO^- + H_2O \rightarrow CNCl + 2\,OH^-$$
$$CNCl + 2\,OH^- \rightarrow CNO^- + Cl^- + H_2O$$
$$\overline{CN^- + ClO^- \rightarrow CNO^- + Cl^-}$$

The detoxification of the less toxic cyanate ions can be accomplished by aeration or by using hypochlorite.

$$2\,CNO^- + 3\,ClO^- + H_2O \rightarrow 2\,CO_2 + N_2 + 3\,Cl^- + 2\,OH^-$$

Also, peroxide can be used for detoxification of cyanide, but this method is much more expensive.

The reduction of Cr^{+6} is commonly made by using SO_2, often in the form of a metabisulphite solution.

$$2\,CrO_4^{2-} + 3\,SO_2 + 2\,H_2O \rightarrow Cr_2(SO_4)_3 + 4\,OH^-$$

At a central treatment plant, receiving large amounts of acid iron pickling baths in oxidization stage 2, the ferrous ions are advantageously used for the reduction process:

$$CrO_4^{2-} + 3\,Fe^{2+} + 8\,H^+ \rightarrow Cr^{3+} + 3\,Fe^{3+} + 4\,H_2O$$

Hardening salt which often contains cyanide can be dissolved and treated as indicated above.

3.3.2 Organic Toxic Wastes

Pesticides containing chlorine can be destroyed by dechlorination processes.

Some methods are based upon reaction with organometallic compounds containing sodium. The end products are dechlorinated (and detoxified) compounds and metal chloride.

Recently a method, based upon reaction between hydrogen and organic chlorine containing compounds, has been described. By this reaction hydrocarbons and hydrogen chloride are formed [12].

3.4 Wastes Containing Heavy Metals in Association with Acids or Alkalis

The waste originating from

- acid pickling baths
- electroplating baths
- corrosive baths
- sludge from electroplating industries.

can contain a number of metals
— iron
— chromium
— nickel
— copper
— zinc
— cadmium
as well as hydrochloric acid, sulfuric acid or nitric acid and alkalies such as sodium hydroxide.

The treatment methods most often in commercial use for these kinds of wastes consist of neutralizing the solutions by adding bases or acids and of immobilization of the metals by precipitation as hydroxides, or stabilization by adding cement compositions or other silicate products (or precipitation, followed by stabilization). The solid products, filter cakes or stablilzed products, are then disposed of under controlled circumstances.

The Fig. 3.2 presents the solubility of various metals often present in electroplating wastes as a function of pH. [13].

Table 3.2 presents a leachate test performed on a fixation product based on a commercial available process. The leachate test has been carried out on sample ground to a powder grain size 0,5–5 mm diameters. The sample is mixed with 10 times it's weight in water at 20 °C and the pH adjusted to approximately 6. The mixture has been shaken for 6 hours and then filtered and centrifuged.

Many companies offer fixation products most often based on adding cement or

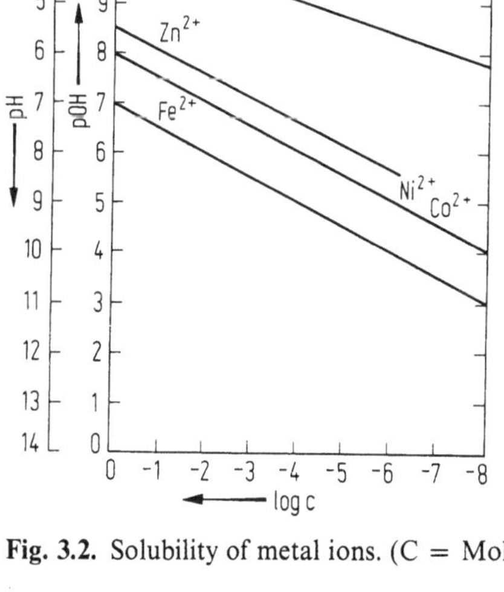

Fig. 3.2. Solubility of metal ions. (C = Mol/l)

Table 3.2. Leaching test results of fixation products from electroplating wastes[a]

	Measured on Electroplating Waste	Measured on Leachate	
Total Solid Matter %	44.7	Cyanide (ppm)	<0.007
Ash on Ignition %	46.5	Cadmium (ppm)	<0.01
pH	<1	Chromium (ppm)	<0.01
Cyanide	0	Copper (ppm)	<0.3
Cadmium (ppm)	5	Lead (ppm)	<0.3
Chromium (ppm)	120,000	Nickel (ppm)	<0.1
Copper (ppm)	11,600	Total Mercury	—
Iron (ppm)	—	Organic Phosphorus	—
Lead (ppm)	220	Arsenic (ppm)	<0.001
Nickel (ppm)	2,200		
Zinc (ppm)	3,200		

a From Stablex company brochure.

puzzolanes and claim leachate values in the same range as presented in the table.

3.5 Waste Which Needs Special Investigation or Sorting Before Treatment

3.5.1 Special Investigations

Wastes which are received at a central treatment plant without proper documentation on their composition and/or origin, must be subjected to detailed examination before being allocated to a specific treatment plant.

Problems of this nature may especially occur when a waste generator starts delivering wastes to a central treatment plant.

When investigating waste of unknown composition one would, from experience often make use of trade names as codes.

The assignment consists of getting the waste analyzed to a degree, that a proper type of treatment can be submitted. At the treatment plant there will often be only two possibilities.

(1) Incineration
(2) Disposal in special deposit (e.g. underground storing).

A proper specific physical-chemical treatment would request such a comprehensive analytical work, that it would be avoided in most cases.

The limiting factors for materials, which are to be incinerated are the following:

(a) Hazardous elements which are released during the incineration and emitted in the smoke phase.
(b) Elements which by stoking can cause explosion.
(c) Elements which, in the processing in the plant, could react, causing risks of explosion.

Under (a) belongs halogen sulphur, Hg and As.

Under (b) belongs the explosives.

Aluminium when heated with carbon in a reducting atmosphere, forms a carbide which when contacted with water (cooling of slag), forms CH_4. Aluminium and carbides should therefore not be loaded into an incinerator. Some other elements are under suspicion for forming similar explosive compounds. For a number of these compounds it is true, however, that they contain elements, which either not at all or only in very limited amounts are to be found in the wastes normally delivered to a treatment plant.

The quantity of waste in a load of unknown origin is obviously a major consideration in deciding the extent of analyses which are necessary before its allocation for treatment. In cases where less than 10 kilograms of burnable waste are involved, a quick examination in the laboratory to ensure no possibility of explosion, is all that is necessary before submitting for incineration.

Solution tests, heating and incineration of a sample in a Bunsen burner will give an indication of whether the element is of organic or inorganic composition, and the way it burns. Tests are also carried out for halogenes and sulphur, mercury and arsenic. In the presence of greater amount of halogenes solid waste will be ground before incineration. If the wastes contain mercury and arsenic it should not be incinerated.

The determination of loss of weight on the ignition will indicate how much organic material is present in the waste. If the waste has a large inorganic content it should normally be disposed of on a controlled landfill (or in salt mines, caves etc.).

At the disposal in salt mines it is a condition that the waste is not liquid, combustible or chemically reactive.

If the disposal is made on a landfill, the composition and the physical-chemical characteristics (solubility) of the compound should be known precisely.

3.5.2 Sorting of Wastes

Wastes delivered in small containers (cans, bottles, etc.) and again placed in drums have to be sorted before treatment.

This is the case for chemical wastes from households delivered to the collection system for hazardous waste in Denmark.

The chemicals are delivered in the original containers which before transportation are packed in steel drums (50 gallon) using stock absorbent materials.

In some of the municipalities a yearly campaign for collecting hazardous wastes from households is arranged. Table 3.3 presents examples of the waste types and quantities collected at such occasions [14].

In the table "units" means cans, bottles, etc.

3.6 Hazardous Waste Which cannot be Treated by Simple Incineration or Detoxification

Even methods for destruction by chemical methods do exist for almost all compounds, some products may arise in such small quantities that the operation of a specific process is not profitable or commercialized. If these products are not destroyed in a safe way by incineration a special controlled disposal may be actual.

Table 3.3. Hazardous wastes from households, collected in Danish municipalities

Municipality	Year	Homes which Participated		Waste Received							Total metric tons approximately
		Nos.	% of Nos. in municipality	Oil & Tar Liters	Paint Residues Units	Medicin Units	Herbicides & Pesticides	Contaminated Packages: Nos. of cans & bottles	Contaminated Packages: Nos. of spray cans	Others	
Holmegård	1977	211	10,7	1950	*)	668	48	1434	154	23	5
	1978	216	11	1400	773	1193	148	1400	184	1558	6
	1979	120	5,7	2100	570	287	*)	23	232	422	10
	1980	134	6,3	none	630	637	343	82	506	70	6
Ravnsborg	1977	154	6,2	7160	*)	1732	416	1673	255	115	10
	1978	98	3,9	1840	201	499	848	53	86	15	4
	1979	94	3,1	5900			Not specified				10,5
Hvidovre	1978	290	1,8				Not specified				3
	1979	551	3,5				Not specified				3,5
	1980	458	2,9				Not specified				3
Kolding	1979	265	1,3	661	1644 + 2 t	1316 + 250 kg	593 + 1,1 t	603 + 440 kg	297 + 170 kg		6

*) not specified separately

An example of these waste categories are small quantities of solid cyanide containing hardening salts or mercury containing batteries. Such waste types might be disposed of in mines or underground caves.

Quite another type of hazardous wastes which need special treatment are explosives or cylinders containing presurized gases. In some countries such waste products are handled by special military staff.

3.7 Wastes Which can be Landfilled

In principle only hazardous waste products which are solid compounds and originate from processing of hazardous wastes, or for which no process exists, which render them more harmless, should be permitted to be landfilled.

Of these types filter cakes from precipitating of metals, and slag and ashes from incineration mentioned.

A condition for landfilling is that the landfill is always properly designed and well controlled.

References

1. Environmental Protection Agency, Hazardous Waste Management System, Federal Register Vol. 45 — No. 98, Book 2 May 19, 1980 (USA).
2. Nato: Challenges of Modern Society, Vol. 4, Hazardous Waste Disposal, Washington DC, USA, June 1982.
3. Environmental Protection Agency. Identification and Listing of Hazardous Wastes, May 19, 1980.
4. Directive No. 78/319 of March 20, 1978, on Toxic and Hazardous Wastes.
5. Pao C. Chan, Daniel Coffey and David F. Ollis. Hazardous Waste Generation and Off-site Disposal Patterns in California. University of California, Davis, September 1981.
6. Verordnung zur Bestimmung von Abfällen, nach § 2, Abs. 2 des Abfallbeseitigungsgesetzes, May 24, 1977.
7. Bekendtgørelse om Kemikalieaffald, July 3rd, 1980.
8. Nyt fra Miljøstyrelsen. Kemikalieaffald, Brancheforteguelse, Oct. 1980.
9. Office of Technology Assessment. Technology and Management Strategies for Hazardous Waste Control, Washington DC, USA, March 1983.
10. Chemcontrol A/S 1st International Symposium, Odense, Denmark, 1982.
11. Feasibility of Destroying Hazardous Wastes in High Temperature Industrial Processes. PedCo Environmental, Inc., U.S. Environmental Protection Agency, Industrial Environmental Research Laboratory, Cincinnati, April 1983.
12. New Scientist, December 1983.
13. Svend Erik Jørgensen, Spildevandsrensning.
14. M. Palmark, Chemical Waste, Association of Danish Engineers, 1980.

Rapid Analysis Methods for Special Wastes

E. Thomanetz and O. Tabasaran

This publication discusses the basic problems encountered in the physical-chemical characterization of special wastes.

For practical purposes, a number of what is called "rapid screening analysis methods" are recommended; these are quite deliberately to be conducted not by means of hi-tec equipment that is prone to malfunctions but by means of simple manual processes (these methods require well informed and experienced laboratory-staff).

Contents

Chemical Waste
Managing Editor: K. R. Müller

1 General Comments Concerning the Analysis of Special Wastes

Special wastes are wastes which, because of their type, composition and/or amount, should not be mixed with household wastes. They are mainly produced by industry.

Special wastes are not, in general, single substances but usually mixtures of different substances, whose identity and composition depend upon the production conditions of the waste producer and the technology of the waste-producing process. The physical and chemical evaluation of a special waste is thus best compared to the quality control of a large-scale, mass-produced product by random sampling, in which variations from the norms laid down are revealed in many of the cases (control of identity).

Since the methods of investigation must, above all, be adapted to the practical realities of the daily routine of a special waste disposal site it is realistic and practicable to use *rapid screening tests*. Such tests can, for instance, simply reveal the presence or absence of an unwanted contaminant and thus allow a yes or no decision concerning the acceptability of the material for disposal. There are also *screening tests* that produce semi-quantitative evidence (high, medium, low, not detectable).

The methods, thus, differ from the usual DEV/DIN[1] regulations, which are generally too labour-intensive, too time-consuming or sometimes not applicable for the purpose of screening special waste before disposal.

The aims of screening analysis are as follows:

— Rapid estimation of the amount of environmentally relevant components in the waste.
— The rapid identification of unauthorized components in the waste.
— The rapid determination of the chemical reactivity of the waste with rainwater.

The rapid estimation of the leachability of environmentally relevant components within the disposal site, by percolating rainwater, is of particular importance. Rapid in this instance often means in the presence of a waiting truck.

At this point it is appropriate to mention the qualifications and motivation of the technical staff responsible (this also applies to the disposal site staff and to the staff of the responsible authorities), without whom the effectivity of such analyses — even in the best equipped of laboratories — is nil. Close personal contact and exchange of information with each other and with the managers of other special waste disposal sites and with the authorities are just as important as the obligation to participate in specialist educational meetings.

2 The Consequences of Special Waste Analysis for the Running of a Disposal Site

All the waste analyses described later have the purpose of identifying the waste that is delivered as being the type of waste which has been authorized or detecting possible variations from the authorized specifications and drawing conclusions from the information obtained. If the acceptance analysis reveals conspicuously higher values than the originally authorized waste, then the following can occur:

— Entry into a "transgression register" (computer) and contact with the waste producer. He is invited to explain the occurrence and requested not to deliver such loads in future. (Normally the site operator should advise on pretreatment).
— Depending on the nature of the offending contaminant it may be necessary to make special arrangements for the material within the disposal site (i.e. separate storage).
— If one or more analysis values are very high, or if the material that has been delivered contains unauthorized substances, or if the delivery of an unacceptable load is repeated, the objectionable delivery is returned and, if necessary, the authorization withdrawn.

3 Authorization Criteria for the Disposal of Special Waste, General and Recommendations

The authorization criteria for the disposal of special waste in the Federal Republic of Germany are not uniformly regulated at the present time. This is mainly because,

[1] DEV/DIN: German standard methods for the examination of water, wastewater and sludge.

in contrast with, for instance, the TA-Luft (technical guidance for the maintenance of clean air) regulations where clear upper limits are laid down for polluting gases, clear limits cannot be laid down for the acceptance of chemical waste. The TA-Luft regulations refer, in general, to clearly defined components of effluent gases whose upper acceptable load is laid down, whereby there is the possibility of staying within these limits by the application of suitable processing techniques. In such a case the establishment of upper limits is very worthwhile and effective. However, the problems are much more complicated in the disposal of special waste.

The establishment of upper limits for environmentally significant substances in the disposal of special waste appears to be impractical. On the one hand, the presence of a pollutant — even in high concentration — is less significant than the possibility of the component being mobilized in the area of the disposal site. (For example, the high heavy metal content of ore beds comes to mind in this context). On the other hand, the absolute amount of a waste product is of importance; a single delivery of a small quantity of waste with a relatively high concentration of a particular noxious substance is to be viewed in a quite different light, as far as its potential danger is concerned, compared with a very large quantity of a continuously produced waste containing small concentrations of the same material. From this it is clear that a classification of chemical wastes according to fixed concentration limits cannot be practicable. It is much better to judge and evaluate each waste material individually according to its potential dangers. The major criteria for evaluation and, thus, acceptance or rejection criteria for special wastes are as follows:

- The amount of waste (per unit time).
- The concentration of environmentally relevant substances in the waste.
- The evaluation of the mobility of these environmentally relevant substances within the disposal site.
- The evaluation of possible chemical reactions between the waste and other wastes in the disposal site.
- The concentration of environmentally relevant substances in the aqueous eluate from the waste.
- Regular analysis of the leach water at the disposal site and the positive or negative trend of slow concentration changes in the leach water composition.

From this it can be seen that an expert committee can only evaluate the environmental effect of a special waste by consideration of a whole range of interconnected criteria. This means that there may be a certain nonquantifiable residual risk that can only be minimized by the competence of the approval body. With the progress of knowledge it is quite possible that a waste material, which is approved at present, may, in the future, not find approval for disposal because of the appearance of environmentally relevant, technological or other reasons. The decision of the expert committee to approve a waste is, in the end, a yes or no decision. If the disposal of the waste is allowed the original analysis of the waste becomes the criterion for evaluating the acceptability of the analyses during the time that it is delivered.

The use of computerized data handling is a great advantage for the optimal running of a chemical waste disposal site. Such a facility is not only important for a great deal of administrational work but also leads to better safety control. The amount and type of information stored should be decided by consultation between the site operators, the relevant authorities and, if necessary, expert consultants. It should be mentioned here that, as well as the data needed for costing and organizational purposes, the relevant information about the physical and chemical properties of the waste should be stored and can be recalled and updated by the disposal site supervisors.

4 Physical and Chemical Characterization of Special Wastes

The disposal site operator should not only make an eluate analysis but also a complete characterization of the waste itself. The clues from these tests then determine whether exact analyses of individual substances are necessary. Such rapid screening tests are described in later sections.

4.1 The Comprehensive Investigation of the Individual Components and the Cumulative Properties of Special Wastes

The comprehensive analysis of special waste should be carried out in the following cases:

— On the first delivery of the waste.
— On obvious deviations from the normal nature of the waste as reported by the deliverer or the disposal site personnel.
— On deviations in the check analyses of deliveries.
— When alterations in the technology of the waste-producing process (changes in additives) are (mandatorily) reported by the waste producer.
— At fixed periods as a check with the aim of confirming the identity of each waste product at "certain intervals".
  Explanation of the meaning of a "certain interval":
  The recharacterization of each waste product according to an individual schedule depending on both mass and *time* criteria has advantages, since, on the one hand, the amounts of the individual wastes that are delivered vary a great deal and, on the other hand, the delivery schedules are variable and it cannot be guaranteed that the agreed delivery times and quantities will be adhered to.

The following procedure is worthy of consideration:

Time criterion: The carrying out of a comprehensive analysis once a year for deliveries of less than $\times$ t/a

Mass criterion: The carrying out of a comprehensive analysis after the delivery of every $\times$ t.

Whichever criterion is fulfilled first assumes priority. The value of x does not have to be rigidly laid down but rather the operator of the disposal site, possibly with the help of technical advisers, must make allowances for the particular situation of the producer, the waste-producing technology and the type of waste. Here the application of computerized data processing is of particular advantage.

4.2 Waste Acceptance Testing

Alongside the periodic characterization of the waste by the analysis of its individual constituents and cumulative properties, the control analysis on delivery, which should be carried out on a sample from every load, constitutes an important part of the whole continuous analytical surveillance. In some cases this control can be limited to a check on the appearance of every truck-load and the removal of a sample, if necessary after tipping, by technically qualified persons capable of judging and evaluating the waste material, the supplier and the technology producing the waste.

The check must be suitably enlarged in cases of deviation. The sample should be stored in a sample archive for several weeks. The point of such sample storage is to assist the reconstruction of possible incidents on the disposal site (gas production, fires, temperature variations, odour production etc.) which may be caused by mixing wastes or by the effect of weathering.

4.3 Official Analysis of Random Samples

The purpose of official random sample analysis is to check on the disposal site's internal waste control system. For this purpose the officers of the controlling authorities should take samples either from the disposal site's sample archive or from the disposal site itself during unloading and report the agreement or disagreement of the analyses thus obtained (for purposes of motivation and self-correction) to the operators of the site.

5 Recommendations for the Execution of Rapid Methods for the Physical and Chemical Characterization of Special Waste

These methods have been particularly chosen because they do not require the use of high technology instruments and apparatus (AAS, ICP, RF, GC, GC/MS, TOC analysers etc.). This has been done deliberately to allow for the personnel and space available to a laboratory at a disposal site. On the one hand, such instruments require enough highly qualified operators and, on the other hand, the preparation and down times of such instruments are normally considerable. It is much better to try to develop manual methods that are simple, reliable and sufficiently sensitive and adapted to the conditions of a waste disposal site.

The methods that are to be described are not yet generally accepted. They have been developed as a result of practical requirements — many of them were developed and tested on the chemical waste disposal site at Malsch in Baden-Württemberg and are used there and at the disposal site which came into use at Billigheim in autumn 1983. There are no well-tried rapid screening tests at all for some problems (e.g. rapid biological toxicity screening). It appears an important task for the future to develop suitable techniques for these and, if necessary, to improve the existing techniques.

5.1 Sampling Special Wastes for Subsequent Investigation

Obtaining a representative sample of a chemical waste that has been deliverd by truck is a difficult task in many cases. Some wastes do appear very homogeneous (e.g. wastes from water and waste water treatment), but many consist — in appearance and/or composition — of inhomogenous refuse. Whereby the enviromentally relevant components are mingled with large amounts of inert material, such as stone, bricks, lumps of concrete or even empty containers. The actual environmentally relevant substances can either be present as small particle interstitial material or as large inhomogeneous lumps or sheets weighing several hundreds of pounds or as glaze-like coatings on fragments or as residues in cans and drums. The sampling techniques developed for use with commercial bulk materials are not applicable to such wastes. The sampling of the contents of drums and other containers is just as problematical, since density gradients often form so that the surface layers consist more or less just of exuded liquid.

Unless special aids are used samples can usually only be taken from the surface layers of an untipped truck-load and this leads to several disadvantages:

- Any solvent present may have been selectively evaporated — helped by the wind created during the journey.
- During rainy weather reactions and elution may have taken place.
- Problem waste which has been deliberately hidden in the middle of the load is not detected.
- Even if such problems do not occur a proper laboratory examination of waste in a waiting truck is often an unrealistic proposition because of the time involved.

Since the heterogenous nature of special wastes precludes the adoption of a patent remedy for sampling problems, several alternative possibilities will be discussed here.

The following alternatives are to be recommended as examples:

- A sample is taken from the loaded truck at the entrance to the disposal site and the sample and the load are checked visually by trained and experienced personnel on the orders or under the supervision of the site manager.
- The contents of the truck are tipped on an area of the disposal site indicated by the site manager.
- Problem-free and well-known wastes are sampled again at the disposal site after tipping, by one of the bulldozer drivers under instructions from the site manager. The bulldozer drivers are instructed to report anything out of the ordinary.
- This sample can be returned to the disposal site laboratory by the truck driver, since the distances within the site do not allow the site manager or other qualified personnel to be present at the tipping of every load (assuming 30 deliveries daily and a one-way distance of 300 m this amounts to 18 km internal travel daily).

This procedure necessitates that the personnel be *well trained* by the site management and that they be continually kept up-to-date concerning the waste itself, the way the waste is produced and, above all, concerning investigational criteria for problematical contaminants.

- With less straight forward wastes, with anything out of the ordinary or under suspicious circumstances the disposal site manager himself should take the sample at the actual dumping site.

Another alternative, in order to allow the removal of as representative a sample as possible from the loaded truck at the entrance to the disposal site, would be to install a sampling rig with servo-assisted sample-taking probes, which penetrate into the mound of waste from above. Suitable assemblies are not commercially available, however, and must be individually designed, built and tested. This type of sample-taking is inapplicable to certain wastes that the sampling probes cannot penetrate or for which none of the material remains in the sampling probe tube.

The third, rather more theoretical, possibility for a precise visual inspection and for representative sampling would be to set up a tipping and reloading station for all delivered waste at the entrance to the disposal site but separate from it. However, apart from the space requirement, there are several dangers inherent in this method (dust, waste water, smell). In addition such measures would require an internal transport fleet and a larger staff.

The preparation of very inhomogeneous waste for sampling can be drastically improved by the use of crushers and mixers (jaw breakers and cement mixers), but the danger of sparks causing fires during crushing has to be considered.

5.2 The Water Content or State of Dryness of the Waste at 105 °C According to DEV/DIN (Loss of Weight at 105 °C)

The water content and residue on drying are important parameters for the evaluation of the suitability of water-containing sludges for disposal. The test also provides information concerning possible offensive odours which the waste could cause. At states of dryness of less than 15% by weight the waste is not usually able to retain its shape. It should be noted that the presence of volatile organic materials (solvents, liquid fuels, sublimable organic solids) can cause considerable errors in the determination. By appropriate indications or when the sample smells of organic solvents etc. further investigations are necessary to determine the content of easily volatile solvents.

5.3 Loss on Ignition and Residue on Ignition of the Waste According to DEV/DIN

These cumulative parameters provide guidance as to the proportion of organic material in the *residue on drying*. (It should be noted that considerable amounts of organic compounds can evaporate even during the drying of the sample.)

False results and false interpretations are possible because of:

- Failure to maintain a uniform ignition temperature of 550 °C (conventional method — different results are obtained under different conditions).
- Evaporation of water of crystallization (above 105 °C) prevention: predry at 180 °C if large amounts of water of crystallization are suspected.
- The decomposition of inorganic compounds (e.g. carbonates with the formation of carbon dioxide) prevention: not possible. Estimate the effect by obtaining information concerning the carbonate content from the producer of the waste.
- With certain materials containing high proportions of organic matter the sample can burn off so rapidly in the muffle furnace that particles are entrained. Result: loss of weight. Prevention: heat the muffle furnace up slowly or reduce the supply of air.
- The organic components of certain materials are only very slowly oxidized at 550 °C and unburned residues remain. Prevention: improved air supply to the muffle furnace or the treatment of the residues, which are usually black, with 1% ammonium nitrate solution.

5.4 Petroleum Ether (or Solvent)-Extractable Materials in Waste (High Boiling Oils and Fats)

This rapid screening test for the cumulative properties is used for wastes from certain origins (the mineral oil industry, oil-contaminated earth etc.):

Method: Extract the waste with low boiling petroleum ether or other solvents (by shaking or in an ultrasonic bath)

↓

Drip a measured amount of extract onto a filter paper to form a circular blot

↓

After evaporation of the petroleum ether (or solvent), if necessary using a hot air drier, measure the diameter of the remaining "grease spot"; this diameter is a measure of the petroleum ether (or solvent)-extractable material

Note: It is necessary to find the right conditions (amounts, time of extraction, paper quality) empirically. The method should be "calibrated" against the precise DEV/DIN method for each individual problem. Experience and skill are required to carry out the test.

5.5 Easily Volatile Solvents in the Waste

Two alternative methods are available for this cumulative parameter screening test. The first method is faster, the second produces more accurate values.

Method 1:

Determine in parallel the residue on drying at 105 °C

Weigh 20 g homogenized sample into a 300 ml screwtopped jar

↓

Add 200 g analytical grade methanol

↓

Place the closed jar in an ultrasonic bath for 10–30 minutes

↓

Determine the water in the filtrate by the Karl Fischer method

↘ ↙

The difference between the residue on drying and the water content is the quantity of solvent

Method 2: (special apparatus required, see Fig. 1)

Weigh out 100 to 200 g homogenized waste into a porcelain boat

↓

Place the boat into a pistol equipped with a jacket heated to 105 °C by means of a heat exchanger fluid

↓

Regulate the rate of flow of the air, which recirculates through the jacketted pistol

↓

The material that slowly evaporates in the heated pistol condenses in the water-cooled coil condenser which follows the pistol and drips into a receiver (separating funnel)

↓

Leave in action for 5–8 hours

↓

Two phases are usually formed in the separating funnel. Separate the phases and weigh each of them

↓

Determine the water content of each phase by the Karl Fischer method (this enables the total water-miscible-solvent content to be measured). The difference between the amount of water present and the total amount of the condensate is the solvent content

(from "Place the boat into a pistol…" ↓) The residue on drying can be obtained by weighing the boat at the end of the experiment

Note: Both methods should be compared with each other at given intervals for particular problems.

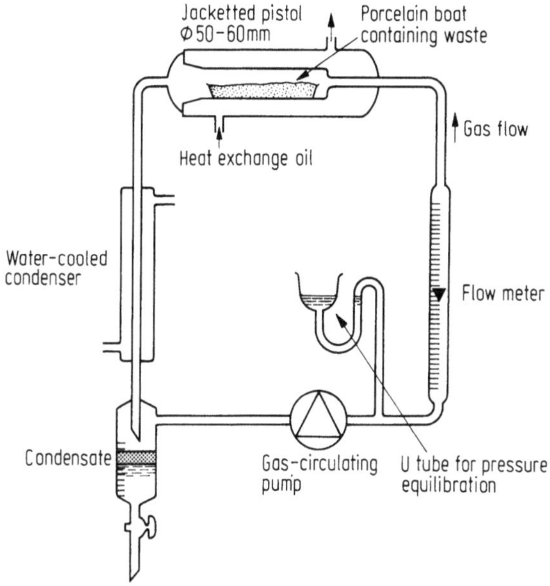

Fig. 5.1. Schematic respresentation of the apparatus for the determination of solvents in chemical wastes (developed at the Malsch special waste disposal site)

5.6 Investigation of the Formation of Potentially Explosive Gas-Air Mixtures above the Waste (Ex Test)

Method: Take a sample, if necessary collecting and mixing pieces of the waste to be investigated. Break up large pieces by hand (investigate odour, colour and other individual characteristics)

↓

Place 1–2 kg sample in a bucket inside a closable plastic bag, leave for 1/2 hour

↓

Admit the gas phase of the bucket into the Ex apparatus (calibrate with nonane). Report values higher than 25% UZG[2] (important for solvent-contaminated earth etc.). Shake the plastic bag before making the measurement, in order to mix any stratified layers of vapour

Consequences: If the Ex test is positive further investigation is necessary, e.g. the flammability and burning behaviour and the solvent content.

[2] UZG: Lower explosions limit

5.7 Flammability and Burning Behaviour and the Behaviour in a Pilot Flame Combined with a Test for the Formation of Acidic Combustion Gases

Method:

Take ca. 100 g homogenized sample

↓

Spread out or place in a heap on aluminium foil

↓

Attempt to ignite with a bunsen flame (fan slightly)

→ The waste ignites easily. Consequences: Further investigations of solvent content Consulation with waste producer Recommendation of pretreatment Withdrawl of permission to tip

↓

The waste is difficult or impossible to ignite

↓

Behaviour in the continuous flame, i.e. the bunsen flame is played continually on the sample

→ Consequences: On certain types of behaviour, e.g. melting or decrepation, the waste must be suitably diluted. In extreme cases withdrawl of permission to tip

↓

Preliminary test for acid formers (HCl, SO_2, NO_x) with moist litmus paper in the gas flow (check the smell simultaneously)

Note: Particular skill and experience are required to carry out this test.

5.8 Test for the Presence of Halogenated Organic Compounds in the Waste (Beilstein's Test)

Method: Prepare as concentrated an extract of the waste as possible using a non-halogenated, water-immiscible solvent, in a Soxleth extractor

↓

Separate cleanly from any aqueous phase that is formed

↓

Dip a piece of copper sheet (size ca. 3×3 cm) or gauze, which has been previously cleaned and heated glowing hot, in the extract

↓

Introduce the sheet or gauze into the nonluminous flame of a gas burner. If organically bound halogens are present the flame is coloured a characteristic green to bluish-green by the evaporation of the copper halides formed

Note: It is not to be recommended that the waste itself be brought into contact with the copper, since the ubiquitous inorganic chloride also gives this reaction. Skill and experience are required to carry out the test. A suitable solvent must be found by trial and error. The sensitivity of the test should be determined by comparison with precise methods for the determination of adsorbable organic halides (AOX) and with gas chromatographic analyses.

5.9 Investigation of the Production of Heat and Gases by the Waste During Hydrolysis

This investigation is particularly suitable for waste from the light metal refining industries (aluminium salt slag, magnesium dross etc.) and assists in determining whether gas formation or heat evolution can occur on the disposal site.

Method: Place ca. 50 g of as finely divided waste as possible in a 300 ml conical flask equipped with an eudiometer and pressure-equalizing dropping funnel

↓

Add dilute sulphuric acid from the cropping funnel (in excess)

↓

Determine the volume of gas evolved (mainly hydrogen). Approximate calculation of the heat of reaction from the amount of hydrogen gas formed using thermodynamic data

Note: The reason for using acid is that — in comparison with water — the reaction is much more rapid and that only the total gas and heat production is of interest. The critical value for the amount of hydrogen, i.e. heat of reaction, depends on the problem and should be determined empirically, in order to decide on special measures for the waste (diluting the waste, removal).

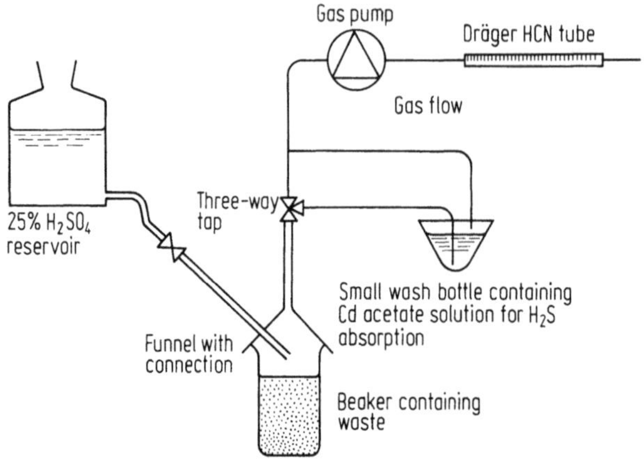

Fig. 5.2. Diagram of arrangement for testing for cyanide in chemical wastes (developed at the Malsch special waste disposal site)

5.10 Test for Cyanide in the Waste

Method: Place ca. 5–10 g waste in a small beaker (10 ml)

Place a glass funnel on it; add enough 25% sulphuric acid to submerge the waste (see Figure)

Suck off the gas phase with a Dräger gas-purging apparatus fitted with a HCN tube (if necessary first pass the gas through a gas-washing bottle filled with cadmium acetate solution to remove hydrogen sulphide)

Note: This rapid screening test constitutes a very fast and sensitive preliminary test for total cyanide in waste. However working with trace gas analysis tubes requires a great deal of experience, since other gases, e.g. hydrogen sulphide and sulphur dioxide, also colour the indicating layer of the tube red (although the shade is rather different, as is the speed of coloration). The preliminary test is not suitable for some complex cyanides.

5.11 Production of a Rapid Eluate from the Waste for Subsequent Eluate Analysis

Method: Weigh out 30 to 50 g waste into a beaker and add water (1:10)

↓

Place the beaker in a thermostatted ultrasonic cleaning bath

↓

Extract for 10 minutes

↓

Filter the eluate

Note: This method is quicker than the DEV/DIN method and (according to the chemical waste disposal site in Malsch where it is used) gives comparable results.

With regard to the preparation of eluates from chemical wastes for later evaluation, it should be considered whether it is appropriate to elute with acids or bases or, if necessary, with leach water from the disposal site as well as with pure water. Such investigations sometimes yield important additional information concerning the mobilizability of the components of the waste which is not obtainable just from eluates produced using pure water.

5.12 Determination of the Water-Soluble Portion of the Waste

Method: The evaporation residue according to DEV/DIN at 105 °C of the rapid eluate is, in my opinion, a good measure of the content of water-solubles in the waste in weight %. Easily volatile organic components are naturally not included and must be determined separately in other investigations.

6 Recommendations for the Carrying out of (Rapid) Tests for the Physical and Chemical Characterization of Eluates from Chemical Wastes

The reason for the preparation and investigation of chemical waste eluates is in order to attempt to obtain information concerning the probable quality of the disposal site leach water. It is true that a laboratory investigation can only approximately simulate the complicated elution processes caused by precipitated water which occur within the body of the tip; however in the absence of better methods there is no alternative. The eluate so obtained (rapid eluate) can then be investigated for all the relevant nonspecific cumulative parameters and solutes. There is no need to determine the following parameters since they do not provide any useful information:

- Loss on ignition of the evaporation residue as a measure of the organic part of the water-solubles (the COD is a much better measure of this parameter)
- Sedimentable materials and loss on ignition of the sedimentable materials (these parameters are variable functions of the amount of comminution of the waste, of the elution technique and the elution conditions)

The remaining determinations are only meaningful if they are carried out on the filtered eluate.

6.1 The Chemical Oxygen Demand (COD) of the Waste Eluate

The COD is one of the most important diagnostic cumulative parameters for evaluating the pollution of the eluate and, hence, of the leach water by organic solutes. It is recommended that the DEV/DIN methods be employed, since large errors are possible using all rapid COD determination techniques. Tests for chloride ions (DEV/DIN) and for nitrite (nitrite stick from Merck) are essential before every COD determination (see the DEV/DIN documentation).

6.2 The Dissolved Organic Carbon (DOC) of the Waste Eluate

In principle DOC is one of the most important cumulative parameters for the quantification of dissolved organics in water. The disadvantage is that a commercial instrument, which generally requires a considerable amount of maintenance, is necessary for the determination of the DOC. This method ought therefore to remain the preserve of analytical chemistry laboratories; the determination of DOC by the disposal site laboratory must be regarded as problematical.

6.3 Substances Capable of Coupling Reactions in the Eluate Waste (Phenols)

Method:

Fill two wells of a spot test plate 2/3 full with eluate

1st well

Add one drop phenol test solution (2 mg/l), in order to detect possible interference with the subsequent reaction

2nd well

Add 1 drop phenyldimethylaminopyrazolone solution (2%) to each well

\+

3–4 drops dil. aqueous ammonia solution

\+

1 drop freshly prepared potassium ferricyanide solution (5%)

If the reaction in the 1st well is negative, i.e. there is interference, then a different detection reaction must be employed

If substances able to couple are present a red to deep red coloration is produced

Note: A very fast preliminary test. If the result is positive further exact investigations should be carried out according to DEV/DIN.

6.4 Solvents in the Waste Eluate (Water-Miscible)

There is no practicable rapid screening test available for this determination. An investigation would be indicated if the investigation of the waste itself (see 5.5) established significant quantities of water-miscible solvents. It is recommended that the sample be sent for gas chromatographical investigation.

6.5 Cyanides in the Eluate

It is only necessary to carry out this test if the cyanide test on the waste is positive or the preliminary test is inconclusive.

Method:

Fill two wells of a spot-test plate 2/3 full with eluate

→ *1st well*: Add one drop standard CN solution (0.1 mg/l)

→ *2nd well*

Add several drops pH 7 buffer solution to each well

+

1 drop 1 % chloramine T solution

+

2 drops reagent solution*

If cyanide is present a red coloration is produced within ca. 1 minute

Detection limits: 0.01 mg/l pale pink colour
0.1 mg/l reddish-violet colour
1 mg/l deep reddish-violet colour

(At a concentration of 5000 mg/l thiocyanates produce a coloration which corresponds to that produced by 0.1 mg/l cyanide)

If the reaction in the 1st well is negative, i.e. there is interference, then apply the the DEV 13 test

* Reagent solution: Suspend 6 g barbituric acid in 10 ml water and bring into solution by the addition of 30 ml pyridine and 50 ml water. Then add 6 ml of 32 % hydrochloric acid

Note: Spot test analysis is well suited for rapid preliminary analyses. The DEV/DIN tests should be carried out if the result is positive.

6.6 Heavy Metals in the Eluate — Preliminary Test for Pb, Cd, Cu, Ni, Zn, Hg

Method: Place 0.06 ml eluate in a stoppered test tube and dilute to 3 ml with distilled water

↓

Add 2 ml pH 7 buffer solution

↓

Add 1 ml dithizone solution (7.5 mg dithizone in 500 ml 1,1,2-trichlorotrifluoroethane)

↓

Agitate for 10 minutes on a rotary mixer

↓

If one or more of the above mentioned metals is present the colour of the green dithizone layer changes:

coloration	detection sensitivity
Zn: brilliant red	1 mg/l
Pb: pale red	5 mg/l
Cd: pink	5 mg/l
Cu: yellowish-brown	5 mg/l
Ni: brownish-violet	10 mg/l
Hg: violet	10 mg/l

(The detection sensitivity has been deliberately decreased by the predilution step, to achieve a practicable sensitivity range.)

Note: If a coloration (or mixed coloration) is obtained corresponding to a total heavy metal content of ca. 10 mg/l then a sample should be sent to an analytical chemical laboratory for the determination of the individual heavy metals (by atomic absorption or ICP). The dithizone method is well suited for preliminary testing. Its sensitivity to zinc is a disadvantage however, since this metal is relatively common. This means that cadmium and mercury can only be detected at what are, for them, relatively high concentrations. In view of the environmental relevance of these metals it would be desirable to be able to detect them at lower concentrations than is possible in the screening test. Another method must be developed or the present method modified so that zinc no longer interferes.

6.7 Chromium VI in the Eluate

Method:

Fill two wells of a spot test plate 2/3 full with eluate

1st well
Add one drop Cr-VI test solution (0.2 mg/l)

2nd well

Add one drop saturated sodium fluoride solution to each well (to mask ferric iron)

+

2 drops 25% H_2SO_4

+

1 drop diphenylcarbazide solution (1% in acetone)

If the reaction in the 1st well is negative, i.e. masked, then the DEV E10 test should be used

When Cr VI is present a red-violet coloration is produced (detection limit 0.1 mg/l)

Note: This screening test is fast and relatively reliable. The DEV method has inherent problems and can only be recommended with reservations.

7 Recommendations and Comments Concerning the Setting Up of a Disposal Site Control Laboratory for Special Wastes

As can be seen from the previous sections, the control laboratory of a special waste disposal site requires not only an extensive sample-taking system but also special equipment and measuring arrangements, which are neither usual nor of use in other types of laboratories. As far as the suggested analytical techniques are concerned it is recommended that a separate analysis station be set up for each screening test and equipped with the required apparatus and reagents — as detailed in the individual descriptions. Since — as is worth re-emphasizing again and again — the experience of the technical staff is of particular importance in the carrying out of these screening tests, the tests must be sufficiently practised and tried out before they are applied. For training purposes it would be useful to work with samples containing defined amounts of the material being tested for (e.g. earth that has been "spiked" with known quantities of kerosine, cyanide, chromium VI etc.). The limits of detection sensitivity and applicable range should thereby be established and checked against standardized laboratory methods. Such control measures should also be applied from time to time during the daily routine. It would also be useful to establish contacts with other suitable laboratories, in order to be able to discuss the problems involved.

References

K. Mangold, E. Barth, H.-P. Kares: Private information; Gesellschaft zur Beseitigung von Sonderabfällen in Baden-Württemberg mbH, Federal Republic of Germany, Fellbacher Str. 47, 7012 Fellbach — 4 (1982)

No other relevant detailed literatur references are available.

C. Handling and Treatment

Transfer, Storage, Shipment

J. Bromley

During the manufacture of chemicals, wastes are produced at the chemical production plant, at plants using chemical products and by the consumers of the products. Storage, handling and transport of wastes occur at many stages throughout the industrial process. The practical measures that need to be taken to handle wastes safely and reduce disposal costs are considered along with the regulations regarding safety at work, storage, labelling, transportation and transfrontier transfer.

When there are accidents during waste movement it may be necessary to seek advice from the chemical emergency centres. Rail and sea transport and incineration at sea are discussed. The specific problems at waste treatment centres are briefly reviewed.

Contents

1 Introduction

The nature of chemical wastes, the different types of wastes and the problems of their analysis have been discussed in the previous chapter. Before going on in the next section to discuss the treatment of chemical wastes it is necessary to consider their transfer, storage and shipment as it is often these aspects that attract most public attention because of accidents, spills, or illegal disposal or transboundary transfer. This public attention can easily turn via political concern into even more complex rapidly enacted regulations. The responsible authorities have been conscious of the need for regulations for the transfer, storage and shipment of pure chemicals, but public concern over waste transport is a more recent phenomenen. In consequence regulations regarding chemical wastes are receiving more attention from political and regulatory authorities. Certain incidents such as the 'disappearance' of the drums from Seveso, have resulted in a considerable increase in government attention to the problems of toxic wastes. The Chemical Industry has always been careful to ensure that transport of raw chemicals is safely carried out. It is now equally concerned to ensure that chemical wastes are safely transported. Figure 1.1 illustrates the various points at which the storage, handling and transport of toxic wastes may arise. The chemical process plant shown inside the dashed area will produce wastes which may be recycled. If the wastes are toxic, then we have, within the plant, the handling (H), the storage (S), and transport (T) of a toxic waste in house. If waste, either from the plant itself, or from the recycle unit is 'toxic' then further storage handling and transport problems will arise as it is

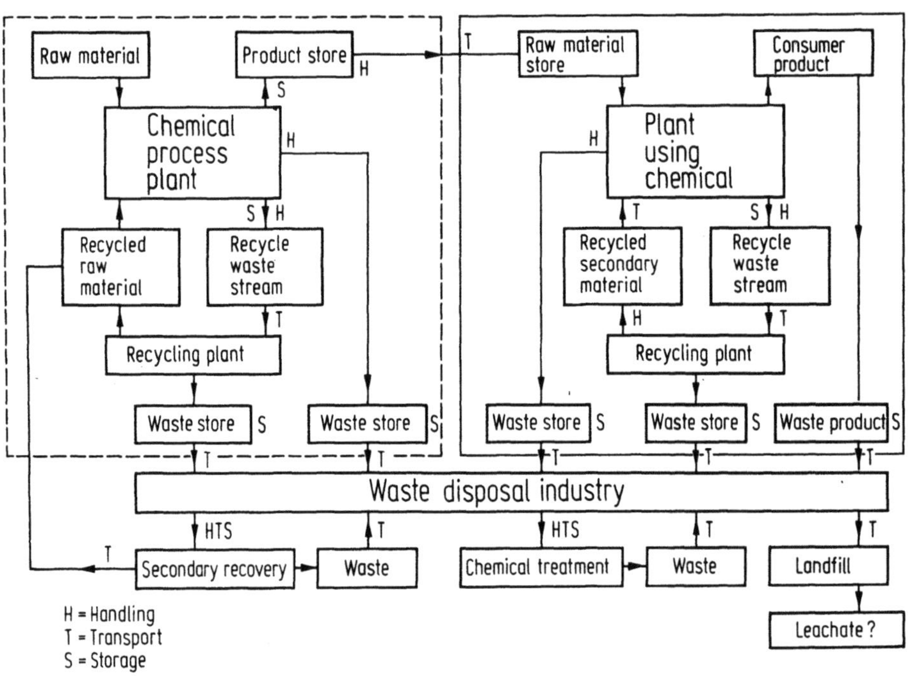

Fig. 1.1. The Storage, Handling and Transport of Toxic Material

transferred to the toxic waste disposal industry as shown on the diagram. Material which may be toxic will also arrive from industries which use toxic chemicals. Such a plant using 'toxic' chemicals is shown inside the outlined area.

The toxic waste industry will, in its turn, have storage, handling and transport problems for the various 'products' it makes which may be of the various kinds shown on the diagram.

This diagram only serves to show that there are many storage, handling and transfer steps possible. Staff within (a) the primary production unit, (b) within a plant using a toxic chemical and (c) staff at a toxic treatment centre must all observe the same precautions and obey appropriate safety regulations for a given chemical waste.

While the 'waste' is within the producing factory or when it is generated at a plant using a chemical as a feedstock the operations will also be subjected to regulations such as "Safety at Work Acts" and also in some circumstances the "Major Hazard Installations Regulations" in the U.K. or equivalent legislation in other countries.

The wastes will be of the many types mentioned in the previous chapter but there are other sources of waste which may arise from the following sources which have not been included in Fig. 1.1. They arise from the following operations:

1. The closing down of laboratories and the removal of old chemical stocks, often including many badly labelled old samples with obscure code numbers of the labels.
2. The removal of contaminated equipment from old plant. (One half of the refuse generated by BASF at Ludwigshafen is understood to be demolition debris which has been removed prior to the construction of new plants!).
3. The decontamination of contaminated land.
4. The washing up of hazardous cargoes onto beaches.
5. The cleaning up of spilled chemicals at factories or on roads.
6. Filter bag wastes.
7. Asbestos wastes.
8. Scrubber wastes from stack effluents.
9. Collections of cadmium nickel batteries for recycling.

Currently wastes from many of the above sources seem to be increasing whilst on the other hand increased recycling has resulted in a decrease in the amounts of electroplating waste, scrap solvents, cutting oils, contaminated oil, and materials with high calorific value.

The spectrum of wastes is therefore changing and the appropriate regulations will have to keep pace with the changing environmental problems that will arise in their transfer, storage and shipment. Currently with the increased activity in the removal of asbestos, its correct transfer at source into double lined marked polythene bags, its safe transport so that no dust is caused and its correct disposal, needs careful surveillance to ensure the safety of those loading it into bags, those transporting it and those operating the landfill sites.

Having indicated the wide range of transfer, storage and shipment problems that can arise, we will look at examples within a chemical plant and then within a factory considering the detailed regulations for transport and the specific problems of disposal.

2 Handling of Wastes at a Chemical Factory

In general the hazards of the chemical waste streams from a chemical plant should be well known but there have unfortunately been many accidents and errors in handling wastes on plants, thus it is worth while listing some of the types of problems that need attention. The German Standards Institute has published a useful guide on the subject which is available in English entitled 'DIN Safety Data Sheet for Chemical Substances and Preparations' (DIN 52,900), W. Hagemann, Burggrafenstraße 4–10, D-1000 Berlin 30, West Germany.

2.1 Transfer of Toxic Wastes — Solids, Liquids, and Gases

Appropriate protective clothing must be worn when wastes are taken from a plant. Appropriate facilities such as showers and cyanide antidotes need to be available depending on the operation. It may be necessary to check on toxic vapour concentrations, inflammable gas concentrations, the possibility of dust clouds and explosion or inhalation when wastes are removed from a unit. At a manufacturing plant the safety precautions will be well known. Checks must be made to ensure the works staff do not become too familiar with the operation and neglect the appropriate safety precautions. The plant managers and scientific staff must ensure plant operatives, cleaners and stores staff are all well aware of the hazards associated with wastes. External contractors must not be allowed to transfer material when there is a potential hazard unless they qualified to do so or are correctly supervised.

When the wastes may have toxicological properties the advice of environmental health physicists and medical experts should be sought. In major companies such advice is usually available on site. Some toxicological or dermatological testing of a simple kind may be required.

At many plants using catalysts or active carbons specialist firms may be called in to remove the material. Fully self-contained breathing equipment may be required in some instances.

Solid Wastes

Solid waste removal may need consideration of the following problems:

(1) The inhalation of powders — special masks or breathing apparatus.
(2) Dust or explosion hazards — professional advice — the use of special non sparking tools.
(3) Resuspension under windy conditions — hood arrangements.
(4) Pyrophoricity — fire precautions.
(5) Adsorbed vapour release — no naked flames — special flame proof electrical gear.
(6) Dermatological effects — protective clothing — barrier creams.
(7) Eye irritants — goggles.
(8) Adherance to clothing — complete clothing change and showers.
(9) Smell — suction hoods with extract and clean up system.

Liquid Wastes

Liquid waste removal may need the following precautions:
(1) Inflammability — special flame proof pumps — no naked flames.
(2) Toxicity — breathing apparatus.
(3) Skin contact — protective clothing.
(4) Change in viscosity as tank is emptied — homogenisation before discharge. Scraping out of the sludge through an opening end door on the tanker may be necessary.
(5) Corrosivity — water flush facilities of any spills to sewer only if it is safe to do so.
(6) Volatility — special refrigerated facilities may be required if a suitable pressurised tanker facility is not suitable.

In the U.K. low flash point materials are controlled by the Petroleum Regulations Act 1928 and 1936 and the ensuing Statutory Instruments and the Highly Flammable Liquids and Liquefied Petrolum Gases Regulations 1972.

Gaseous Wastes

The removal of gas wastes from a plant is rare. However, a frequent problem is that of faulty valves on gas cylinders. Cylinders are often left exposed to the elements for lengthy periods. They may be too dangerous to open and release the gas. In the United Kingdom the Chemical Emergency Centre specialise in dealing with this problem and run a service for the safe disposal of cylinders. If in doubt do not try and open difficult toxic gas cylinder valves that are stuck. Call in specialist advice. Oxygen cylinders that are suddenly opened can under exceptional circumstances set fire to any bronze fittings.

Much of the above advice will be in house knowledge to chemical producing companies. They have a responsibility to see that the hazards of handling wastes are appreciated by their own staff and also that any waste disposal operator is also fully aware of any potential hazards of materials he is asked to dispose of. Preferably waste disposal contractors for toxic wastes should be members of the national accredited waste disposal contractors Association. In the U.K. this is NAWDC — The National Association of Waste Disposal Contractors.

2.2 Storage of Wastes

The appropriate warning labels giving appropriate safety precautions should be firmly affixed to the containers and preferably painted on them. Good quality containers should be used if any spillage of the waste would be hazardous.

The wastes should be stored prior to recycling or disposal in safe conditions. The quantities should not be allowed to accumulate until they represent a hazard. For some substances the amount that can be stored may be limited by the Major Hazards Legislation or appropriate national response to the E. E. C. Seveso directive (82/501/EEC).

Following an incident at a transport company's warehouse in Salford, Manchester, U.K. in which some 600 large drums of sodium chlorate (29.5 tons) exploded, concern has been expressed in the U.K. about the storage of chemicals at warehouses on their way to either a consumer or possibly a disposal operator. The U.K. Chemical Industries Association has published guidelines for operators and users of warehouses to cover this problem. ("Guidelines for Safe Warehousing", CIA, Alembic House, 93 Albert Embankment, London. SE1 7TU. U.K. Guardian and Daily Telegraph, 9. 9. 83).

This incident does not concern toxic waste which in the normal U.K. circumstances should leave the waste generators premises with documentation indicating its destination at a disposal site or treatment. Toxic wastes should not spend any time in warehouses as a normal matter of course if handled by a responsible waste disposal contractor.

The Salford incident has brought pressure on the Health and Safety Executive (HSE) to ensure that the local authorities, the HSE, the fire and police services are notified of operations involving handling, storing or transporting hazardous substances. Health and Safety at Work. 6 (1) p11, Sept. 1983. (The full report on the incident is available from HMSO entitled 'Fire and Explosions at B & R Hauliers, Salford. 25 Sept. 1982. ISBN 011-8837028).

Some countries have regulations in operation regarding the storage of wastes. Duphar has been involved in a court case in Holland over the storage since 1968 of, what the prosecutor maintains, was hexachlorocyclohexane waste (European Chemical News, *41* [1110], p22, 5 Dec. 1983).

It is therefore important to ensure that if any wastes are stored at a plant that they do not have to be declared to the public authorities.

In general a company will not wish to use space for storage of waste materials which it has to eventually dispose of. However storage on site may occur if:

(1) A firm cannot afford the high waste disposal charges.
(2) There is plenty of space and the company proposes to operate a disposal facility on its own premises when the license has been obtained.

Generally however most companies do not wish to be involved in waste disposal and have not the space for on site storage which may present some hazard.

However Izvestiea has reported that a 600 km section of the Dniestr River was contaminated with raw potassium salts when a waste dam wall at a fertiliser plant collapsed on September 15 1983. 2000 tons of fish were killed. 8 persons had to face trial over the incident. (World Env. Rept. *9* [22], p2–3, 30 Nov. 1983).

The U.S. House of Representatives has voted to prohibit the owner of any storage disposal facility operating under interim status from expanding the capacity by more than 10 per cent unless the storage is in tanks, containers or enclosed piles. (Haz. Mat. Int. Rep. *IV* [44], 1–2, 11 Nov 1983).

Various figures are given for the amount of toxic waste recycled in house in the U.K. It has been suggested that some 8 million tons is recycled compared with the 4M tons that is sent for disposal. If the recycle process is a batch one, appreciable amounts of material may accumulate before a recycle run is carried out.

If the material is a solid and is not drummed but just bunkered, it must not be allowed to spread or dissolve and it must be clearly labelled. If it is liquid it must be stored in suitable tanks with suitable fire precautions if it is flammable. Problems of smell have to be overcome.

The segregation of the different streams of toxic waste in a storage area may be advisable of safety grounds. The cost of waste disposal can be appreciably increased if a toxic waste is allowed to mix with other inert or toxic materials. This is particularly true for chlorinated solvents. They should be kept in a separate container and not mixed with general solvent wastes. The highly dangerous practice of mixing polychlorinated biphenyls with waste oil must be strictly prohibited as it may result in highly toxic vapours being created when the waste oil is burned. Confusion over labelling can cause problems. A situation is known to the author where an HCN generating waste was confused with another very similar one with somewhat disastrous eventual consequences. Waste disposal storage is a responsible job, it cannot be left to the scrap and disposals section with no knowledge of the hazards involved. Training is required.

2.3 Transport of Wastes

Before wastes can be transported off site in the U.K. the sending company has to be sure that the disposal company are going to transport the waste to a landfill site or treatment centre that is authorised to receive it. When this has been ascertained then the various consignment notification procedures will have to be completed for the 'cradle to the grave' transport of the waste. In the U.K. the procedure falls under Section 17 of the Control of Pollution Act. Some materials which may not be subject to this particular clause for the more highly toxic wastes, may require clearance with the landfill site operators to ensure that he is allowed to receive the waste under the detailed terms written into his landfill site license.

The details of regulations regarding the transboundary transfer, packaging, labelling, transport vehicle design, road, rail and sea transport and arrangements to deal with accidents are dealt with in detail in later sections.

The House of Lords Gregson report on hazardous waste urged the introduction of licensing for waste handlers and registration of waste producers. Registration, the report argued, would also require producers of hazardous waste to make quarterly returns of their waste and disposal routes. Whilst local authorities are keen on such regulations there is some opposition from other quarters. (ENDS, p16, July 1983).

2.4 On-site Disposal of Waste

The disposal of wastes by on site processing, be it incineration, water treatment, borehole injection or biological treatment, is increasing in the larger companies because of some governments policies that set out to make the landfill waste disposal process expensive or difficult. The eventual storage of the incinerated ash or sludges on site must be in an environmentally acceptable manner. In some countries the exact details of companies in house disposal practices are subject to full inspection and approval. Assurance may have to be provided that the detailed arrangements for on site storage of the wastes is a long term solution which is environmentally and hydrogeologically acceptable. In the United States it has been estimated that 1% of U.S. waste generators generated 90% of the 40M tons of waste each year and that only 5% was shipped off site for treatment. (HMIR Vol. IV [35] 9. 9. 83).

At Milford Haven refinery they have a waste incinerator on the plant site.

In August 1983 soot particles from the incinerator's 240 ft. high flare stack started a fire in an oil tank. 460,000 tons of oil was set alight and 160 firemen were called to put out the blaze. The operation of waste disposal facilities on one's own premises can lead to disastrous consequences. (Guardian, p2, 2 Nov. 1983).

In the U.K. 78% of the wastes from chemical industry are landfilled (40% in house and 38% off site), 13% is dumped at sea, 5% is chemically treated and 2% subjected to cement encapsulation. ('We care about waste'. Chemical Industries Association, Alembic House, 93 Albert Embankment, SE1 7TU. U.K.).

3 Handling of Wastes at Plants Using Chemicals

Many plants use solvents, acids, organic chemicals, photographic chemicals and generate similar waste streams. Generally the hazards of the waste streams are known but the degree of chemical control over the process may not be as close as at a major chemical plant where the raw material was produced.

3.1 Collection of Wastes

In order to reduce toxic waste disposal costs it is important to segregate the toxic waste and concentrate it if possible to reduce charges for its disposal or recycling. Careful training of staff at all levels on the hazards of the toxic wastes and of rules for its collection are required.

3.2 Storage of Wastes

The same criteria apply as at a chemical plant. However, more care may be required to ensure that staff are protected from exposure to the waste before it is removed.

3.3 Transport of Wastes

If the material is toxic it is best removed and transported from a small company by a waste disposal company familiar with the problems of transporting toxic chemicals. The material may be suitable for recycling in which case the recycling company may arrange to collect.

4 Regulations Controlling the Transfer of Wastes

4.1 The Comprehensive Investigation of the Individual Components and the Cumulative Properties of Special Wastes

Wastes must be notified under the Article 16 of Directive 78/319/EEC if they are on the inclusive lists given in the introductory chapter. The U.K. Section 17 Regulations which cover the transport of the waste from its source to the landfill site and the details of the landfill site license conditions should ensure that the public are protected from chemicals which could be dangerous to a child if he were to ingest 5 ccs of the chemical and that the aquifers from which we draw our drinking water are also protected. However, it is the obedience to the law that ensures our safety and not just the law itself. Lord Gregson in the House of Lords enquiry into hazardous wastes was concerned that it was too easy for someone to set up as a waste disposal operator with little experience and unsuitable vehicles. He suggested that we should have an inspectorate to strengthen the county waste disposal officers in ensuring that the law was obeyed. Such an inspectorate has now been set up to help ensure that the 'cradle to grave' legislation is working satisfactorily.

In the United States this concern is paralleled by the U.S. Environmental Protection Agency's Office of Solid Waste and Emergency Responses decision to deploy its resources in 1984 with priority given to enforcement and compliance inspectors. Permits have been issued in 1983 to all major hazardous waste handlers and facilities. 3% of generators and transporters will be inspected in 1984. (Cathy H. Bombrowski, World Wastes, p34, October 1983).

One of the problems that faces different countries to different extents is that there is money to be made out of illegal dumping of wastes. The fines that are administered are often derisory and the cost of the legal procedures and the collecting of satisfactory evidence is expensive and difficult. In some countries there is a tendency for the population to obey laws more carefully than in others. There is also a marked difference in the way the regulatory authorities have been staffed. The EEC has a very small staff indeed. The U.K. has a staff at the Department of the Environment that is mainly scientific. The legislative action is left to the 43 county waste disposal officers in England and Wales and their counterparts in Scotland and Northern Ireland. In very sharp contrast the U.S.A. with many hundreds of sites needing remedial measures and no site licensing procedure and transportation documentation yet fully operational has a very large staff trained in law with only a few scientific staff at the U.S. Environmental Protection Agency in Washington.

The USEPA have now extended the full Resource Conservation Recovery Act Regulations to include companies generating 25 kg of hazardous wastes per month. They will have to use the manifest procedures which requires (1) the name of company generating the waste; (2) a description of the waste and its hazardous class; (3) the number and types of hazardous waste containers; (4) the quantity of waste; (5) an EPA identification number. (HMIR, 4 Nov. 1983, p5).

Whilst individual countries may have systems in operation for controlling the transport of toxic wastes Mr. Narjes was only able to report on 13 June 1983 that the Federal Republic of Germany, the United Kingdom and Luxemburg had produced reports to the EEC on the disposal of toxic and dangerous wastes. No doubt this situation will improve and the data from other countries within the EEC will be forthcoming soon. (Official Journal of European Committees 26 [C212/31], 8. 8. 83).

Although the EEC Directive on Transfrontier Shipment of Toxic Wastes is still under detailed discussion systems of forms for transferring waste from country to country have been operating satisfactorily for some time. For example, toxic wastes which cannot be satisfactorily treated in the U.K. have been accumulated at Harwell Laboratory from various U.K. concerns and the necessary clearance documents have been obtained for them to be accepted at the Kali and Salz mine in the

Federal Republic of Germany. Once permission for them to be accepted has been received the other necessary forms for rail transport and customs clearance and the Section 17 notification procedures are completed together with the appropriate authorisation for deposition in Germany. Packaging has to be acceptable to Kali and Salz, the British and Continental Railways and the shipping firms. The appropriate multi lingual labelling on each package is affixed and the appropriate warning notices for each country through which the material passes are arranged. These various aspects will be discussed in detail later in the chapter.

4.2 Transfrontier Shipment of Hazardous Wastes

On 17 January 1983 a proposal for a directive on the supervision of control of transfrontier shipment of hazardous wastes was submitted by the Commission to the European Council. Since that date there has been continuing discussion of the exact details of the legislation. The situation was not helped by the disappearance of 41 drums of waste containing a small amount of dioxin from Seveso, Italy in March 1983.

Mr. Narjes was questioned on May 4th 1983 regarding progress with the directive. He replied on June 15th 1983 to say it was still pending. By July the decision had been made to convert the directive into a regulation. This would mean it had to be obeyed immediately. There was also opposition by the Commission to the European Parliament's call for waste to be processed in its country of origin. (Eur. Chem. News. 25. 7. 83, *41*, 1092, p13).

The Germans criticised the directive as not being as tough as their own existing legislation. The French published in the Official Journal (Aug. 2 1983) rules to control toxic waste movement into or across France. Meanwhile, in the U.K., the House of Lords Select Committee on the European Communities opposed the EEC plan to turn the directive into a regulation. The problem of the definition of toxic wastes would have to be solved first so the regulation could work. There is also much controversy about wastes that are going for 'recycling'. This was particularly an issue because of wastes received by a U.K. company, Riafield, from Holland which subsequently turned out not to be recyclable.

The European ministers who discussed the proposal in June 1983 discussed the matter again in November. At this stage discussions commenced on the use of special routes which offered maximum guarantee of public safety. They avoided busy routes, built up areas and peak traffic periods. Only certain border crossing points equipped with special staff would be used. A situation report on the implementation of the regulation would be required every year instead of once every three years.

OECD have already been discussing the issue. They first met in November 1982 and again in July 1983 and in October 1983. They were anxious to harmonize national regulations, to have a prenotification system, to avoid informing all local areas through which the waste passed and to place responsibility on the waste producer.

In November 1983 the U.K. House of Lords Select Committee considered that the EEC should not be panicked into a regulation just because of the issue of the Seveso drums. A regulation would need to be precise and accurate and it would take time to draft it. (Transfrontier Shipment of Hazardous Wastes, Lords Paper 50. HMSO, High Holborn, London WC1).

The U.K. Chemical Industries Association were concerned that wastes were being subjected to tighter control than toxic chemicals used as feedstocks by industry.

The debate still continues. In the U.K. House of Commons Mr. Waldegrave said the U.K. remained unconvinced that recycleable materials should be subject to this kind of legislation. (Mat. Rec. Weekly, *143* [8], p11, 3 Dec. 1983).

The EEC meeting on November 26 1983 went on into the night and ended in deadlock.

The U.K. Government had announced proposals to control imported waste at the end of 1981. The Institute of Public Health Engineers commenting in early 1982, welcomed these proposals and called for the necessary legislative changes to be brought into effect as soon as possible. (Env. Health. Rept. 1982, pp39–40).

Now the French preside over the European Community they will push the legislation on transfrontier shipments of toxic wastes. A compromise between the German and French positions will be required. At the December 1983 meeting the French pressed for a system that would make the producer liable for anything that happened to the waste. The Germans opposed this approach. At the October 1983 OECD meeting the Germans and other nations had approved of a draft set of rules which would place moral but not strict legal liability on waste producers. (Inter. Environ. Rept. *7* [1] 11. 1. 84).

However it now seems that views in Germany have changed and it is reported that Germany is trying to make any new EEC legislation in the field as tough as possible. Currently the Dutch favour a simple notification system for the 50,000 tons of waste shipped each year from Holland through West Germany to East Germany. (Inter. Environ. Rep. *7* [1], p3, 11. 1. 84).

Currently the EEC is opposed to West German suggestions, as are leading waste exporters in Netherlands and Italy. Further discussion will take place on March 1st 1984.

Unfortunately at the present time (March 1984) the details of the transfrontier shipment of waste legislation are not yet clear. From the above account of the events leading up to an EEC decision the reader will be able to gauge the issues that are likely to come out in the final directive or regulation. It is very clear that all nations require some control. It is also clear that transfrontier shipment is currently taking place with government consent and suitable documentation. The harmonisation of these policies has yet to be achieved.

Hardy Wing of the Ontario Ministry of the Environment has said that all the Canadian provinces agree on the importance of a national hazardous waste manifest to account for the safe transportation of hazardous wastes across provincial lines. A group of federal provincial and local waste management and regulatory officials has designed a draft manifest form which is currently being reviewed by all the provincial environmental agencies. (Hazardous Materials Intelligence Report *IV* [40], p4, 14 Oct. 1983).

In West Germany the Bundesrat (upper House of Parliament) has given its approval to new rules which propose a permitting system for foreign waste shipments through the country.

The European Environmental Bureau, the voice of environmentalist groups in Brussels, has criticised the delay by the common market's environment ministers of

EEC, to produce rules on the transfrontier transport of toxic wastes. (European Chemical News, *41* (1107), p26, 14 Nov. 1983).

Within the United Kingdom it has been found that the practice of transporting wastes from the Republic of Ireland through Northern Ireland via the shortest sea route from Larne to Stranraer in Scotland and then to disposal facilities in Great Britain has placed a great pressure on the district council for the area at the land frontier in trying to comply with the Section 17 Special Waste Regulations. Similar and possibly more onerous administrative duties will fall on these authorities if the EEC proposals come into operation. (NAWDC News, p19, December 1983).

Patricia Shaw has reviewed the deliberations of the United Nations and the OECD expert committee on the transport of dangerous goods. Whilst they have been primarily concerned with pure chemicals their deliberations do cover the problems of wastes when they may present a problem. (P. M. Shaw, International legislation and the transport of hazardous wastes UNEP Industry and the Environment. Special Issue 1983).

The following organisations may issue documents with information relevant to toxic waste transfer, storage and shipment.

IMO — International Marine Organisation
ICAO — International Civil Aviation Organisation
IAEA — International Atomic Energy Agency
OCTI — Central Office for International Rail Transport
IATA — International Air Transport Association
ICC — International Chamber of Commerce
ICS — International Chamber of Shipping
IRU — International Road Transport Union
CEFIC — European Council of Chemical Manufacturers Federation
SEFA — European Syndicate for Steel Drums

4.3 Labelling and Packaging of Wastes

Cumberland and Feates have reviewed the various schemes for the rapid identification of chemicals and their hazards in an emergency.

(Chapter 2. Transport of Hazardous Materials. The Institute of Civil Engineers, London, 1978. Proceedings of a symposium held in London on 15 Dec. 1977).

The hazard information systems developed for pure chemicals can be applied to wastes to indicate the nature of the hazard within certain limits.

The UN hazard classification system classifies pure chemicals in 9 classes. 1. Explosive; 2. Gases; 3. Inflammable liquids; 4. Flammable solids; 5. Oxidising agents; 6. Poisonous substances; 7. Radioactive substances; 8. Corrosive substances; 9. Miscellaneous dangerous substances; and uses a system of diamond symbols to indicate the hazard.

The Agreement concerning the International Carriage of Goods by Road (ADR) and the International Regulations concerning the carriage of Dangerous Goods by Rail (RID) have developed a combined information panel system using a different set of numbers. This European Hazard Identification Number (KEMLER) is a code of properties which needs to be interpreted by the fire brigade. An example is shown in Fig. 4.1.

The HAZCHEM scheme is an action code telling the brigade what to do. The U.K. as a signatory to this agreement must use this system for goods sent to Europe. However, within the U.K. there is a system known as U.K. Hazard Information System (UKHIS). A typical label is shown in Fig. 4.2.

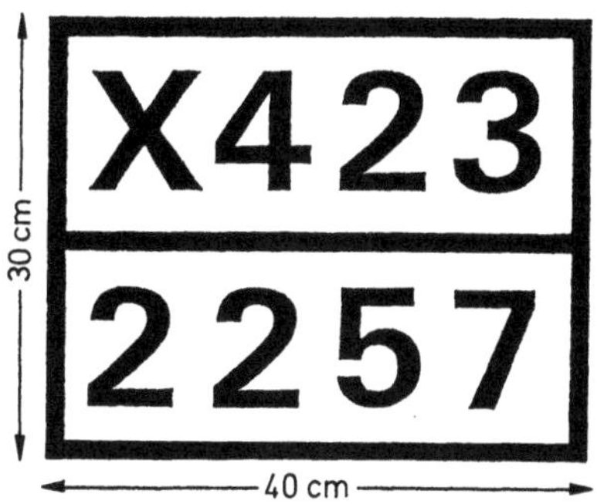

Fig. 4.1. ADR/RID hazard information panel

Fig. 4.2. UKHIS hazard information panel

From the actual number in the code the fire and emergency services know exactly how to deal with the hazards. The code is described in Fig. 4.3.

There are therefore established systems for marking tanker loads of hazardous pure chemicals. These same symbols can be used on vehicles of waste to indicate the hazards of the waste. The U.K. HAZCHEM scale card has the advantage that it tells the emergency service what to do.

In addition to having the HAZCHEM sign vehicles may also carry a 'Tremcard' (or transport emergency card) which gives more detailed information than the simple HAZCHEM sign.

A similar system for labelling pure chemicals is in operation in the United States known as the MCA CHEM-CARD — Transport Emergency Guide.

The extent to which these various systems should be used for waste chemicals has not been finally agreed by the various organisations concerned. Moreover these systems are in general designed for use on tankers full of chemicals. Waste chemicals may be in smaller volumes and in drums. A vehicle may contain a mixed load of chemical wastes. The labelling problem to cover a transport emergency then becomes a matter for very careful consideration by the company producing the waste who will know its composition in more detail than the company which is transporting the waste.

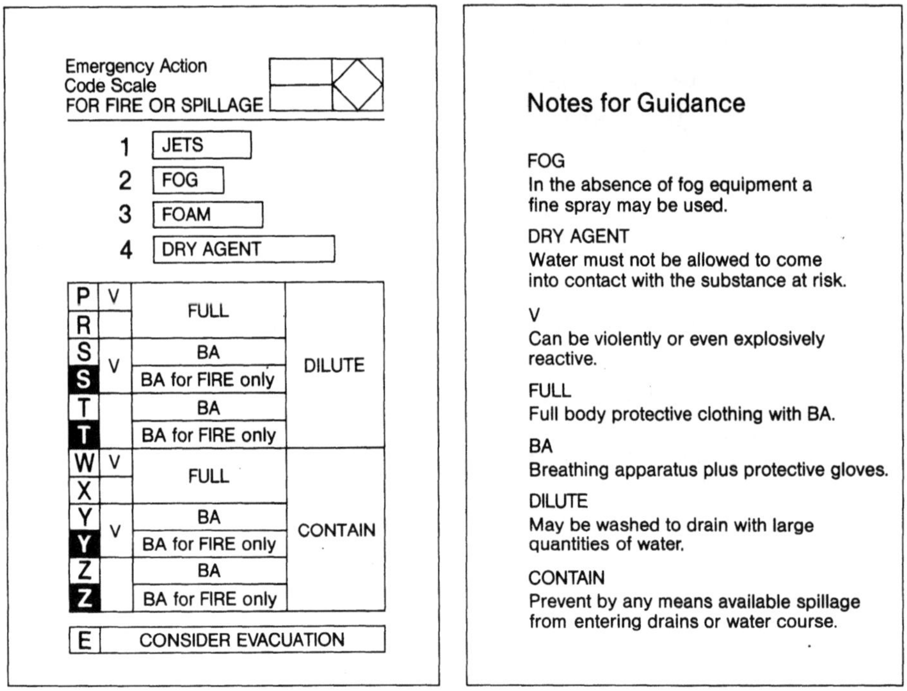

Fig. 4.3. Hazchem scale card

The problems that can arise during the conveyance of hazardous waste have been discussed by L. E. Baker. (L. E. Baker. The Conveyance of Hazardous Waste — The problems that arise. Public Works Congress 1970. Under auspices of the U.K. National Association of Waste Disposal Contractors).

Baker commented that in 1978 the legislators had been quilty of largely ignoring the problems of waste disposal. The international agreement for hazardous substances although detailed and extensive, made no obvious attempt to recognise the problems of waste disposal.

One approach suggested at that time was to substitute for the U.N. number on the so called HAZCHEM plate the words 'DISPOSABLE WASTE' and allocate the most severe HAZCHEM code 4WE to all wastes. This suggestion was quickly dropped and a Working Party was set up which decided on 18 groups of waste shown below.

1. Hazardous waste liquid containing acid.
2. Hazardous waste solid or sludge containing acid.
3. Hazardous waste liquid containing alkali.
4. Hazardous waste solid or sludge containing alkali.
5. Hazardous waste inflammable liquid flash point less than 23 °C.
6. Hazardous waste inflammable liquid flash point 23 °C to 61 °C.
7. Hazardous waste inflammable solid or sludge, n.o.s.*
8. Hazardous waste miscellaneous packaged, n.o.s.
9. Hazardous waste solid or sludge, n.o.s.
10. Hazardous waste liquid, n.o.s.
11. Hazardous waste solid containing inorganic cyanides.
12. Hazardous waste liquid containing inorganic cyanides.
13. Hazardous waste solid or sludge, agrochemicals, toxic, n.o.s.
14. Hazardous waste liquid, containing inorganic cyanides.
15. Hazardous waste solid or sludge, agrochemicals, toxic, n.o.s.
16. Hazardous waste liquid, agrochemicals, toxic, no.o.s.
17. Hazardous waste containing isocyanates, n.o.s.
18. Hazardous waste containing organo-lead compounds, n.o.s.

* n.o.s. — not otherwise specified

These groups decided in the U.K., were allocated 'quasi;' U.N. numbers and given appropriate HAZCHEM codes.

Since this time progress has been made. On February 9th 1984 the Health and Safety Commission in the U.K. have issued a revised edition of the approved substance identification numbers, emergency action codes and classification for dangerous substances conveyed in road tankers and tank containers. It supercedes an earlier listing published in July 1981. (British Business, *13* [7], p286, 17 Feb. 1984).

The EEC has also produced the 5th update of the 1967 Directive on Classification, Packaging and Labelling of Dangerous Substances which will have to be complied with by January 1st 1985. (Int. Env. Reporter, *6* [10], p453, 12 Oct. 1983).

These documents do not however give much attention to the problem of the labelling of waste.

However the following U.N. numbers have now been allocated in the 7000 series for wastes (see Table 4.1).

4.4 Regulations Regarding Road Transport

The regulations regarding the transport of materials by road are very extensive. In the United Kingdom there are special regulations covering the following:

Explosives —	The Conveyance Explosives Bylaws 1958 (SI 1958, No. 230)
Petroleum Spirits —	The Petroleum Spirit (Conveyance by Road) Regulations 1966 (SI 1966, No. 1190)
Carbon Disulphide —	The Petroleum (Carbon Disulphide) Order 1968 (SI 1968, No. 571)

Other Dangerous Substances —	The Petroleum (Liquid Methane) Order 1957 (SI 1957, No. 859)
Dangerous Substances —	(Conveyance by Road in Road Tankers and Tank Containers) Regulations 1981 (SI 1059/81)

There are no specific regulations regarding wastes but obviously if the waste is similar to one of the above classifications the appropriate regulations are a good guide as to the requirements that need to be followed.

However the transporter of wastes is faced with a difficult problem in that his waste will probably vary in composition from load to load and he needs to purchase a multipurpose vehicle that can varsify a wide range of regulatory conditions and also be able to stand the corrosive and other problems placed by a wide range of wastes.

Table 4.1.

Substance Identification Number	Substance	Emergency Action Code (See Fig. 4.3)	Additional Advice on Personnel Protection
7006	Hazardous waste, liquid containing acid	2 WE	
7007	Hazardous waste, solid or sludge, containing acid	2 WE	
7008	Hazardous waste, liquid containing alkali	2 WE	
7009	Hazardous waste, solid or sludge, containing alkali	2 WE	
7010	Hazardous waste, flammable liquid flash point less than 23°′ C	3 WE	
7011	Hazardous waste, flammable liquid flash point more than 23 °C but less than 61 °C	3 W	
7012	Hazardous waste, flammable, solid or sludge, N.O.S.*	3 WE	
7013	Hazardous waste, including water reactive substances, and packages, N.O.S. (relates to drums, jerricans, boxes and other receptacles)	4 WE	A*
7014	Hazardous waste, solid or sludge, N.O.S.	2 X	
7015	Hazardous waste, liquid, N.O.S.	2 X	
7016	Hazardous waste, solid or sludge N.O.S.	2 X	B
7017	Hazardous waste, liquid, toxic, N.O.S.	2 X	B
7018	Hazardous waste, solid containing inorganic cyanides	4 X	
7019	Hazardous waste, liquid containing inorganic cyanides	4 X	
7020	Hazardous waste, solid or sludge agrochemicals, toxic, N.O.S.	4 WE	B
7021	Hazardous waste, liquid, agrochemicals, toxic, N.O.S.	4 WE	B
7022	Hazardous waste, containing isocyanates, N.O.S.	4 WE	C
7023	Hazardous waste, containing organo — lead compounds, N.O.S.	4 WE	BC

* Not otherwise specified

The Code of Practice published by the U.K. National Association of Waste Disposal Contractors gives an extensive list of the regulations that control the transport and movement of waste in the U.K.

Apart from the construction of the vehicle to fulfil the regulatory requirements and the labelling aspects mentioned in the previous section, the following points need to be considered:

— The labelling of packages and drums carried on vehicles.
— The training/instructions to driver operators. This is to some extent covered in the U.K. by the Health and Safety at Work Act and the provisions associated with the Heavy Goods Vehicle License.
— The nature of the wastes and the problems of letting sludges drain out.
— The cleaning of vehicles between collection runs. Serious accidents have occurred when acids have been pumped into vehicles previously containing organic wastes that have not been completely removed.
— Many tankers have a fully opening rear door so that sludges can be pulled out using long scrapers.

The United Kingdom Health and Safety Executive have given consideration to the specific problems of specifications for the design and use of vehicles for the collection of wastes but as yet no formal guidance document is available.

The routes that vehicles travel when carrying waste loads may have to be planned to avoid adverse reactions from the public. The hours at which the vehicles operate may likewise be said by some to cause a public nuisance. Often local opposition to waste disposal operations is based on vehicle movements so this matter may have to be given careful consideration.

The training aspect of drivers has been well covered by L. Baker (loc. cit.) and also in a document by the U.K. Chemical Industries Association. (Drivers and Hazardous Loads — A training manual for road tanker drivers).

The points that need to be covered in training will include:

(a) The hazards and physical and chemical properties of the wastes.
(b) Emergency action — fire fighting, first aid, spillage.
(c) The use of protective clothing and breathing apparatus.
(d) The proper use of valves, pumps and vehicle equipment.
(e) Proper securement of loads, particularly packages.
(f) Communications in emergencies.
(g) A knowledge of regulations.
(h) Vehicle stability and the effect of contents.

4.5 Accidents During Transport

During road, rail or sea transport chemical waste may be involved in accidents which may involve spillage, fire, or members of the public may come in direct contact with the waste.

For road transport there are a number of well developed systems for dealing with accidents involving chemicals which may be applicable to some extent to wastes.

The system of labelling of wastes that has already been described is often sufficient to give the emergency service the information they require in order to render the situation safe. However, the applicability of the labelling system to waste is not as universally accepted as it might be. The UN 7000 series numbers adopted in the U.K. which have already been mentioned is not yet in use in Europe.

The European Parliament and the European Commission have agreed that the U.K. Hazchem Code should be extended to apply to all hazardous waste materials. The emergency action code has the merit of instructing firemen, police and rescue services on exactly what should be done in the event of an accident. The Commission plan to issue draft proposals early in 1984 so that a regulation can come into effect in summer 1984. (Hazard. Cargo. Bull *5* [1] p4–5, Jan. 1984).

In the United States the Manufacturing Chemists Association have a free phone telephone system which will link any enquirer with the Chemtrec Centre in Washington where data is held on a very wide number of trade named and chemically identified chemicals. The service will pass on to the enquirer the relevant information from a set of data files supplied by the chemical manufacturing companies. The system operates 24 hours a day but does not have any personnel available to deal with incidents. This is left to the local emergency services. Each year they answer many thousands of calls, some of which may refer to wastes.

In the United Kingdom the Chemical Industries Association and the Department of the Environment funded the setting up of a National Chemical Emergency Centre at Harwell which is manned 24 hours a day. It is only available on an ex directory number to the emergency services and other chemical emergency centres throughout the U.K. Originally the data supplied by the Chemical Industries Association was accessed by fire brigades directly on a dedicated main-frame computer. However, this system has now been augmented by the CHEMDATA system to which most U.K. fire brigades belong. The CHEMDATA system supplies the brigades with a 4 monthly updated hard disc (or a set of floppy discs) which contain emergency response information on some 60,000 trade named chemicals. No specific wastes have yet been included on this data base but the 24 hours service will enable fire brigades to obtain advice from a qualified chemical accident expert on the action that should be taken for accidents involving a waste.

The data files can be supplied with a translation file which will enable the user to obtain the information in German, Italian, French or Spanish. The system is already in use in Eire and Australia and shortly is to be installed in a number of other countries.

In the United States some four billion tons of hazardous waste are transported each year. The National Academy of Sciences Committee has reported that the safety record for transporting the roughly 2400 hazardous materials is good in comparison with that for general transportation. Nevertheless 6,115 accidents involving hazardous materials in 1980 caused 19 human deaths. They view the current piecemeal regulation of hazardous material transport as unnecessarily complex. They also view one major area of neglect is in the preparedness for emergencies involving transported hazardous wastes and suggests that the U.S. Department of Trade should develop a master plan to train both public servants and private personnel. Courses should be held by certificated state instructors. (Ind. Res. & Dev., *25*, Sept. 1983).

The concern to improve safety standards and reduce accidents during transport of wastes is illustrated by Chemical Waste Management Inc. of Oakbrook, Illinois who announced that the company's 1982 safety record of 0.755 accidents per million miles of transport of hazardous waste was 41 % better than the national average of 1,289 accidents per million miles of transport for similar sized private carriers of similar size in the U.S.A. (Haz. Intelligence Materials Report, p7, 29. 7. 83 *4* [29]).

In the United States the public attitude towards hazardous wastes hauliers has been very militant. Carl Smiley reports that U.S. Pollution Control Inc. ensures that its drivers have 10 years truck driving experience. They are given full training and the company has a full time safety director. When trucks of polychlorinated biphenyls are transported they have two drivers. Especially dangerous loads are accompanied by an escort. The drivers carry a complete set of safety gear in their cabs. (C. Smiley, WSPCI Operates Wastes Haling Division. World Wastes, Sept. 1983, p30).

4.6 Transport of Hazardous Wastes by Sea

The transportation of wastes by sea falls into two categories:

(A) The transportation from one country to another.
(B) The transport and incineration of hazardous wastes at sea.

(A) *The Transportation from One Country to Another*

The transport of waste chemicals by sea is subject to new regulations published by the International Maritime Organisation (IMO). In 1983. They replace earlier suggestions produced by (IMO) in 1972. The regulations apply to dangerous goods (wastes and pure chemicals) packed in packages, boxes and drums, portable tanks, tank vehicles, freight containers and other units.

Particular attention is given to improving the reporting of incidents involving the loss of chemicals overboard (Lloyds list, 28. 1. 84).

Some ports have their own detailed regulations regarding the carriage of 'dangerous' goods. The Port of London Authority (PLA) has a Schedule of Dangerous Goods first published in 1975 which controls goods within the docks premises of the PLA. Chemicals are divided into ten classes. Each class is further divided into three sub groups. Group I may be deposited in sheds for up to 2 weeks. Group II must be passed direct from land transport to the ship. Group III materials may only be brought into PLA premises with special permission. Whilst the document contains no specific reference to waste it is obvious that any wastes must conform to the instructions relative to its most dangerous constituent. Similar regulations exist at other ports and enquiries should be made before wastes are consigned to ensure no infringement of regulations will occur.

There was considerable concern expressed in the U.K. parliament when chemical wastes from Holland were shipped into the U.K. as a 'recyclable' material. The company that received the wastes was unable to recycle them. Since that incident close cooperation between the Dutch and U.K. authorities have ensured that such irregular practices are now far less likely to occur.

(B) *The Incineration of Wastes at Sea*

The technology of burning wastes at sea has been studied in some detail. In 1984 the U.S. will again authorise the incineration at an EPA designated site 150 miles off shore in ships 375 feet long filled with custom engineered liquid injection and incineration systems. The burning rate of 28 tons per hour is equivalent to one 55 gallon drum per minute.

Toxic wastes and hazardous chemicals are now being placed in landfills. The alternative of ship borne incineration is viewed as a 'stop gap' since no one knows for sure how long existing landfills will be effective in isolating their poisonous contents according to World Wastes (Jan. 1984, p. 6.).

There has been difficulty with some of the previous ship borne incinerator runs. The boats have become polluted. The new equipment will ensure that the previous problems do not arise.

4.7 Transport of Hazardous Wastes by Rail

The transport of wastes to a suitable disposal or treatment facility can often constitute the major part of the disposal costs. In the United States a very much higher proportion of the country's pure chemicals are transported by rail than is the case in the countries of Europe. It is also probably true that there is a higher proportion of rail transport of hazardous wastes in the United States than in Europe.

Peirce and Pierson however have recently analysed the problems of rail transfer for ordinary solid wastes in North Carolina, U.S.A. (J. J. Peirce and B. A. Pierson. Waste Management and Research, 1983, 1, 127–8).

As a result of a survey of the literature and personal communications they concluded that there was little understanding of the available designs and associated costs of rail haul transfer systems for ordinary municipal solid wastes. They make no mention of hazardous wastes.

The regulations that apply to the rail transfer of hazardous chemicals are well developed in many countries. When wastes are transferred it is normal for them to have to comply with the same regulations as those that apply to pure chemicals which present the same hazards. However few railway companies have rolling stock dedicated to the task of hazardous waste disposal. The cleaning out, filling and emptying of the rolling stock would be carried out by experts familiar with the hazards presented by the wastes. When the waste is drummed or packaged it has to be accompanied in most countries by the same type of documentation as is required for pure chemicals.

5 Handling and Storage of Wastes at a Treatment Plant or Disposal Site

5.1 Reception of Wastes

The appropriate documentation covering the load of chemical wastes may need to be produced by the driver of the vehicle to the staff at the reception point for the waste. This will have to tie up in the U.K. with an advance agreement with the recipient that the waste is acceptable at the treatment plant or disposal site.

Samples may need to be taken in order to decide on the exact treatment to be given to the waste. The problem of obtaining a representative sample of wastes is a difficult one which will not be dealt with here.

Once the waste has been accepted the driver will be told when and where to unload drums, tip solids or discharge liquids.

Considerable technical supervision will be required at this stage to ensure that no accidents arise through the admixture of incompatible wastes. The tragic accident that occurred in the U.K. when a driver was killed as a result of hydrogen sulphide released when he discharged his load of liquid into a mixing tank has made U.K. operators very conscious of the care that is required at this stage of waste handling.

The careful labelling of loads is crucial during the unloading so that the identity of each drum is not lost.

5.2 Storage of Wastes

Considerable attention needs to be given to the temporary storage of wastes at a treatment (or even a disposal) site. The different type of wastes, acids, alkalis, cyanides, solvents, solid organics, chlorinated solvents, will all need different storage areas prior to treatment.

The drums, sacks, boxes, bottles and other containers need to be examined to ensure that they are in good condition if it is intended to store them for some time before treatment. If there is any doubt as to the integrity of the packing the container should be placed in a larger good quality container which should then be appropriately labbelled.

The storage areas should be fitted with bunds and drainage systems and suitable fire fighting and first aid facilities should be available in storage areas.

Staff operating in storage areas should be given good training in the hazards of the wastes and in emergency and first aid procedures.

Some of the worst hazardous waste incidents have occurred in storage areas at waste treatment centres. The extent of the fire damage has been exacerbated by the large amount of waste stored under poor conditions at the premises.

It should be the aim of the plant management to keep the total inventory of toxic wastes present on his site as low as possible.

Storage of wastes for which there is no suitable alternative disposal route is allowed in the deep underground salt mine of Kali and Salz at Heufe Neuroda in West Germany. The packaging of the wastes, the vapour pressure of any constituents of the waste have to conform to tight specifications.

6 Conclusion

The chemical industry has developed excellent labelling systems for pure chemicals. The system has been extended to cover waste chemicals. The universal acceptance of such systems to waste labelling during both handling, storage and shipemnt will do much to allay public concern over chemical wastes. Whilst this concern has previously concentrated on chemical waste disposal, it is slowly moving into concern over waste transport. The continued co-operation by the chemical and waste disposal industry to improve standards and training will do much to reduce any potential accidents. It will allay public concern and will probably result in well thought out governmental regulations based on existing practices which have been already proven to be satisfactory.

References

1. "Guidelines for Safe Warehousing", CIA, Alembic House, 93 Albert Embankment, London, UK. Guardian and Daily Telegraph, 9. 9. (1983).
2. HMSO: Fire and Fire Explosions at B & R Hauliers, Salford, 25. 9. (1982).
3. European Chemical News, *41*, 22 (1983).
4. World Env. Rept. *9*, 2–3 (1983).
5. Haz. Mat. Int. Rep. *IV*, 1–2 (1983).
6. ENDS, p 16 (1983).
7. HMIR, Vol. IV (35) (1983).
8. Guardian, p 2, 2. 11. (1983).
9. "We Care About Waste, Chemical Industries Association, Alembic House, 93 Albert Embankment, London, UK.
10. C. H. Bombrowski, World Wastes, p 34 (1983).
11. HMIR, p 5 (1983).
12. Official Journal of European Committees 26 (C212/13) (1983).
13. Ener. Chem. News *41*, 1092, p 13 (1983).
14. Transfrontier Shipment of Hazardous Wastes Lords Paper 50. HMSO, High Holborn, London.
15. Mat. Rec. Weekly *143*, 11 (1983).
16. Env. Health Rep. 39–40 (1982).
17. Inter. Environ. Rep. *7* (1984).
18. Inter. Environ. Rep. *7*, 3 (1984).
19. Hazardous Materials Intelligence Report *IV*, 4 (1983).
20. European Chemical News *41*, 26 (1983).
21. Nawdc News, p 19 (1983).
22. P. M. Shaw, International legislation and the transport of hazardous wastes UNEP Industry and the Environment. Special issue (1983).
23. Chapter 2. Transport of Hazardous Materials. The Institute of Civil Engineers, London (1978).
24. L. E. Baker, The Conveyance of Hazardous Waste — The Problems that Arise. Public Works Congress. Under auspices of the U.K. National Association of Waste Disposal Contractors (1970).
25. British Business *13*, 286 (1984).
26. Int. Env. Reporter *6*, 286 (1983).
27. Hazard. Cargo Bull. *5*, 4–5 (1984).
28. Ind. Res. & Dev., *25* (1983).
29. Haz. Intelligence Materials Report *4*, 7 (1983).

21. C. Smiley, WSPCI Operates Wastes Hauling Division. World Wastes, p 30 (1983).
22. J. J. Peirce, B. A. Pierson, Waste Management and Research, *1*, 127–128 (1983).

Detoxification and Decomposition

D. Martinetz

Toxic by-products and residues, which may not be released untreated into the environment, are produced by industry and research in the form of solids, liquids, and gases.

First, general possibilities of detoxification of industrial exhaust gases are to be considered; distinction is made between procedures which separate toxic gases from the rest of the exhaust and methods by which toxins are destroyed. Detoxification of laboratory waste gases is referred to in tabular form.

Further, the most important thermal and chemical detoxification methods for industrial solid waste are discussed. Possibilities to rapidly detoxify solid laboratory wastes are also presented in tabular form.

Industrial waste liquors and waste waters are discussed in respect of mechanical and physico-chemical separation procedures of toxins; if recycling of the separated chemicals is not possible, an additional detoxification step must follow. A variety of chemical transformation reactions into less toxic or nontoxic compounds are feasible and lead to either reusable, disposable, or biologically transformable substances; these chemical procedures are discussed and evaluated. Biological degradation is only briefly considered in this chapter.

Contents

1 The Detoxification of Industrial Waste Gases

The following processes are available for the detoxification of waste gases by removing the toxic components or destroying them:

- Condensation from fogs and vapours.
- Removal by adsorption on the surface of solid phases such as active carbon or silica gel.

- Absorption by chemical reaction at solid surfaces.
- Absorption in a washing liquid by physical or chemical sorption.
- Oxidation by thermal combustion; catalytic oxidation by combustion far below the normal oxidation temperature.
- The reduction of the toxic components of waste gases.

Recently reports have been published of the application of micro-organisms to waste gas purification [e.g. 64].

When choosing a process for waste gas purification it is necessary to distinguish in principle between low concentrations of various polluting compounds [1] from open processes or high concentrations of pollutants consisting mainly of one or a few compounds from a closed process; it is only in the later case that it is possible to justify economically a requirement to hermatize the process.

1.1 Process for Removal of Toxic Components from Waste Gases

1.1.1 Waste Gas Condensation

Condensation processes constitute detoxification processes in so far as they can be used to separate dangerous solvent vapours from exhaust gases, as for example with the solvents methylene chloride, ethyl acetate and ethanol which are used in the manufacture of plastic films [45]. Condensation is, in general, employed as a preliminary purification process for high concentrations and is combined with a subsequent adsorption process. Further applications lie in the condensation of waste gases from oil refineries, of odorous waste gases having a high water content in the foodstuff and allied industries and in the condensation of ammonia, chlorine and SO_2 under pressure and with refrigeration.

1.1.2 Adsorptive Waste Gas Purification

The principle of this method consists of the concentration and fixing of the gaseous or vaporous substance at the gas/solid interface.

The principle areas of application are:

- Recovery of chemicals, particularly solvent vapours, from extraction plants, paintshops, foil manufacture, dry cleaning etc.
- The separation of toxic gases and vapours from industrial exhaust gases or from the surrounding air by means of breathing mask filters.
- Air filtration in ventilation and air-conditioning plants for domestic accommodation, working and storage areas and air filtration in the ventilation plants of industrial production areas, laboratories etc.
- Specialized processes for the separation of gases and vapours.

In general substances with a rigid pore structure (active carbon, aluminium oxide, silica gels, zeolites) are employed as adsorbents) in particular cases substances with flexible pore volumes (tetracalcium aluminate hydrate, swellable layered silicates) find application.

Various types of active carbon and silica gel which are characterized by high porosity and hence a high surface area are frequently employed for the adsorption of gases and vapours. Thus, the internal surface area of active carbon generally amounts to ca. 1000 m^2/g, the most common pore diameter is 1–2 nm [2]. Since the particle sizes of the gases and vapours to be removed lie between 0.0001 and 0.01 μm, they are easily able to penetrate to the site of adsorption. The criteria for the choice of adsorbent are reported in [3]. When organic molecules are being adsorbed there is the additional effect of capillary condensation brought about by adhesive action, which finds useful application in the elimination of toxic or revolizable vapours from waste gases. Since active carbon possesses hydrophobic properties it can be applied to the adsorption of many solvent vapours from air. Adsorption is, in principle, a discontinuous process; it can, however, be employed continuously, for example by the utilization of two adsorption units with changeover (see Fig. 1.1). Modern active carbon air filters for the destruction of pollutants work on the fluidized bed principle with thermal regeneration [4] and catalytic combustion of the odorous or noxious material (see Fig. 2 for example). In some cases the adsorption or the subsequent desorption is accompanied by a chemical reaction.

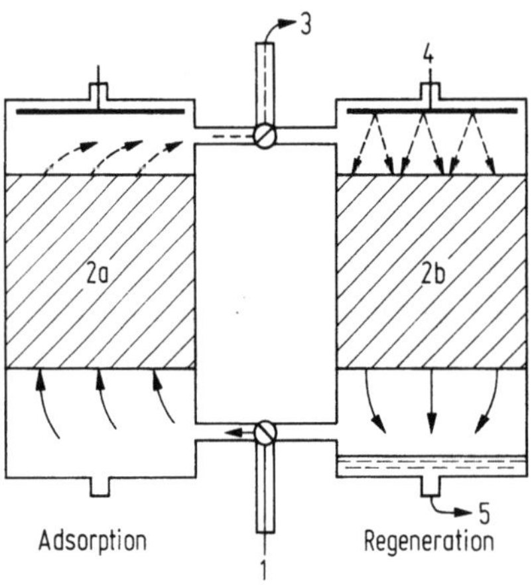

Fig. 1.1. The principle of active carbon adsorption by the solid bed process. 1 Waste gas entrance, 2a Active carbon, 2b Apent carbon, 3 Purified gas, 4 Steam, 5 Eluate containing impurities

Thus in the presence of atmospheric oxygen H_2S is accumulated as elementary sulphur which can be reutilized after elution [5]. In the presence of atmospheric oxygen and water vapour sulphur dioxide yields sulphuric acid which can also be utilized. In general active carbon offers many possibilities for recycling, as is indicated by the many practical examples (recovery of solvents from the exhaust air in mechanized dry cleaning [6], carbon disulphide from the effluent air from artificial silk, viscose and rayon production etc) that have been documented.

The amount and concentration in which a gaseous or vaporous substance is produced is decisive for whether it should be recovered or destroyed. Recycling is uneconomic at concentrations of less than 3 g/m^3 waste gas; this limit is displaced to higher concentrations for mixtures and for water-soluble materials.

Filters for protective facemasks for the removal of toxic substances (especially for accidents) constitute a quite different example of application, since the adsorption capacity is different for individual gases and solvents — depending on their hydrophobic nature — for example their capacity for n-butanol is 20-fold that for methanol. Larger molecules are preferentially adsorbed, small molecules, on the other hand, such as oxygen, nitrogen or carbon monoxide are not adsorbed, so that they pass through the filter unhindered. For this reason, for the elimination of CO for example, special filter combinations are necessary consisting of a dry filter followed by a catalyst (hopcalite), which bring about its oxidation to CO_2.

On the other hand, if the molecules are too large they cannot infiltrate into the pores of the active carbon and thus "penetrate" the filter. According to TRAUBE's rule long-chain molecules are adsorbed better than branched-chain ones. The dew point of the substances also plays a role. The higher it is the more easily condensation takes place in the larger capillaries of the carbon.

A high partial pressure at normal temperatures can displace the adsorption equilibrium in the direction of desorption. Water vapour coats the active surface of the carbon and can even displace already adsorbed molecules. Atmospheric humidity, therefore, greatly limits the working life of mask filters. The adsorption capacity depends primarily on the following factors:

It increases with

- increasing C content, } of the substance to be adsorbed
- increasing molecular mass, } of the substance to be adsorbed
- increasing boiling point, } of the substance to be adsorbed
- increasing pressure,
- and decreasing working temperature,

and can break down if large particles block the pores (hence a dust prefilter) or if the concentrations are too high. Since the pore size distribution of active carbons varies greatly depending on their method of manufacture so too does their loading capacity. Once its loading capacity is reached the carbon must be regenerated by either the thermal or the elution technique. In some cases this can be accomplished by nearly saturated water vapour. The adsorbed substances are displaced from the pores by the capillary condensation effect. Desorption by steam is preferentially employed for recycling purposes, while inert or exhaust gas is usually used for the desorption of noxious or odorous substances, which are then catalytically oxidized. Substances that have been desorbed using steam can be condensed without losses, but are recovered containing more or less water, which can extend to complete water miscibility (alcohol), so that a dehydration step must be undertaken before the recovered material can be reused [7]. Elution techniques are to be recommended if the adsorbed molecules have undergone chemical reactions with the adsorbent surface.

In general, adsorption processes using active carbon alone are only employed for the vapours of substances whose boiling points (for room temperature adsorption) are greater than 25 °C. Otherwise the carbon is impregnated with chemically active substances. This leads to a combination with the large surface area with chemical reactivity, so that various gases and vapours (e.g. H_2S, HCN, Hg vapour) can be bound in large quantities (see Table 1.1). When working at high temperatures in the presence of atmospheric oxygen the possibility of fire should be taken into account, particularly during the summer months, if the adsorption is extremely exothermic. As was mentioned previously the removal of SO_2 is an important appli-

Table 1.1. Adsorption data for various gases and vapours at a concentration of 0.1 vol. % in air on nonimpregnated active carbon and active carbon impregnated with heavy metal compounds

Compound	Molecular weight	Uptake $g \cdot l^{-1}$ filter volume
Nonimpregnated active carbon		
Ammonia	17	1.2
Hydrogen cyanide	27	1.9
Propylamine	59.1	6.5
Chlorine	71	7.7
Hydrogen sulphide	34	7.7
Methanol	32	25
Carbon disulphide	76.1	69
Benzene	78	96
Ethyl chloroformate	108.5	130
Chloropicrin	164.4	135
Tetrachloromethane	153.8	173
Bromine	158	173
Active carbon impregnated with heavy metal compounds		
Ammonia	17	15.4
Hydrogen cyanide	27	17.3
Chlorine	71	27
Cyanogen chloride	61.5	38
Phosgene	98.9	50
Hydrogen sulphide	34	50

cation. Adsorption is only capable of removing a more or less large proportion of the SO_2 in a waste gas.

Some techniques involve blowing alkaline dust (calcined dolomite or lignite fly ash) into the combustion zone. In England and the United States processes are in use involving alkaline aluminium hydroxide at 330 °C [9]. Regeneration is performed reductively using producer gas at 650 °C. The H_2S-containing gas produced is processed to yield elementary sulphur. Other processes use active carbon. In these the cooled and purified raw gas is separated from the active carbon, which is not consumed, as sulphuric acid in the presence of O_2 and water vapour [10]. This principle is applied in the BABCOCK BF process. In this other pollutants such as chlorine, fluorine and SO_2 are adsorbed simultaneously. The exhaust gases flow through a vertical silo-shaped adsorber from the centre to the outside. The active carbon flows through the adsorber from top to bottom over sheets arranged like venetian blinds (see Fig. 1.2) so that a continual removal of the loaded coke is guaranteed. Desorption is carried out at 600 °C. During this process the bound sulphuric acid is decomposed and reduced to SO_2 by the oxidation of carbon, expelled and further processed to H_2SO_4, liquid SO_2 or elementary sulphur. The effluent gas has the composition 25% SO_2, 15% CO_2 and 60% H_2O.

The methods of treating H_2S and CS_2-containing waste gases from the manufacture

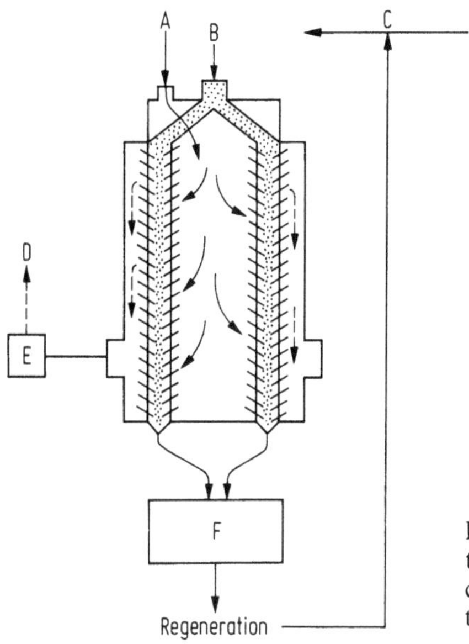

Fig. 1.2. Example of active carbon adsorption by the fluidized bed process. A Waste gas, B Active carbon, C Make up carbon, D Purified air, E Suction fan, F Spent carbon

of viscose products include two adsorption processes. In the sulphosorbon process the H_2S is oxidized to sulphur on wide-pore, KI-impregnated active carbon, the CS_2 is adsorbed on small-pore active carbon. The regeneration of the sulphur-loaded active carbon is accomplished with liquid CS_2, the sulphuric acid (small quantities) is washed out with water. The CS_2 is desorbed with water vapour. The thiocarb process involves the oxidation of the H_2S on heavy metal-poor carbon. The sulphuric acid which is produced in small quantities is neutralized with NH_3.

During the electrolytic production of aluminium fluorine-containing compounds are produced as dusts or gases (chemically either as almost water-insoluble cryolite [Na_3AlF_6] and aluminium fluoride or as hydrogen fluoride). The dry adsorption on aluminium oxide is regarded as an economic option for elimination of the gaseous components, since then the almost quantitative fluorine removal can be accompanied by its simultaneous recycling to the production process with the Al_2O_3 raw material. In practice fluidized beds are often utilized [11]. One disadvantage is that other volatiles are adsorbed and recycled along with the fluorine compounds, which can, under some circumstances, affect the quality of the aluminium that is produced. For this reason attempts have been made to separate out the accompanying elements (mainly iron, vanadium, phosphorus, carbon and sulphur), either by means of an electrostatic filter before or after the adsorption step or by recovering the fluorine content by means of a pyrohydrolysis technique, which then, however, increases the consumption of the fresh oxide since the aluminium hydroxide produced during pyrohydrolysis is unsuitable for aluminium production [12].

Highly porous active carbons can also be used in the laboratory for the adsorption of various pollutants (gases or vapours) such as halogens, halogenated hydrocarbons, phenols, mercaptans, acryl monomers and also finely dispersed mercury or mercury vapour as well as radioactive vapours (e.g. iodine or methyl iodide).

The natural and synthetic zeolites constitute a special group of active adsorbents.

Chemically these are hydrated alkali or alkaline earth aluminosilicates with an atomic ratio O/(AL + Si) = 2, whose structure is a three-dimensional anionic network with numerous cavities, that are in turn connected together by rigid pores of constant radius (ca. 3,4, or 5 Å etc. depending on the mineral). Charge neutrality is maintained by the presence of loosely bound cations within the cavities which are capable of base exchange. If the water present in the cavities is expelled, then an extremely active adsorbent is obtained that is selective because of the constant pore size. Since the molecules to be adsorbed are selected according to size these substances are known as "molecular sieves". Molecular sieves are under test for various applications. Thus, laboratory and pilot plant investigations have been described for the removal of ammonia [13]. On the laboratory scale two columns in series filled with klinoptilolite, a natural zeolite, can achieve an ammonia-removal rate of better than 99%. Regeneration is performed using lime. The use of zeolites as catalysts has also been described [14]. Their main action is as acid base catalysts, but other catalytic effects are known.

1.1.3 Absorption Processes

A distinction must be made between absorption at the solid phase boundary and absoption in liquids, whereby the former has the advantage that no waste water problems are created. An example will make this clear.

The removal of SO_2 from power station and other flue gases is a very significant problem for both dry and wet absorption, whereby each process reveals its own advantages and disadvantages. Scrubbing techniques have the disadvantage that the exhaust gases are reduced in temperature during the absorption process and this removes the necessary driving force for expelling them into the atmosphere, so that after desulphurization a supplementary heating step in necessary. In addition the high water content after scrubbing causes the formation of unwanted fogs at the chimney. The liquid scrubber effluent contains sulphites which pollute the waste water. The dry processes which are based on catalysis, on the other hand, require absolutely dust-free gas, so that their economic use is limited here in another manner. It should also be considered what is the most suitable form in which to remove the sulphur, i.e. which recycling product industry can best utilize. Dilute sulphuric acid will probably not find much demand; other sectors already produce sufficient. Neither is ammonium sulphate likely to be in great demand as a fertilizer because of its limited nutrient content. Liquified SO_2 finds a market under specialized conditions, e.g in cellulose processing or oil refining and for refrigeration, but does not have widespread application. The processes used for the desulphurization of flue gases are well reviewed in [15]. With the absorption processes the removal also constitutes a detoxification when chemical reactions are utilized.

1.1.3.1 Dry Processes

The technical developments in this area have been mainly concerned with the elimination of SO_2 from exhaust gases (compare Table 1.2). The removal of SO_3 is achieved with great succes using alkaline ferric oxide with the formation of sulfates [17] which cannot, however, be achieved with sulphur dioxide since the decomposition temperature of sulphites is lower than the temperature of the flue gases to be treated.

Table 1.2. Examples of the application of absorbents [16]

Absorbent	Pollutant	Absorption temperature (°C)	Amount absorbed (wt %)	Transformation to	Regeneration with	Remarks
Alkaline ferric hydroxide	SO_2 SO_3	100–300		sulphate		
Alkaline aluminium oxide gel (60% Al_2O_3, 30% Na_2O)	SO_2	130–330	19–24	sodium sulphate	H_2, CO, producer gas to H_2S and COS; reduction to S in Claus kiln	catalytic oxidation occurs in parallel to absorption
Manganic oxide (vacuum dried)	SO_2		25–27	manganese sulphate	NaOH to MnO	the waste gas must be dust-free
Ferric oxide (Fe_2O_3, finely divided)	SO_2	1100–1300	20–30 with the addition of 1.2 times the amount of absorbent compared to the stoichiometric amount	1. SO_3 by catalytic oxidation with Fe_2O_3 2. solid sulphase because of the chalk 3. normal flue gas dust removal		fuel gas desulphurization during oil and coal gasification; absorbent injected directly into the combustion chamber
Calcium carbonate	SO_2	1000–1200 ($CaCO_3 \rightarrow CaO + CO_2$)	20–30 with the addition of twice the amount of absorbent compared to the stoichiometric amount	calcium sulfate magnesium sulphate (not industrially utilizable)		
Aluminium oxide	fluorine compounds	300	1.7	aluminium fluoride	electrolysis	

One possibility of utilizing this technique is a preliminary catalytic oxidation to SO_2 which, however, involves high costs. The processes for the dry absorption of SO_2 are not so technically perfected that they can be economically applied on an industrial scale [16], even though various processes exist.

A Japanese process works at 130–150 °C with manganese oxides in a fluidized bed process with absorption of the SO_2 as manganese sulphate [9] followed by regeneration with water, ammonia and air. The manganese oxides so produced can be re-employed, after drying and grinding. Solid ammonium sulphate for fertilization is obtained from the solution remaining. Absorption processes employing managanese-containing absorbents, which are regenerated at 900 °C with powdered coal [9], are also in use in the Federal Republic of Germany.

The most significant hurdle for the practical application of a dry nonregenerative process for sulphur dioxide removal — based on the injection of the solid absorbents — is the fact that the reaction is not quantitative and a degree of utilization of only about 30% is achieved for the absorbent. Recently laboratory investigations have revealed that the reaction of SO_2 and SO_3 with calcium and magnesium compounds is positively influenced both in the degree of utilization of the absorbent and in the rate of binding by certain additives [18]. An example of such an ingredient is $CaCl_2$, which can either be utilized by impregnation to the extent of ca. 2 mol % with respect to $CaCO_3$ as an absorbent or as a simple dry mixture at a concentration of about 5 mol % as a promoter. Other substances that have been investigated are NH_4Cl, NaBr, NaOH, $NaHCO_3$, CH_3COOH, H_2SO_4 [18]. The real disadvantage of the wet processes (low flue gas temperature, the formation of waste water-polluting sulphites) would thus be avoided; however, a preliminary dust-removal step would be necessary. These additives seem to open up the possibility of a nonregenerative process at moderate temperature (about 500 °C) [18]. Good results have been obtained in investigations of the absorption of fluorine on aluminium oxide, sodium aluminate and alkali-containing aluminium hydroxide [9].

1.1.3.2 Wet Processes (Waste Gas Scrubbing)

The main technological solutions for washing out unwanted components (Fig. 1.3 illustrates the principle of an industrial installation) are [19]:

- Scrubbing towers with or without packing, working on the countercurrent principle.
- Scrubbers with water bath and eddy zone.
- Scrubbers with liquid spray (radial or rotating spray).
- High performance velocity scrubbers (injector or venturi scrubbers).

The scrubbing effect depends on the specific amount of scrubbing fluid, the surface area (i.e. contact area) of the washing liquid and the relative velocities of the two media [19]. In contrast to dry absorption, in which only a surface absorption or a reaction with uptake occurs, in the wet process there is a complete penetration and intimate mixing of the gas and liquid, whereby the solubility or thermal reactivity of the gas or vapour plays a role. The wettability may be improved by the addition of detergents. Hot exhaust gases must be cooled before coming into contact with the absorbing agent.

In principle the gas can be held back in part by purely physical adsorptive forces, whereby the amount of gas dissolved is proportional to the partial pressure and a

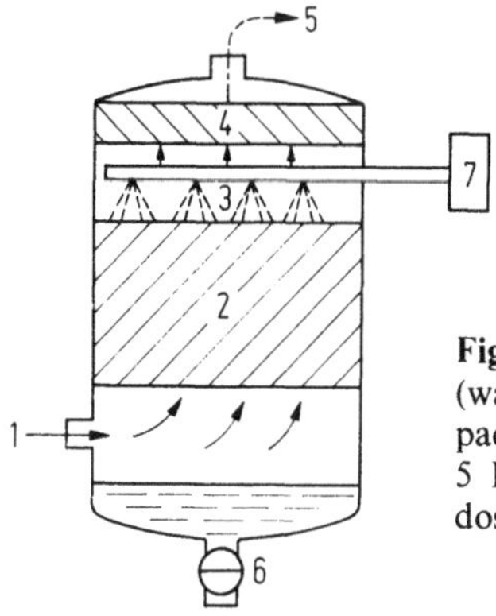

Fig. 1.3. The principle of a one-stage chemical absorption (waste gas scrubbing plant). 1 Wyste gas entry, 2 Ceramic packing, 3 Distribution jets, 4 Droplet remover (packing), 5 Purified gas, 6 Sludge valve, 7 Water + detoxifying agent dosing equipment

vapour pressure equilibrium is set up, when the temperature is raised the gas solubility is reduced and increasing partial pressure of the gas accelerates absorption (saturation of the absorbent). The simplest absorbent is water. Various organic solvents also find application.

It is only in a few cases that the purely physical solution process is adequate for the absorptive removal of noxious contaminants (e.g. for highly hydrophilic solvent vapours such as methanol, ethanol, acetone and N,N-dimethylformamide). It is usually preferable if a chemical reaction occurs between the gas or vapour to be absorbed and the absorbent (chemisorption). The use of chemical reactions allows the removal of larger quantities of gas and the reactions of the most various types, extending from precipitations to oxidations, can be utilized. Aqueous solutions of acids, bases and salts are frequently employed. Simple examples are the introduction of ammonia gas into water with the formation of NH_4OH and the removal of amines as amine sulphates using dilute sulphuric acid [22]. The use of catalytic solutions constitutes a specialized possibility. Thus a patent [20] describes the simultaneous introduction of hydrogen cyanide and oxygen or an excess of air into a catalyst solution consisting of water, acetic acid and 1.25% copper nitrate, when oxamide is produced and precipitates.

$$2\,HCN + 1/2\,O_2 + H_2O \rightarrow H_2NCOCONH_2$$

Similarly cyanogen can be bound in the form of an ester of cyanoformimidothioic acid by the introduction of appropriate quantities of a mercaptan-containing inert organic solvent in the presence of a metal compound as catalyst [21].

$$(CN)_2 + HSR \rightarrow NC\text{-}\underset{\displaystyle SR}{\underset{|}{C}}\text{=}NH$$

In practice, however, such reaction are scarcely suitable for detoxification, unless the noxious gases are to find a secondary application. Favourable reaction velocities (removal of the gases by the end of the absorption) are achieved if low viscosity absorbents are employed and the greatest possible surface area is created. Further a large drop in concentration from the gas to the absorbent also has a favourable effect.

Table 1.3. Examples of the application of liquid absorbents (according to [16])

Absorbent	Pollutant	Pollutant source	Conversion to	Absorbed portion (% absorbent)	Regeneration with	Remarks
Water	chlorine	chlorination and other processes	$HCl + HClO$	100 (30% HCl)	HClO eliminated by active carbon	hot absorption; exhaust gas, HCl-free
Caustic soda	chlorine		bleaching lye	high		for exhaust gases containing much Cl_2
Milk of lime	chlorine		calcium hypochlorite and chloride	high		
Carbon tetrachloride	chlorine			satisfactory	rectification	absorption takes place with cooling
Caustic soda	hydrogen sulphide	in many processes	sodium hydrosulphide	small amounts		circulating absorbent
Alkazid caustic (K salts of amino acids)	hydrogen sulphide		potassium hydrosulphide	relatively high	stripping with water	HCN, SO_2 and other acid components must be removed previously; hot absorption
Ferric hydroxide suspension + small amounts Na_2CO_3	hydrogen sulphide	viscose manufacture	sulphur thiosulphate	ca. 85 (elementary sulphur)		sodium carbonate accelerates the reaction; high gas velocities necessary
Water + silicate packing (trickle tower)	fluorine compounds (HF, SiF_4)	metallurgy, chemical industry, ceramic industry	hexafluorosilicic acid	ca. 90	further processing to salts	mainly large amounts of exhaust gas with low fluorine compound content

Aqueous alkali	SO_x (SO_2)	flue gases	sulphites sulphates	good		
Calcium carbonate suspension	SO_x (SO_2)		calcium sulphate	good	sedimentation	large quantities of limestone required
Magnesium oxide manganese dioxide suspension (manganite)	SO_x (SO_2)		sulphates sulphites bisulphites	17–22	reduction with the addition of carbon, sulphur and oil; S cleavage on roasting at 900 °C	highly active and easily regenerated absorbent
Water containing chlorine and chlorine dioxide (coke filled scrubbing tower)	odours (H_2S)	chemical industry, metal refining		good, odourless		necessary residual chlorine content of 10 mg/l in basic scrubbing water
Water + chlorine (limestone packed tower)	odours (H_2S, mercaptans, amines)	chemical industry, slaughterhouse wastes		odourless		chlorine is added to the entering gases

The pressure and temperature also have to be optimized. In choosing an absorbent care should also be taken that the absorbate is easily regenerable in order to facilitate continual operation. At the present state of the technological art trace amounts of toxic substances can often only be removed by economically unwarrantable expenditure (strive for a circulatory specification). Technical applications are summarized in Table 1.3. The waste water produced is often a source of problems.

Wet processes also find application in the removal of SO_2. Here the already mentioned diminution in the upward draught of the purified gases because of the cooling occurring during absorption constitutes a disadvantage. They can rapidly sink into atmospheric layers close to the ground, when precipitation causes the formation of acid rain. In practice this can be avoided by reheating the exhaust gas after absorption has taken place, so that the warmed gas is adequately diluted by vortex action during its climb into the upper layers of the atmosphere. Dry processes are better in this respect, since they operate at higher temperatures anyway, even if the residual pollutants are not so effectively eliminated. The first technical process for the removal of SO_2 from power station flue gases was operated in England. It involves the use of suspensions of chalk with the formation of calcium sulphite/sulphate suspensions. Even today about 80% of the world's flue gas desulphurization plants operate using chalk, for example the Knauf Research-Cotrell process with the production of gypsum [23].

Further development utilize chalk with the addition of small amounts of HCl (to form $CaCl_2$) and secret water-soluble catalysts [25] or milk of lime containing formic or nitric acids; calcium hydrogen sulfite is produced that can be transformed into gypsum [24].

The WELLMAN-LORD process [26] operates with sodium sulphite solution, which has a high capacity for SO_2 and allows its recovery as a high percentage gas on regeneration. The absorption capacity is more than 90% for most plants. In order to regenerate the bisulphite liquor formed its solution is concentrated, when SO_2-containing vapours are released and Na_2SO_3 crystallizes out.

$$SO_2 + Na_2SO_3 + H_2O \rightarrow 2\,NaHSO_3$$

$$2\,NaHSO_3 \xrightarrow{T} Na_2SO_3 + SO_2 + H_2O$$

The SO_2 released is processed further.

Other weakly basic absorbents have been investigated, e.g. sodium citrate and sodium phosphate solutions or ammonia in the form of an ammonium sulfite solution (whose pH is so chosen that no gaseous NH_3 is expelled but SO_2 is absorbed). If oxidation catalysts are introduced, then conversion occurs to ammonium sulphate which can be recovered by evaporation. If phosphate solutions are employed they can be regenerated with H_2S and the separation of a suspension of sulphur [9]. The WALTER process, where the dust-freed flue gas is treated with the calculated ammount of ammonia solution, is of interest [27]. Ammonium sulphite is formed, which is washed out an reacted with atmospheric oxygen to form utilizable ammonium sulphate fertilizer.

The Peracidox process works according to the principle of oxidative gas scrubbing [10] to purify exhaust gases of SO_2. The SO_2 is removed by the addition of a chemical oxidizing agent such as Caro's acid or H_2O_2 to the scrubbing solution. The dissolved SO_2 is oxidized to H_2SO_4. In all the wet processes described a mixture of sulphites and sulphates is usually obtained. This causes problems for a meaningful further utilization; tipping is often the only answer. Absorption by the melt is a special case of absorption and has also been investigated as a process for the removal of SO_2. Thus a molten mixture of sodium, potassium and lithium carbonates can be utilized at about 400 °C. It is regenerated using fuel gas with the formation of H_2S [9].

Hydrogen sulphide is another example of an acid exhaust gas component which can be removed by alkaline absorption liquids (often bases whose salts are easily regenerated by high temperature hydrolysis), when the CLAUS process is not suitable because of low concentrations. Weakly basic nonvolatile amines (such as triethanolamine or amino acids) are mainly used. Absorption takes place in the cold; desorption occurs on heating.

During the removal of the oxides of nitrogen from the exhaust gases from nitric acid manufacture (NH_3 combustion) the use of pressures of 10 atm allows achievement of residual concentrations of 0.1–0.2 vol. % (in comparison to 0.3–0.4 vol. % at normal pressure) [9].

Fluorides in exhaust gases are captured in water or brine and bound as flurosilicic acid or salt, which can be converted to utilizable aluminium fluoride or cryolite. This is accomplished by treating the prewarmed flurosilicic acid with aluminium hydroxide, centrifuging off the silicic acid formed and crystallizing the aluminium fluoride trihydrate out of the remaining solution, centrifuging it off, drying it in a fluidized bed and calcining it.

$$3\,SiF_4 + 2\,H_2O \rightarrow 2\,H_2SiF_6 + SiO_2$$

$$H_2SiF_6 + 2\,Al(OH)_3 \rightarrow 2\,AlF_3 + SiO_2 + 4\,H_2O$$

The flurosilicic acid can also be transformed to cryolite. The first stage is the reaction with sodium carbonate to form sodium fluoride and finely crystalline silicic acid, which is separated. In the second stage the NaF so produced is reacted with aluminium fluoride, the precipitated cryolite is then filtered off and calcined.

$$H_2SiF_6 + 3\,Na_2CO_3 \rightarrow 6\,NaF + SiO_2 + 3\,CO_2 + H_2O$$

$$3\,NaF + AlF_3 \rightarrow Na_3AlF_6$$

Basic pollutants, e.g. amines, can be washed out with acidic scrubbing solutions — in general dilute solutions of mineral acids (beware! risk of corrosion). Often — particularly under laboratory conditions and for the purification of large volumes of exhaust air from small concentrations of pollutants (e.g. ≤ 10 mg/m^3 org. substances) — powerfully oxidizing scrubbing liquids (concentrated sulphuric acid, H_2O_2, potassium permanganate, ClO_2, chlorite or hypochlorite solutions) are employed or oxidizing agents are added [28] since in such cases thermal or catalytic combustion is not appropriate.

DEGUSSA has developed a pilot plant for the quantitative oxidation of SO_2 to

H_2SO_4. The hydrogen peroxide concentration can be adjusted at will by means of the redox potential. Low concentrations in the ppm range can be oxidized just as successfully as higher ones in the gram and kilogram range [29].

The industrial disposal of hydrocarbons in exhaust gases can be achieved using sodium chlorite ($NaClO_2$) in hydrochloric acid solution in a one-stage washing process (see Fig. 1.4) [28].

$$4\,C_mH_n + (4\,m + n)\,NaClO_2 \xrightarrow{(HCl)} (4\,m + n)\,NaCl + 2n\,H_2O + m\,CO_2$$

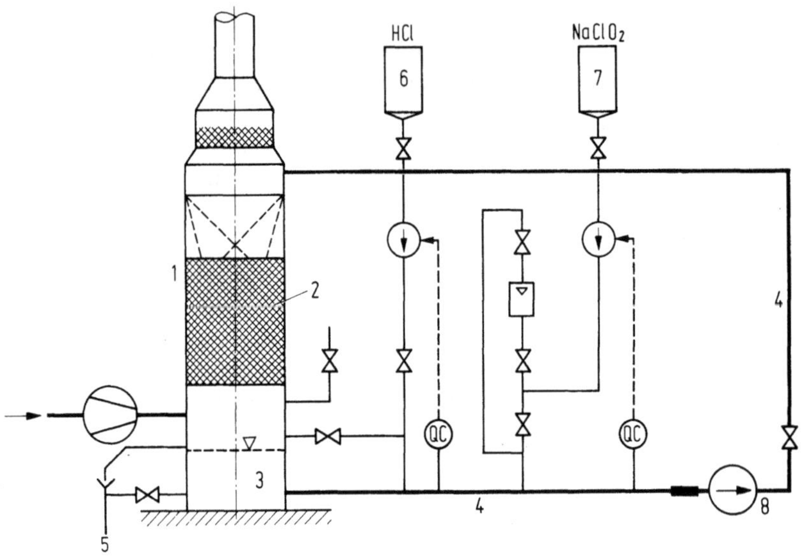

Fig. 1.4. Adsorptive waste gas purification plant. Chemical oxidation with sodium chlorite [28]. 1 Scrubbing tower, 2 Packing, 3/4 Scrubbing liquid circulation, 5 Removal and neutralization of scrubbing liquid, 6 Automatic pH adjustment to between 1 and 3, 7 Automatic dosage of $NaClO_2$, 8 Circulating pump

The chemical reaction takes place via the formation of chloric acid. The decontamination process is controlled by measurement of the redox potential.

Further examples are the oxidation of H_2S to H_2SO_4, of RSH to RSO_3H, of RSR to RSO_2R, of RCHO to RCOOH and HCN to CO_2 and NH_3. The exact course of the reaction is unknown in most cases.

Older processes involve scrubbing with acids and/or bases with the simultaneous injection of chlorine or ozone, which, however, requires expensive plant and does not always lead to optimal gas purification. Often the pollutants are only partially oxidized by ozone which under certain conditions can cause the formation of more toxic or more odorous products. The possibility of the formation of toxic organohalogen compounds by reaction with chlorine is well known. Nevertheless the possibility of gas cleansing with ozone is worth pointing out since it has the advantage that, of itself, it does not leave behind any undesired products. There are pilot plants in operation where, in certain cases, SO_2 employed for bottle sterilization, where relatively high gas concentrations occur, is immediately oxidized by ozone to sulphate on absorption in alkaline solution [30]. The plant consists of two packed spray towers and two separate wash liquid circulations, that can be connected if desired. Injectors allow the addition of ozone to each stage. It was found that the oxidative effect of ozone was better in sodium carbonate solution than in dilute caustic soda solution; this is because the lower pH of the sodium carbonate solution results in a slower decomposition of the ozone. The alkalinity is, however, sufficient for an efficient absorption of the SO_2. Optimization studies [30] revealed that for an exhaust stream of 3200 m^3/h with a SO_2 concentration of 1975 mg/m^3 the best reaction conditions were obtained by the utilization of 5–10% sodium carbonate solution and 150 g/h ozone. On this basis a fully automatic bottle sterilizing plant with an hourly capacity of 5000 bottles was built.

Investigations have been undertaken of the treatment of other odorous and noxious gaseous components based on the experience obtained using ozone in water purification, because the reaction between ozone and the pollutants is very slow in the gas phase and the degree of completion is unsatisfactory. It has been reported that ozone concentrations of 0.5–1 ppm in aqueous solutions yield redox potentials of more than +900 mV [31]. The reaction can be performed in both acid or neutral solutions. However, the most favourable oxidation velocities are obtained in the alkaline range at about pH 12. At higher pH the rapid ozone decomposition competes with the oxidation. An example [31] is the absorption and oxidation of H_2S in 5% sodium carbonate solution at concentrations of 30–600 ppm H_2S (pilot plant with a throughput of 500 m^3/h, exhaust gas; 0.2–0.8 ppm O_3 in 5% sodium carbonate solution). The alkaline wash solution was found to contain: — sulphide traces <1 ppm; a few ppm elementary sulphur; ca. 10% iodine-oxidizable sulphur compounds such as sulphite and thiosulphate; 90% sulphate ions. The oxidation can be so controlled that the ozone content of the purified air is less than 0.1 ppm. H_2S can no longer be detected (<0.01 ppm).

Phenol and formaldehyde-containing exhausts can be washed well with alkalis at pH 14. However, the oxidation products of phenol with ozone have not been individually studied.

Now an example for the use of chlorine as an oxidizing agent, since for all its disadvantages it is an attractive and economical oxidizing agent. In the limestone tower process [32] chlorine is employed for the elimination of the decomposition products of proteins, fats and carbohydrates, i.e. a conglomeration of nitrogen-containing compounds (particularly ammonia and amines), sulphur-containing compounds (H_2S, mercaptans, disulfides), alcohols, aldehydes, ketones and carboxylic acids (e.g. acetaldehyde, acrolein, butyric acid). For this purpose columns filled with limestone dolomite or other alkaline earth carbonates are employed in which the gas to be purified is passed countercurrent to a stream of percolating water. At the same time chlorine which has been injected into the gas stream reacts in the limestone bed to form chloride and hypochlorite. The hypochlorite oxidizes the odoriferous materials, while the acid components are neutralized by the limestone. The degrees of purification listed in Table 1.4 were obtained in a plant for the processing of poultry carcasse waste and feathers. We would like to treat the reactions which include a precipitation, which are less employed in gas purification, as a special case of the wet absorption process. But it is naturally possible that in such processes the

Table 1.4. The efficiency of the limestone tower process in a plant for the processing of poultry carcase refuse and feathers [32]

Components	Concentrations		
	raw gas (mg/m^3)	purified gas (mg/m^3)	efficiency (%)
H_2S	2.20	0.007	99.7
NH_3	12	0.7	ca. 95
Organic nitrogen (e.g. amines)			
— very volatile	1.6	0.15	$\geqq$90
— not very volatile	2.7	0.8	70–75
Total C	$\geqq$50	10–12	75–80
Chlorine	—	<1–1.5	—

chemical reactions take place with the formation of solid immobilization products. For example, laboratory quantities of hydrogen sulphide can be disposed by passing them into a solution of heavy metal salts, most economically a suspension of ferric hydroxide.

1.2 Processes for the Detoxification of Poisonous Exhaust Gas Components

1.2.1 Oxidation of Exhaust Gases

Under the heading of oxidative processes for the detoxification of poisonous exhaust we wish to deal exclusively with thermal and catalytic combustion in industry and the combustion of small amounts of surplus gases or exhausts on the laboratory scale. Wet oxidation techniques have already been treated in the previous section.

1.2.1.1 Thermal Processes

In principle thermal destruction can take place in furnaces (for difficult to burn gases with large amounts of air and preheating to the combustion temperature by burners or heat exchangers) or flaring (requirement: stable combustion; large quantities of industrially produced combustable gases or vapours that cannot be used or laboratory-produced gases that require destruction). In industry the flame temperatures in the furnace average 1400 °C, the mean combustion chamber temperatures ca. 800 °C and the exit temperatures ca. 450 °C. The purifying effect can amount to up to 97%. The reasons prohibiting the universal application of purely thermal processes for large amounts of gas are economic (fuel consumption). The substances that can be disposed of in furnaces include solvent vapours, organochlorine compounds and hydrocarbons, as well as the exhaust gases from waste incinerators. Flaring in particularly employed in the petrochemical industry and in coke production, as well as for residual gases in the laboratory. Explosive gases are burnt in special chambers [33].

1.2.1.2. Catalytic Oxidation

In the presence of certain catalysts (in particular heavy metals and the oxides of the heavy metal iron, chromium, cobalt, copper, molybdenum, nickel and vanadium and also certain salts on inert carriers) the oxidation of gases is so accelerated that they can be oxidized at temperatures way below the combustion temperature. This means that the fuel consumption of the process is much less than that for purely thermal combustion. If the calorific value is below 30–40 kcal/Nm^3 (126–167 kJ/Nm^3) then additional energy has to be supplied even for catalytic combustion. For calorific values in the range 40 to 400 kcal/Nm^3 (167–1670 kJ/Nm^3) the necessary heat can be supplied by heat exchange [34].

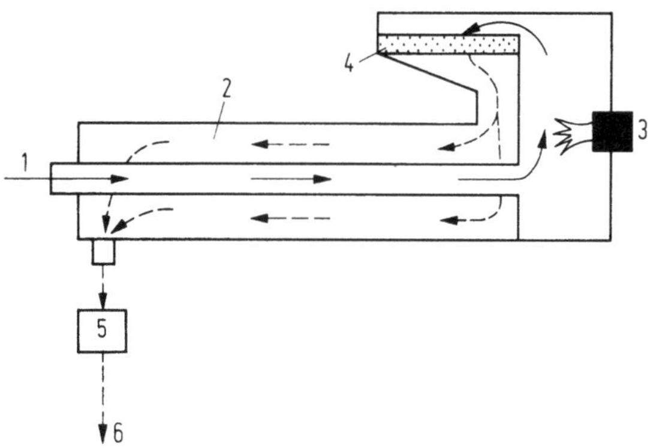

Fig. 1.5. The principles of a plant for the catalytic oxidation of waste gases. 1 Waste gas entrance, 2 Heat exchanger, 3 Supplementary burner, 4 Oxidation catalyst, 5 Suction fan, 6 Purified gas exit

Nevertheless, the thermal combustion process is preferred in practice, since there is no diminution of the effectivity or the degree of conversion because of catalyst poisoning. It should also be noted that certain unwanted side reactions can lead to the formation of more toxic compounds in the catalytic process. There is reported to be a danger of phosgene formation in the catalytic combustion of gases containing chlorinated hydrocarbons [35].

Like all catalytic processes catalytic oxidation depends on the production of a sufficiently large surface area by finely dividing the particles of catalyst on a carrier material (for an example of a plant see Fig. 1.5). The metallic components are usually platinum or palladium or their oxides; the carriers are usually chromium-nickel steel nets or honeycombs or various ceramic materials [36]. The latter have the advantage of a large specific surface area, which endows them with high activity and a certain resistance to contact poisons.

The average working temperatures lie between 200 and 400 °C, where a purification to the extent of 90% is attained at 300 °C and of 99% at 350–400 °C, depending on the type of catalyst. The gases are heated to the reaction temperature by burners or heat exchangers before they are led into the catalyst chamber. The catalysers must be so constructed that they can withstand temperatures of up to

Table 1.5. Examples of the application of catalytic waste gas oxidation ([16] with additions)

Source of pollution	Pollutant	Reaction temperature (°C)	Catalyst, working life	Remarks
Electroplating	hydrogen cyanide	300	platinum	Cyan-cat process
Chemical industry, dry cleaning	chlorinated hydrocarbons	420–500	platinum/Al_2O_3	laboratory study
Motor vehicles (gasoline engine exhausts)	carbon monoxide, fuel residues, aldehydes, cracking products	200–250	noble metals on ceramic carriers; 3–6 mm spheres; 2000 hours	oxides of nitrogen are not destroyed
Chemical industry, asphalt melters, fish meal plants, etc.	inorganic and organic odours	ca. 650	perforated metal sheets; 3000 hours	partial decomposition to CO_2, H_2O, N_2
Used-oil refineries	exhaust gas	370		
Metal smelters	hydrogen sulphide		bauxite catalyst	formation of SO_2 and sulphur
Flue gases	sulphur dioxide	80	carbon-containing catalyst	wet catalysis

600 °C for short periods of time, since under some circumstances localized overheating can occur during the reaction. Depending on its character the contact surface can be regenerated by blowing out, washing, sublimation or burning off, where the cost of regeneration (after 3–4 years operation) with decreasing activity of the noble metal can amount to 30–50% of the purchase price. For trouble-free operation it is necessary to remove dust and above all catalytic poisons (e.g. lead, arsenic, cadmium and chlorine compounds) as well as encrusting substances (e.g. silicones, phosphorus compounds) before catalytic treatment, in order to guarantee a working life of 15000 to 40000 hours [9]. It is necessary to find the optimum catalyst for each application experimentally (see also Table 1.5). Catalytic oxidation has also been used as a means of SO_2 removal, since detailed information is available concerning the oxidation of SO_2 from the contact process for sulphuric acid production. A process has been developed in the USA that involves the oxidation of the SO_2 content of flue gas with the aid of vanadium catalysts. The fog of SO_3 so formed can be converted to solid ammonium sulphate by the injection of NH_3; but its electrostatic

precipitation as sulphuric acid was also investigated [9]. Further industrial applications of catalytic combustion are [31, 37]:

- The combustion of waste gases from the aerial oxidation of paraffins and olefins (ethylene oxide, propylene oxide, acrolein), of aromatics (phthalic anhydride, maleic anhydride), of naphthalene, of alcohols and of formaldehyde.
- The destruction of the exhaust from hydrogenations, decarboxylations, polymerizations and pyrolyses.
- The combustion of odiferous or toxic exhaust gas contaminants e.g. from mercaptan and sulphide production or acrylate monomers.

Good results have been obtained in the purification of waters contaminated with hydrocarbons by stripping the hydrocarbons and then burning the solvent-containing air. Because of the high calorific value of such vapours an autothermal energy supply can be ensured by the use of heat exchangers. So much energy is released during adiabatic combustion that heat-resistant materials must be employed. For this reason the reaction is conducted on two contact layers, the reaction at the first being controlled by means of the temperature [36].

If nitrogen-containing exhaust gases are to be processed, then at an oxygen concentration of ca. 15% between 30 and 70% of the nitrogen is converted to oxides of nitrogen, so that such gases should not be processed by catalytic oxidation [37]; in contrast the concentration of oxides of nitrogen does not exceed 50 ppm even very high inlet concentrations in the thermal combustion process.

It has been verified under laboratory conditions that a whole range of aliphatic chlorinated hydrocarbons can be very effectively oxidized to HCl and CO_2 at 450 °C over a catalyst consisting of platinum γ-Al_2O_3 [38]. With catalysts containing 0.8% or more platinum the conversion rate is generally over 95%. Catalysts containing less platinum produce lower conversion rates (e.g. 0.2% platinum yields 32 ± 2%). The yields are almost independent of the vapour pressure of the chlorinated hydrocarbons and also exhibit little temperature dependence between 420 and 500 °C. Catalysts consisting of $CuO—Cr_2O_3$ on Al_2O_3 at 550 °C are particularly suitable for the total oxidation of chlorinated hydrocarbons with olefinic, paraffinic and aromatic skeletons [39].

The oxidation of HCN (Cyan-cat process) will be considered in more detail as a practically important example of catalytic oxidation. Cost calculations [40] indicate that the catalytic burning of hydrogen cyanide is one of the most economical processes for the disposal of cyanides or cyanide-containing concentrates (>10 g CN^-/l); its hypochlorite oxidation to cyanate costs six times as much, to CO_2 and N_2 ca. twelve times as much. HCN is liberated from these by the addition of waste sulphuric acid; this takes place at relatively high pH values in the case of the alkali metal cyanides. At pH 7 the equilibrium of the reaction

$$2\,NaCN + H_2SO_4 \rightleftharpoons 2\,HCN + Na_2SO_4$$

lies to 90% on the product side.

Complex cyanides release <2 hydrocyanic acids quantitatively at suitably low pH. The addition of the acid and the consequent dilution has, because of its exothermic character, simultaneously the advantage of aiding the driving off of the HCN because of the increase in temperature produced. This process is performed in packed trickle towers with a countercurrent of air. At pH 2 the injection of 1 m^3 of air per litre of cyanide solution removes more than 99% of the cyanide. The HCN-air mixture is burnt in a reactor over a platinum catalyst at a temperature of ca. 300 °C to CO_2 and nitrogen [40].

$$4\,HCN + 5\,O_2 \rightarrow 4\,CO_2 + 2\,N_2 + 2\,H_2O$$

Since the process is exothermic the entering HCN/air can be preheated by heat exchange.

Post-combustion processes find particular application [41]:

- If economic recycling is impossible (low value residue, highly contamined or complex residue, no suitable technology).
- If the contaminants are of a type that allows their oxidation to nonpolluting substances (e.g. aliphatic, alicyclic, aromatic and heterocyclic compounds, alcohols, terpenes etc.).
- If the exhaust gases do not contain catalyst poisons (mechanical inactivation by dust, soot, polymer-forming materials or chemical inactivation by Si, P, As, Se etc.).

In general higher conversion rates are achieved by thermal processes, but significantly more energy must be expended.

1.2.2 Thermal and Catalytic Reduction of Waste Gases

It sometimes occurs that a reductive processing of waste gases is necessary. One of the most important examples is the reduction of the oxides of nitrogen [37], which occur in the waste gases from nitrate manufacture, from the decomposition of nitrates to produce oxides, from the manufacture of nitric acid, from the oxidation of organic substances with nitric acid and the nitration of organic compounds. The reduction to nitrogen can be performed thermally by employing a specialized burner that operates sub-stoichiometrically; gaseous or liquid fuels are employed as the reducing agent. The excess reducing agent is then either thermally or catalytically oxidized in a subsequent oxidation unit. Or noble metal catalysts (on Al_2O_3 for example) are employed [42]. The important reducing agents are hydrogen, carbon monoxide, hydrocarbons and ammonia [43]. The catalytic reduction of the oxides of nitrogen takes place at adequate flow rates at temperatures of between 100 and 500 °C depending on the reducing agent and the catalyst. When hydrogen is employed Pt or Pd catalysts are used and temperatures of only 100 °C are necessary, while temperatures of about 300 °C are required with ruthenium catalysts under otherwise identical conditions. The situation is reversed with carbon monoxide. Here a ruthenium catalyst achieves a 99% reduction in the oxides of nitrogen at temperatures of 250 °C, while temperatures of 450 °C are not sufficient for Pt catalysts. These results are independent of the carrier and, to a degree, of the noble metal content. Much higher working temperatures are required for non-noble metal catalysts. The composition of the products is also influenced by the type of catalyst [42]. The reaction product with Pt and Pd catalysts is mainly ammonia, with ruthenium catalysts this is an unimportant product. When CO is employed as the reducing agent enough H_2O is synthesized by carbon monoxide reforming for the production of NH_3. Another route proceeds via isocyanate which yields NH_3 and CO_2 hydrolytically. Such reduction processes are particularly applied if no nitric acid recovery is attempted, if cheap reducing agents are available, if the exhaust gases contain little or no oxygen and relatively constant concentrations of the oxides of nitrogen. In favourable cases the final concentrations of nitrogen oxides can be held at about 10 ppm [42, 44]. Problems occur with catalyst poisoning and residual emissions (with excess reducing agent these include unconsumed hydrocarbons, CO, NH_3, HCN).

As previously mentioned, not all recyclizable products are required in the quantities that they could be produced from waste materials, as can be the case, for instance, with ammonium sulphate production from waste NH_3 (e.g. from coke-oven gas). Another alternative to combustion is the catalytic cleavage of ammonia in a reducing atmosphere at Ni catalysts [46]. Here any HCN that is present is also reduced and at the same time hydrocarbons are converted by reaction with water vapour so that almost no soot is produced.

$$2\,NH_3 \rightleftharpoons N_2 + 3\,H_2$$

$$2\,HCN + H_2O \rightleftharpoons N_2 + 2\,CO + 3\,H_2$$

$$C_nH_m + m/2\,H_2O \rightarrow CO + m\,H_2$$

1.2.3 Radiation-Chemical Treatment of Waste Gases

It is probable that in the future radiation-chemical detoxification will find application for particular exhaust gas problems [59, 60]. Austrian authors discovered that H_2S and CS_2 in exhaust gases from viscose production could be reduced in a zeroth order reaction on radiation with 500 keV electrons [61]. Japanese workers were able to demonstrate that NO_x and SO_2 could be removed from exhaust gases by energetic radiation. The required dosage rates for a 100% elimination of 80 ppm NO_x and a 80% removal of 600–800 ppm SO_2 were 1 and 4 Mrad respectively. The process was carried out in a flow-through pilot system and an electrostatic eliminator was installed for the removal of the liquid droplets and solid particles formed. The water-soluble portion of the separated solid liquid mixture consisted of sulphuric acid and probably metal sulphates [62, 63].

1.3 Detoxification and Destruction of Toxic Gases in the Laboratory

The destruction of toxic laboratory gases and liquids that boil at room temperature is often very difficult technically, particularly with regard to worker protection and the safety of the surrounding region. Several important examples will be mentioned.

It may be stated in general that reducing gases can be detoxified by means of Ca hypochlorite solutions or other oxidizing agents. Oxidizing gases can be led into reducing agents (such as Na_2SO_3). Chlorine (and bromine) form the hypohalogenite on being led into dilute alkali. This can be utilized as bleaching solution or be detoxified by reaction with sodium thiosulphate. The reaction solution can be washed down the drain with an adequate supply of water.

$$Cl_2 + 2\,NaOH \rightarrow NaOCl + NaCl + H_2O$$
$$NaOCl \xrightarrow{Na_2S_2O_3} 2\,Na_2SO_4 + NaCl$$

Neutralization with sulphuric acid would regenerate chlorine.

Fluorine gas can reacted with solid bicarbonate, soda lime or granulated wood charcoal in a solid gas purifier. The toxic hydrogen fluoride (and its aqueous solutions) can be immobilized with milk of lime at pH 12 as the insoluble calcium fluoride. The strictest security precautions should be taken and polyethylene bottles used as the reaction vessels.

Phosgene, which is of great synthetic importance, can be destroyed by leading into 15% caustic soda solution (phosgene solutions should be treated dropwise with an excess of 15% caustic soda) and stirring for one hour at room temperature. After neutralization with hydrochloric acid the reaction mixture can be disposed of in small amounts down the drain.

$$COCl_2 + 2\,NaOH \rightarrow CO_2 + 2\,NaCl + H_2O$$

According to the patent literature [47] hydrolysis with a suspension of active carbon is a very simple and economic method of disposal. It takes place about twice as fast as the normal hydrolysis. In general 1 volume of active carbon is capable of the destruction of 40–70 volumes of phosgene. Enough water must be added that the concentration of the HCl formed on hydrolysis does not exceed 10%. Higher concentrations inhibit the complete hydrolysis of the phosgene. A detoxification reaction which allows a secondary utilization occurs on leading the phosgene into tetrahydrofuran in the presence of DMF [48] or $ZnCl_2$ [49]; when ring opening of the tetrahydrofuran yields 1,4-dichlorobutane which is available for further use as a dihalide.

At high temperatures and in the presence of active carbon or metal catalysts [50–55] phosgene disproportionates to CO_2 and carbon tetrachloride.

$$2\,COCl_2 \xrightarrow{\text{cat. } 250-450\,°C} CCl_4 + CO_2$$

In the absence of water phosgene is quantitively decomposed by solutions of NaI or LiBr in acetone with the formation of CO.

$$COCl_2 + 2\,NaI \rightarrow CO + I_2 + 2\,NaCl$$

The reaction of phosgene with ammonia or certain amines leading to the formation of urea or urea derivatives and the combination with urotropine discovered by Willstätter, which have found use as, amongst other things, gas mask packing $[COCl_2 \cdot 2\,(CH_2)_6N_4]$, are well-known.

Gaseous sulphur trioxide is best led into concentrated sulphuric acid. The oleum so formed is transformed into dilute sulphuric acid by dropwise addition to ice water.

This can then be carefully neutralized with caustic soda and poured down the drain with plenty of water.

$$SO_3 + H_2SO_4 \rightarrow H_2SO_4 \cdot SO_3 \xrightarrow{(H_2O)} H_2SO_4 \xrightarrow{(NaOH)} Na_2SO_4 + H_2O$$

Sulphur dioxide can be absorbed in caustic soda and washed down the drain with plenty of water.

Hydrogen cyanide (as hydrocyanic acid solutions) is easily attacked by oxidizing agents. Thus the harmless oxidation product can be obtained by appropriate reaction with hydrogen peroxide, otherwise cyanic acid is formed. Javelle water is also a suitable oxidizing agent.

$$2\,HCN + H_2O_2 \rightarrow (CONH_2)_2$$

Slow hydrolysis to formic acid occurs in water. Chlorine reacts with aqueous HCN to form the toxic cyanogen chloride, with alcoholic HCN solutions to form chloroethyldiurethane. HCN is reduced to methylamine by nascent hydrogen. In the presence of atmospheric oxygen sodium thiosulphate yields thiocyanic acid.

$$HCN + Na_2S_2O_3 + 1/2\,O_2 \rightarrow HCNS + Na_2SO_4$$

Usual methods of disposing of HCN and its solutions include reaction with alkaline hypochlorites or with alkalis and ferrous salts.

Cyanogen chloride, which boils at 13 °C, can be trimerized by reaction, under the correct conditions, with water, chlorine and HCl to form the crystalline, physiologically nontoxic cyanuric chloride. The detoxification with hypochlorite is described on p. 231. It should be destroyed in the laboratory by adding it to dilute caustic soda (cyanate formation). Ammonia reacts to cause the quantitative formation of cyanamide.

A further example of the disposal of laboratory gases depends on the sorption of excess carbon monoxide onto a mixture of active carbon and hopcalite (manganese dioxide and copper oxide). The oxidizing agent in the mixture ensures the transformation to CO_2. Recent investigations describe the use of clusters $[Rh_6(CO)_{16}]$ as catalysts for the oxidation of CO to CO_2 [56].

The destruction of gaseous hydrocarbons, such as methane, ethane, propane, ethylene and acetylene, is necessarily performed by burning. Particular precautions are necessary because of the dangers of explosion and to ensure proper control of the combustion process. The detoxification of hydrogen sulphide can also be performed by combustion; it must, however, be performed with particular care — even with laboratory quantities. Hydrogen sulphide in small containers (e.g. gas jars) is better destroyed with a heavy metal salt solution (a suspension of ferric hydroxide is the cheapest source). The detoxification and deodorization of the lower mercaptans, such as methylmercaptan, in a mixture of formaldehyde and a secondary amine such as morpholine has been described [57].

$$CH_3SH + CH_2O + NHR_2 \rightarrow CH_3SCH_2NR_2 + H_2O$$

Hydrogen selenide and telluride can be passed into 15 % caustic potash and then worked up as the elements. As a powerful reducing agent hydrogen telluride is attacked by virtually all oxidizing agents and yields tellurium, tellurium dioxide, tellurite or tellurate depending on the conditions. Aqueous solutions of hydrogen selenide react with atmospheric oxygen to yield selenium, with nitric acid, ferric chloride and ferricyanide ions to yield selenium, with acidic $KMnO_4$ solution to yield selenic acid, with chlorine or bromine water to yield selenic and selenous acids.

Arsine burns with a pale blue flame to yield arsenic which is also poisonous. Aqueous solutions of arsine rapidly decompose to arsenic compounds, which can be precipitated as calcium arsenate. The immobilization of small amounts in the laboratory is best performed by oxidation to arsenious acid with silver nitrate or iodine.

$$AsH_3 + 6\,AgNO_3 + 3\,H_2O \rightarrow 6\,Ag + As(OH)_3 + 6\,HNO_3$$

$$AsH_3 + 3\,I_2 + 3\,H_2O \rightarrow H_3AsO_3 + 6\,HI$$

Passing into adequate amounts of bromine water or alkaline hypobromite solution is also to be recommended. Chlorine decomposes arsine with combustion to arsenic and arsenic trichloride, chlorine water to arsenic acid, nitric acid to arsenious and arsenic acids, hypochlorite and neutral and acidic $KMnO_4$ solutions to arsenic acid. Stibine can also be destroyed by passing it into strong oxidizing agents (hypohalogenites, bromine water, $KMnO_4$). Its reaction with chlorine or nitric acid is explosive, as it is with ozone at 90 °C. Acidic dichromate solutions and H_2O_2 react vigorously. Phosphine is oxidized to phosphate by passing it into hypochlorite solutions. It burns in chlorine to form phosphorus pentachloride. Phosphite and phosphate are formed with permangante. Phosphine is resistant to H_2O_2 and dichromate.

Monosilane (SiH_4) is detoxified by passing it into solutions or suspensions of powerful oxidizing agents such as chloride of lime, hypochlorite, or $KMnO_4$ to form silicic acid. Care is recommended with halogens which react with it explosively.

Diborane can be decomposed to boric acid and hydrogen (!) by passing it into water. Vigorous reaction occurs with oxidizing agents; e.g. it explodes with chlorine. Hypoborate is formed with alkalis and is decomposed by acids.

In principle, the removal of mercury vapour from gas streams is possible by adsorption, but normal adsorbents are unsatisfactory; impregnated or loaded adsorbents must be employed. Thus active carbons impregnated with mineral acids or iodine find application, as do those loaded with sulphur, Na_2S, NaHS, Na_2S_x, thiocyanates and thiosemicarbazides; loads of up to 40 % by wt. are employed. Removal rates of up to 99 % are achieved from air and hydrogen by these means [58]. Impregnation with certain metal halides has also been recommended. Treatment of the active carbon with chlorinated lime or chlorine is said to lead to 100 % removal of mercury from Hg-saturated air [58]. Silicon dioxide (kieselguhr, silica gel) can take up mercury vapour if, for example, it is impregnated with a metal which is more electropositive than mercury. The mercury reduces the metal ion to the metal; in the process more than the stoichiometric amount of mercury is taken up indicating the formation of amalgams. The use of silica gel loaded with various metal sulphides, metal iodides and $KMnO_4$ has also been suggested. Other examples of active adsorbents are impregnated zeolites and manganese dioxide, silver dust and ion exchange resins containing sulphide or sulphhydryl residues. The simplest variant

Table 1.6. The rapid laboratory scale detoxification of important toxic gases

Toxic gas	Disposal
Arsine	pass into hypohalite solution; or react to form As_2O_3 with aqueous $AgNO_3$ solution or KOH and then treat further
Carbon monoxide	pass through a gas purifier filled with a mixture of active carbon and hopcalite
Chlorine	pass into dilute caustic soda and decompose the hypochlorite so formed with $Na_2S_2O_3$ solution
Cyanogen chloride	pass into 15% NaOH
Diborane	pass into water when boric acid and hydrogen (!) are formed; hypoborate is formed in alkali, which can then be decomposed by acidification
Ethylene oxide	pass into cold dilute HCl (glycol formation)
Fluorine	pass through a gas purifier filled with $NaHCO_3$ or soda lime
Formaldehyde gas	pass into ethanol and burn the solution
Hydrocarbons	burn using the appropriate safety precautions
Hydrogen bromide	pass into 15% NaOH
Hydrogen chloride	pass into 15% NaOH
Hydrogen cyanide	pass into dilute NaOH, oxidize the cyanide formed with $Ca(OCl)_2$ solution, dilute with plenty of water
Hydrogen fluoride	pass into milk of lime at pH 12 (use a polyethylene flask)
Hydrogen selenide	pass into 15% KOH followed by reduction to element
Hydrogen sulphide	pass into ferric hydroxide suspension
Hydrogen telluride	pass into 15% KOH followed by reduction to element
Mercury vapour	adsorb on iodine-loaded active carbon
Monosilane	pass into hypohalite solution
Oxidizing gases	pass into Na_2SO_3 or $Na_2S_2O_3$ solutions
Phosgene	pass into 15% NaOH or active carbon suspension
Phosphine	pass into hypohalite or $AgNO_3$ solution
Reducing gases	pass into $Ca(OCl)_2$ solution
Solvent vapours	adsorb onto active carbon or condense
Stibine	pass into hyophalite solution
Sulphur dioxide	pass into 15% NaOH
Sulphur trioxide	pass into conc. H_2SO_4 and add the oleum formed dropwise to ice water and neutralize with NaOH

is probably adsorption onto iodine-loaded active carbon. Table 1.6 reviews the rapid detoxification of important laboratory gases.

References

1. Rafflenbeul, R.: Wasser, Luft u. Betrieb 20 (1976) 600.
2. Storp, K.: VDI-Berichte 124 (1968) 11.
3. Wirth, H.: Staub-Reinhalt.-Luft 36 (1976) 288.
4. Rolke, D.: Umwelt 1979 (3), 169.
5. Stöcker, U.: Chemie-Ing.-Tech. 48 (1976) 833.
6. Altmann, W.: Chemiker-Ztg. 99 (1975) 204.
7. Ruhl, E.: Umwelt 1976 (5), 349.
8. Franke, S.: Lehrbuch der Militärchemie Bd. 1 und 2 Berlin: Militärverlag der DDR 1977.
9. Leithe, W.: Umweltschutz aus der Sicht der Chemie. Stuttgart: Wissenschaftliche Verlagsgesellschaft mbH. 1975.
10. Sander, U.: Dechema-Monographien 80 (1976) 2, 379.
11. Wilde, G., Reh, L.: Aluminium 48 (1972) 738.

12. Fluorrückführung in der Aluminiumindustrie durch neue Verfahren. Berichte des Umweltbundesamtes 6/1976 (Berlin/West).
13. Mercer, B. W.: J. Water Pollut. Control. Feder. (Washington) 42 (1970) 2, Pt 2S, R95–R107.
14. Schirmer, W.: Z. Chem. 13 (1973) 122.
15. Schneider, G.: Chem. Rundsch. 28 (1975) 29, 1–3.
16. Stief, E.: Luftreinhaltung. Berlin: VEB Verlag Technik 1975.
17. Knop, W. et al.: Technik der Luftverunreinigung. Frankfurt/Main: Otto Krausskopf Verlag 1972.
18. Von Houte, G., Dehmon, B.: Chemie-Ing.-Tech. 48 (1976) 863.
19. Grimm, H.: VGB Kraftwerkstechnik 57 (1977) 2, 121.
20. BRD-Pat. 2 402 354 (1975), Riemenschneider, W., Hoechst AG.
21. BRD-Pat. 1 543 424 (1974), Gruber, W., Quis, P., Röhm GmbH.
22. Das techn. Umweltmagazin. April 1978, 39.
23. Hamm, H. et al.: Energie 31 (1979) Nr. 12.
24. Esche, M., Igelbüscher, H.: Das Umweltmagazin. Dezember 1980, 33.
25. Esche, M.: Das techn. Umweltmagazin. Dezember 1977, 36.
26. Helmer, L., Vangala, R.: Dechema-Monographien 80 (1976) 2, 365.
27. Bechthold, H.: Das techn. Umweltmagazin. August 1978, 34.
28. Geipel, H.: Dechema-Monographien 80 (1976) 2, 435.
29. Die Beseitigung von SO_2 in Abluft mit H_2O_2. DEGUSSA company brochure.
30. Schwarzbach, E. et al.: Wasser, Luft u. Betrieb 20 (1976) 235.
31. Schneider, W.: Dechema-Monographien 80 (1976) 2, 459.
32. Berndt, J.: Wasser, Luft u. Betrieb 21 (1977) 353.
33. Kümmel, J., Wiese, D.: Chemie-Anlagen und Verfahren 1977 (7), 56.
34. Hermann, E.: Erdöl u. Kohle, Erdgas, Petrochem. 19 (1966) 426.
35. Wicke, E.: Chemie-Ing.-Tech. 37 (1965) 892.
36. Brand, H.-D.: Chemiker-Ztg. 95 (1971) 458.
37. Quillmann, H.: Chem. Industrie 1973 (8), 497.
38. Bond, C. G., Sadeghi, N.: J. Appl. Chem. Biotechnol. 25 (1975) 241.
39. Laidig, G. et al.: Erdöl u. Kohle, Erdgas, Petrochem. 34 (1981) 329.
40. Jola, M.: Galvanotechnik 61 (1970) 1003.
41. Wenske, R.: Techn. u. Umweltschutz 7 (1974) 115.
42. Berndt, M. et al.: Dechema-Monographien 80 (1976) 2, 545.
43. Meier, H., Gut, G.: Chimia 31 (1977) 19.
44. Pohle, H., Vahrenholt, F.: Umwelt 1978 (4), 296.
45. Kromer, E.: Chimia 36 (1982) 87.
46. Tippmer, K.: Chemie-Ing.-Tech. 46 (1974) 628.
47. US.-Pat. 2 832 670 (1958), Wollthan, H., Farbenfabriken Bayer.
48. BRD-Pat. 1 188 570 (1965), Zeigenbein, W. et al., Hüls AG.
49. Lutkova, V. I. et al.: Ž. prikl. Chim. 32 (1959) 1635.
50. US.-Pat. 3 069 481 (1962), Hazeldine, R. N., Pennsalt Chem. Co.
51. Franz. Pat. 1 499 031 (1967), Ager, J. W., FMC Corp.
52. Niederl. Pat. 6 400 645 (1964), Imperial Chem. Ind. Ltd.
53. US.-Pat. 2 892 875 (1959), Kung, F. E., Columbia Southern Chem. Co.
54. BRD-Pat. 1 112 727 (1961), Glemser, O., Hoechst AG.
55. Winter, J. H.: Z. physik. Chem. 51 (1966) 136.
56. Mercer, G. D. et al.: J. Am. chem. Soc. 97 (1975) 1967.
57. Voronkov, M. G. et al.: Ž. prikl. Chim. 49 (1976) 2577.
58. Bergk, K.-H. et al.: Z. Chem. 17 (1977) 85.
59. Drawe, H.: Chemiker-Ztg. 101 (1977) 425.
60. Drawe, H.: Chemiker-Ztg. 105 (1981) 39.
61. Wiesböck, R., Proksch, E.: Kerntechnik 18 (1976) 390.
62. Kawamura, K. et al.: Treatment of Exhaust Gases by Irradiation in Radiation for a Clean Environment. Wien: IAEA 1974, S. 621.
63. Wahino, M. et al.: Kinetic Study on the Irradiation of Exhaust Gases in Radiation for a Clean Environment. Wien: IAEA 1975, S. 633.
64. Gust, M. et al.: Staub-Reinhalt. Luft 39 (1979) 308.

2 The Detoxification of Industrial Chemical Wastes

The disposal of the chemical wastes produced in the most various industrial processes has to be regarded as a complicated problem, that means that the question is not simply one of disposal, incineration or special treatment, but of sensible combination or complementation. The simultaneous disposal with domestic wastes of those residues from chemical waste, which cannot otherwise be disposed of, must be borne in mind, since special chemical waste disposal sites are expensive and underground disposal sites should be reserved for specific cases. Apart from the toxicological evaluation of the material, its odour and the dangers of fire or explosion, the water content of the waste is extremely important. High moisture contents of tipped material can greatly affect the stability and the suitability of the site for vehicular traffic. Additionally the amount of leach water produced is increased, which can, amongst other things, cause dangerous materials to be washed out. When waste of known properties is delivered it can be tipped on the disposal site, if none of its properties makes this inadvisable. Unknown wastes should be chemically analysed and, if necessary, treated to biological testing (cf. section on disposal sites).

When carrying out incineration, the incineration facility must be operated in conjunction with a disposal site where the incineration residues can be stored; account should be taken of the fact that such wastes have a lower water retention capacity than normal household wastes. Solid and paste-like wastes forming unobjectionable combustion products of low toxicity can generally be incinerated alongside domestic wastes. Some countries maintain incineration ships for particular chemical wastes, especially chlorinated hydrocarbons, which incinerate such materials on the high seas. Composting of wastes from the material-processing industries is only possible in rare cases; furthermore even here combination with a residue-incineration plant and waste disposal site is necessary. A review summarizing the basic possibilities for the disposal of industrial wastes can be found in [1]. Examples for the disposal of some important product-specific wastes are quoted in Table 2.1.

If incineration or landfilling alongside domestic waste or a specialized or underground disposal site are not feasible because of the properties mentioned previously then the waste must be treated in a separate plant and by a separate process to transform it to a safe, nontoxic form, so that it can be disposed of (see Fig. 2.1). Since such facilities are very expensive, they must be laid out with a suitably large capacity; i.e. centralized detoxification and chemical waste treatment facilites, where highly concentrated liquid waste can also be treated, must be planned and built for future requirements. In combination with this a system must be set up for the comprehensive collection of all types of waste coupled with information concerning other branches of industry in order to make use of potentials and markets for the reutilization and further processing of wastes.

For those wastes that cannot be reutilized on site the system of regional collection stations, possibly with pretreatment plants (neutralization, preliminary detoxification, dewatering to reduce transport costs to the central plant) together with the planned inclusion of particular waste components in household waste disposal sites must be extended (see Fig. 2.1). The pretreated wastes can then be transported to the supraregional central plants for treatment, disposal and recycling [3].

Table 2.1. Examples of (A) the content and (B) the possibilities for disposal of wastes from particular processes (according to [2])

Lacquer and paint wastes including unusable production batches
A. Pigments, chlorinated hydrocarbons, heavy metals Consistency: liquid, sludgy or dried out B. Dried out, water-insoluble residues can be disposed of by landfill alongside domestic waste. Liquid and sludgy wastes must — if there is no possibility of drying them by driving off the solvent — be burned in a chemical waste incineration plant.
Phenol-containing wastes from coal and crude oil processing
A. Phenols, ammonia Consistency: liquid, sludgy or solid; powerful odour and flavour, water-soluble B. Small quantities of solid waste may be dumped along with domestic refuse, but as a general rule all wastes should be processed by a chemical waste incineration plant.
Solvents and sludges of various origins
A. Many types of substance, very frequently chlorinated solvents Consistency: liquid or sludgy B. The main process is distillative recovery (separate collection, storage and transport). Mixtures and highly contaminated residues should be processed by a chemical waste incineration plant, as should distillation residues and sludges. Because of the possible emission of pollutants on incineration some states practise incineration on specially equipped ships at sea.
Used oils and oil sludges
A. Petroleum spirit and mineral oils Consistency: liquid or sludgy B. They are usually re-refined Unprocessable residues are sent to a chemical waste incineration plant (possibly a suitably equipped domestic waste incineration plant), when the flammability of the residues must be taken into account, since most of them are mixed with gasoline and organic solvents.
Acid resins and tars from oil refining
A. Mineral oils Consistency: viscous or sludgy B. Dumping is scarcely possible since SO_2 is evolved and high temperatures are reached on admixture with alkaline wastes. Incineration in suitably equipped plants is recommended.
Oil emulsions from metal machining
A. Mineral oils, emulsifiers Consistency: liquid B. If their calorific value is sufficient they can be incinerated in suitable installations. Another method is to treat them by breaking the emulsion. The water produced is treated as waste water, the high calorific value oil phase can either be incinerated or re-refined

Table 2.1 (cont.)

Acids, bases and corrosives from the chemical industry and the metal surface-treating industries
A. Acids, bases, heavy metal salts Consistency: liquid, corrosive B. Neutralization and heavy metal precipitation either on site or at a central facility
Salt slags from aluminium refineries
A. Salts that contain NaCl, KCl, phosphates, nitrites and other components Consistency: solid B. Deposition in a chemical waste depot
Electroplating and metal hydroxide sludges from electroplating and pickling plants
A. Chromium and cyanide-containing, metal-containing Consistency: sludge B. Chromium(VI) and cyanide-containing sludges must be detoxified; acid or basic sludges neutralized. Valuable metal compounds can be recovered by various recycling methods. The sludges then remaining are dewatered and deposited at a chemical waste depot.
Hardening salt residues from metal manufacture
A. Cyanides, nitrates, nitrites, barium Consistency: solid B. Underground disposal is an economic possibility
Pesticide residues produced during manufacture and use
A. Various pesticides B. Pesticides not containing metal or semi-metal components can be burnt in a chemical waste incineration facility with appropriate exhaust gas purification. All other pesticide residues should be properly deposited; the following method has proved to be of value: The residues are admixed with sulphate-resistant cement and made up as in normal concrete production; the mixture is then cast into prepared pits, so that the water-wettable surface area is considerably reduced. The blocks can also be deposited at a household waste disposal site.

Such a plant would consist of:

- Preliminary storage facilities.
- Detoxification and pretreatment plants (e.g. oxidation, reduction, neutralization, precipitation, dewatering, decanting, emulsion breaking etc.). A special disposal site for solid and defined — mainly inorganic — wastes.
- A special incineration plant for liquid and paste-like wastes — mainly organic — with subsequent adsorptive or absorptive purification of the exhaust gases.

— A recycling plant for the recovery of valuable products (e.g. solvent recovery, recovery of metals from electroplating sludges etc.).

Thus, the main task of a chemical waste treatment facility is the transformation of wastes into a reusable, disposable or incineratable form. Particular waste can be temporarily stored and underground disposal and other special treatment facilities maintained.

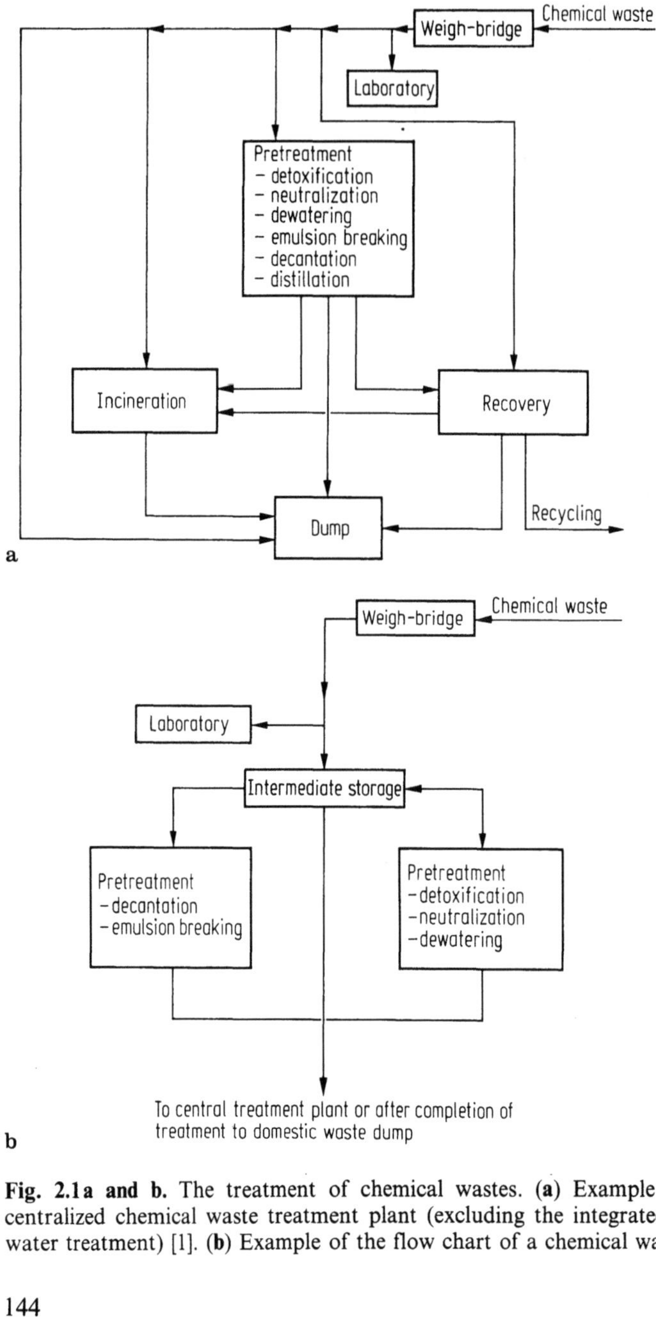

Fig. 2.1a and b. The treatment of chemical wastes. (**a**) Example of the flow chart of a centralized chemical waste treatment plant (excluding the integrated exhaust gas and waste water treatment) [1]. (**b**) Example of the flow chart of a chemical waste collecting site [1]

2.1 Thermal Detoxification Processes

2.1.1 Incineration (see also the section on Incineration of Chlorinated Hydrocarbons)

Incineration is the most highly developed of the thermal processes (cf. Table 2.2) [4, 5]. It is a very complex process, in which many steps occur, in part simultaneously, in part consecutively: drying — degassing — gasification — burning off. The individual processes can be assigned to particular temperature ranges (see Table 2.3). Each phase can be carried out as a process either alone or together with the ones prior to it.

Table 2.2. Classification of thermal waste treatment processes [4]

Process	Drying	Pyrolysis	Gasification	Incineration
Stage I	drying	drying	drying	drying
Stage II		pyrolysis (carbonization)	pyrolysis (carbonization)	pyrolysis (carbonization)
Stage III			gasification	gasification
Stage IV				incineration
Example	drying sewage sludge	waste pyrolysis	waste gasification	waste incineration

Table 2.3. Temperature ranges for thermal waste disposal processes [4]

Main range (°C)	Drying	Pyrolysis (carbonization)	Gasification or oxidative gas production	Incineration
<200	drying			
400–700		charring		
800–1200		coking	gasification	incineration
>1200			high temp. gasification	high temp. incineration

2.1.2 Waste Pyrolysis and Waste Gasification

Originally the term pyrolysis [6] was applied to the thermal decomposition of solids, liquids and gases (also known as coking, dry distillation and carbonization) with the exclusion of oxygen, where large molecules are split into smaller ones under the influence of high temperatures to leave a residue of coke. The smaller molecules can then be cracked at higher temperatures. The final products are hydrogen-poor carbonization residues and hydrogen-rich carbonization products. The following processes occur in the low and medium temperature ranges [6], between 100 and 120 °C drying occurs, espace of hydroscopic water; up to 250 °C desoxidation and desulphurization with visible signs of decomposition, the evolution of water of constitution and carbon dioxide; above 250 °C depolymerization takes place, as does the evolution of hydrogen sulphide; about 340 °C aliphatic compounds split off, methane and other hydrocarbons are produced; at 380 °C what is known as the carbonization phase occurs (the enrichment of the material being pyrolysed with carbon); at 400 °C carbon and oxygen compounds as well as carbon nitrogen compounds are evolved; between 400 and 420 °C bitumenic material is transformed into low temperature carbonization oil and tar; by 600 °C the bitumens have been cracked to form gaseous short-chain hydrocarbons and coke; above 600 °C aromatics are produced, the following reaction scheme is assumed [6]: olefines (ethylene) are dimerized to butylene, which is dehydrogenated to butadiene, this reacts by the diene reaction with ethylene to cyclohexene, which is aromatized to benzene and higher condensed aromatics.

Some authors include the gasification of wastes in the pyrolysis processes [7], others reserve the term pyrolysis exclusively for the dry distillation process. We prefer to employ this latter method.

According to the literature [8] research, pilot plant investigations and practical applications have been concentrated on three process combinations; however, the most various types of retorts have been employed.

(1) Dry distillation and/or gasification of domestic waste in the main with the production of fuel, mostly in the form of gas. Heterogeneous mixtures of waste with a low organic content at 800–1000 °C are treated. Old tyres and sewage sludge etc. can also be included. The main advantages are that little or no atmospheric pollution is caused, more energy is produced than consumed, the products are storable and transportable gas and coke. Disadvantages include the fact that the wastes must be broken up and prepared before treatment and high residence times are necessary.

(2) Dry distillation and/or gasification of industrial wastes mainly with the recovery of usable substances. Largely homogeneous specific industrial wastes with a high organic content are treated (very often plastics, old tyres or acid resins) at 500–700 °C with the production of utilizable liquid or solid products.

(3) Combinations of dry distillation (and/or gasification) and incineration of, in the main, domestic waste with the production of energy. Heterogeneous mixtures of wastes (including added sewage sludge and old tyres) are treated. Firstly a thermal decomposition of performed, then incineration. Preparation of the waste is not necessary. Smaller quantities of flue gases are produced than in conventional incineration and less pollution is caused. The treatment plant is relatively simple (mainly vertical shaft furnaces); a usable slag granulate is produced. The large fuel requirement is a significant disadvantage. Pyrolysis processes (dry distillation processes) can serve, on the one hand, for the processing of wastes with the production of energy. That is the energy is produced in a storable form. On the other hand, these processes are accompanied by less environmental pollution than other pure incineration processes. The amount of gas produced is significantly less, so that gas purification costs are correspondingly less. It is a disadvantage that the wastes have to be comminuted and prepared. More ambitious pyrolysis processes demand sorting steps, since not all "mixed" wastes produce effective results on pyrolysis, even though a range of processes has been developed for the production of energy from unsorted wastes.

Fluidized bed pyrolysis is a very modern technique which is undergoing investigation, whereby a connection has been found between the particle size of the material fluidized and the range of products produced. The fluidized bed consists of a material such as quartz sand at ca. 800 °C and the fluidizing gas is pyrolysis gas — preheated in a heat exchanger. The fluidizing gas and the pyrolysis gas produced are freed from solid matter in a cyclone after leaving the reactor (this is particularly necessary when pyrolysing old tyres, when large quantities of zinc oxide and carbon black are released) and then passed into a condenser; the noncondensables are stored in a gos-holder [9] from which they can be withdrawn as fuel or fluidizing gas.

Plastics and old tyres are particularly important examples of pyrolytically separable waste products [9–12]. A whole range of pilot scale projects and preliminary technical applications for the pyrolysis of sorted plastic wastes have been reported in the literature, whereby particular precautions are necessary for wastes which contain up to 20% PVC [9]. These must be subjected to a preliminary thermal dehydrochlorination before the actual pyrolysation process [9].

Other investigations have been devoted to the problem of old tyres. Old tyres consist of ca. 47% elastomers, 31% carbon black (and inorganic components, ZnO, S), 7% cord and 15% steel [13]. The cord and steel content is relatively easily separated out after mechanical comminution, only the 47% elastomer content actually undergoes pyrolysis. The carbon black content is usually unaffected. The decomposition produces a very complex mixture of saturated and unsaturated aliphatic hydrocarbons and aromatics. The composition of the products obtained by pyrolysis of used tyre material in a fluidized bed at 740 °C is reproduced in Table 2.4. In a pilot plant that operates without the use of supplementary fuel, it is, for example, possible to recover 50% weight of the starting material as 4 liquid fractions [15]:

Table 2.4. The composition of the products of the fluidized bed pyrolysis of worn tyres at 740 °C [14]

Component	wt. %	Component	wt. %
Hydrogen	0.78	Naphthalene	0.90
Methane	10.20	Methylnaphthalene	0.68
Ethylene	2.58	Diphenyl	0.34
Ethane	1.21	Acenaphthylene	0.15
Propene	0.73	Fluorene	0.11
Isobutene	0.18	Phenanthrene	0.08
1,3-butadiene	0.26	Carbon	42.80
2-butene	0.09	Fillers	7.85
Isoprene	0.12	Water	0.35
Cyclopentadiene	0.08	Acid gases (as H_2S)	1.55
Benzene	3.81	C_1–C_4HC, H_2	16.03
Xylene, ethylbenzene	1.93	Pyrolysis oil	30.18
Styrene	2.34		
Indene	0.78		
Total			98.76

— Low boilers (C_4 and C_5 hydrocarbons, a little benzene).
— Benzene fraction (cyclohexene and benzene).
— Toluene fraction (benzene, mainly toluene, a little xylene, traces of alkanes).
— High boilers (naphthalene and methylnaphthalenes).

Unexpectedly it was discovered that the fluidized bed that serves as the reaction medium is insensitive to the size of the material fed into it. This means that a fluidized bed with a diameter of ca. 1 m is able to pyrolyse one whole automobile tyre every 3 min. Investigations are underway to modify the range of products by the addition of catalysts. Laboratory investigations point to very far ranging possibilities.

Many investigations are also being made into pyrolysis under specialized conditions such as in the presence of a protective atmosphere, under vacuum, in hydrogen etc. [13]. Such processes are not, however, of great industrial significance.

Kiener-Goldshöfe's system will be described as a typical example of a waste pyrolysis process [16]. The first stage, decomposition at temperatures of 400–500 °C, takes place in an externally heated drum. With the exlusion of oxygen a complete carbonization of the material occurs with the liberation of volatile heavy hydrocarbons together with phenols, oils and tars. The residue consists of carbonized wood and coal residues, ash and any metal or glass that was present in the waste. After preliminary purification the dust-freed pyrolysis gas passes to a gas converter in which two reactions occur. The first is thermal cracking at 1100–1200 °C whereby long-chain hydrocarbons are cracked to methane, hydrogen and small amounts of the oxides of carbon and simple hydrocarbons. The second, which occurs in the same reactor but down stream from the cracking, is the "dissociation" or cleavage of a certain proportion of CO_2 and H_2O when thermal energy can be recovered. Heating is accomplished in part using the pyrolysis gas. The gas finally obtained contains H_2, CO and short-chain hydrocarbons togethet with CO_2, N_2, NH_4Cl, small quantities of oxygen, traces of HCl and higher hydrocarbons and water vapour [16]. It is utilized in gas engines after purification.

During conventional incineration all the constituent processes occur in one and the same combustion chamber (see also Table 2.2). The hot ash remaining at 1200 °C partially sinters, when nonferrous metals melt out or evaporate so that they cannot be recovered. The excess air necessary for complete combustion also brings disadvantages, even at relatively low incineration temperatures the formation of NO_x is promoted. HCl and SO_2 have already been mentioned several times. In contrast waste pyrolysis (Goldshöfe) is performed in the absence of air relatively low temperatures (see Table 2.3). Any HCl formed can combine with the ammonia that is also released to form NH_4Cl, that is relatively easy to wash out [16]. All the metals remain behind in the pyrolysis drum and can be recovered. Only about one fifth of the amount of gas is produced as compared to conventional incineration and this allows of more economical purification. The components of the pyrolysis gas can be burned without causing pollution. The aqueous effluent from gas scrubbing is slightly alkaline with an average pH of 7.4. More than 30% of the nitrogen is obtained in the form of ammonia. It is stated that the waters can be led into bio-mechanical waste water purification plants.

The Destrugas waste pyrolysis process can be regarded as being ripe for large scale use [17, 18]; it operates with an indirectly heated vertical retort. The comminuted waste is fed in from above, raw gas and coke are withdrawn at the bottom.

The heavy hydrocarbons, tars and other high molecular weight substances produced by carbonization are cracked to low molecular weight gases by the glowing coke in the lower portion of the retort. The coke finally produced contains only small quantities of soluble materials and can be disposed of or utilized for the production of water gas, for heating the pyrolysis plant.

The second basic possibility for thermal decomposition is gasification. This process only begins when the first carbonization stage is already complete; the carbon remaining is then attacked by the oxygen contained in water or in carbon dioxide or by hydrogen. The already completed pyrolysis is reflected in the products obtained from the whole process; the typical pyrolytic products are neutral oils, phenols, organic acids, tars and fuel gas with a varying content of CO, H_2, hydrocarbons, H_2S, NH_3 and HCN. The products produced in the reducing milieu of the pyrolysis process are particularly characteristic. The Andco-Torrax process may be cited as a typical example, which utilizes a vertical shaft furnace with liquid slag removal [4]. The wastes can be introduced overhead without prior preparation. Air heated to 1000 °C is injected below the bed of waste, the following reaction zones are set up from bottom to top:

— Combustion with melting of the slag at up to 1600 °C
  Main reaction: $C + O_2 \rightarrow CO_2$.
— Gasification
  Main reaction: $C + CO_2 \rightarrow 2\,CO$
  $C + 1/2\,O_2 \rightarrow CO$
  $C + H_2O \rightarrow CO + H_2$.
— Pyrolysis
  Release of hydrocarbons.

The produced gas so formed can — if it is not to be utilized — be burned in a subsequent combustion chamber and used to heat the incoming gases in a heat exchanger. The flue gases are purified by a wet process.

The effects of combustion and pyrolysis processes on the environment can include pollution of the atmosphere, water and land surface. Atmospheric pollution consists of dust, exhaust gases, rejected heat (cf. [6]). The major pollutants formed on the pyrolysis of wastes, plastic residues and old tyres are principally sulphur compounds (mainly H_2S), nitrogen compounds (NH_3 and HCN, the oxides of nitrogen are only produced to a limited extent during pure pyrolysis), HCl from chlorine-containing compounds, HF (similar quantities to those produced on incineration) and heavy metal vapours [6]. Carbonization and gasification processes are mainly characterized by the fact that they produce smaller amounts of waste gases that require scrubbing than do incineration processes; they also produce less dust.

Water pollution consists of pollution caused by process water, scrubbing water and silo-seepage water. The gas scrubbing water, which accounts for the major portion of the waste water, accomplishes a transfer of the polluting materials from the gaseous to the aqueous phase (see Table 2.5). Objectionable organic and inorganic compounds and rejected heat are all pollutants; the major material ones are sulphates, chlorides and cyanides, some of the heavy metals contained in the exhaust gases and organic compounds originating in the pyrolytic decomposition (such as neutral oils, alcohols, aldehydes, ketones, acids, phenols, amines and higher aromatics). The pollution of the land's surface is best understood in terms of the necessity for disposal and of the

Table 2.5. The volumes of waste water produced in gas scrubbing and the extent of its pollution for some thermal waste disposal processes [20]

Process	Waste water volume (m^3/t waste)	Emission components enriched in the scrubber water
Waste incineration	2–3[a]	HCl, HF, SO_2, NO_x, dust
Destrugas	0.3	HCl, HF, H_2S, NH_3, dust, phenols, cyanides
Kiener-Goldshöfe	0.3	HCl, HF, H_2S, $SO_4^{2-}NH_3$, tar, phenols, cyanides, dust
Garrett	0.05	salts, tar, oil, organic acids, sludge (highly polluted)
Landgard	2–3[b]	HF, HCl, SO_2, NO_x, dust
Pyrogas	0.06	salts, tar, oil, organic acids, sludge (highly polluted)
Andco-Torrax	3^2	HF, HCl, SO_2, NO_x, dust
Purox	0.3	HF, HCl, H_2SNH_3, tar, phenols, organic acids, alcohols, oils, dust

[a] With scrubbing water recirculation
[b] Estimated as being similar to that in the incineration process

leachability of the residues. The behaviour of residues from low and medium temperature processes on the tip ought to be similar to that from waste incineration, while that from the high temperature processes is better (for instance heavy metals are driven off as vapours).

2.1.3 Specific Thermal Methods

The Flame Chamber (FLC) Process [19] is a modern process by which wastes are "melted down" at 1600 °C. This achieves a reduction of up to 97 % in the volume of the waste. A solid, harmless granulate suitable for building purposes is obtained. The flue gases produced contain relatively little dust and pollutants. In comparison the combustion temperatures obtainable in a roasting kiln are limited to about 1100 °C to avoid forming molten slag which would block the air entry channels. Rotary kilns can be operated at temperatures up to about 1400 °C and wastes of various consistencies can be treated, but the mechanical and chemical wear on the fireproof linings is very great and clinker deposition can occur, which dislocates the whole incineration process by narrowing the kiln's cross-section. A special form of furnace construction was developed for the FLC process. The design is based on the concept of the continual production and removal of liquid slag without continually exposing the lining of the furnace to liquid slag and without having to fear the solidification of the molten slag In the FLC process the waste itself constitutes the combustion chamber and lining. The furnace is insensitive to wastes of varying composition (e.g. sewage sludge, waste oil, industrial wastes). The previously comminuted waste is fed, by means of a suitable conveyor system, into the annulus between an inner and outer cylinder. The combustion chamber is formed at the bottom, between the waste slope and the bell-shaped base of the inner cylinder at whose apex the burner is located

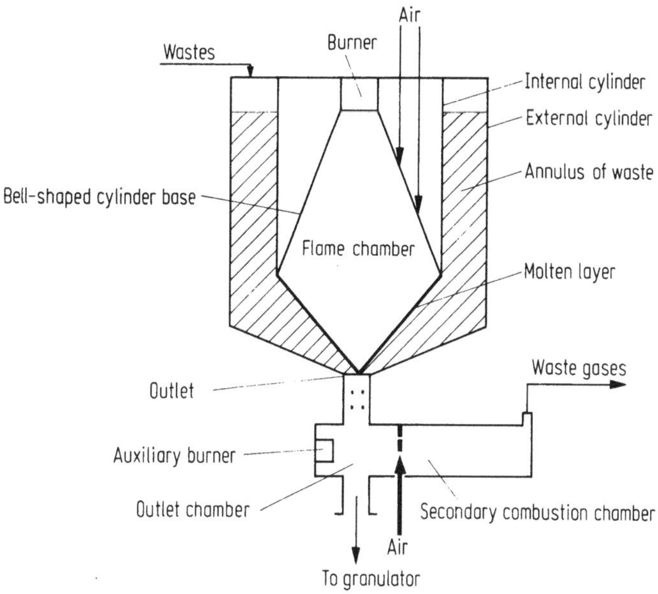

Fig. 2.2. Diagram of the flame chamber process

(see Fig. 2.2). During combustion a relatively thin molten layer is formed on the conical slope of the waste and runs down towards the exit hole. As the liquid slag flows down the reaction surface a corresponding quantity of new waste is fed in through the loading slit. By this method an almost stable horizontal temperature profile is built up between the slag film and the outer wall of the furnace, yielding an outer wall temperature of only 25–45 °C. The sensitive and expensive linings of normal furnaces become superfluous in this process. The material beneath the film of slag is carbonized and gasified; the gas produced penetrates the film of slag and is burnt in the combustion chamber. The liquid slag flows out of the exit into a water bath and is granulated there, while secondary air is admitted to the fuel gases before their entry into the secondary combustion chamber to cool them and to enable them to burn. In order to maintain the necessary high temperatures the system is operated with a small excess of air of ca. 30–60 % (calcination and rotary furnaces require from 80–200 %). The conditions obtaining in the furnace ensure that a high degree of combustion completeness is achieved in spite of the small excess of air [19]. For comminuted wastes it is reported that supplementary fuels can be dispensed with at calorific values as low as 2000 kcal/kg (8374 kJ/kg) [19].

Other high temperature processes have also been investigated with whose assistance a virtually complete decomposition of compounds to their constituent elements can be achieved. Such plants consist, in general, of a porous graphite cylinder, that is electrically heated to 2200 °C. The waste that is led through the tube is decomposed by infra-red radiation. The stability of the reactor tube is ensured by forcing nitrogen through its wall from the outside, thus maintaining a relatively cool mantle of inert gas at its internal wall, which protects the internal surface from oxidation and destruction. Because of the high temperatures employed no troublesome intermediate decomposition products are released in the exhaust gases. Alumina and lime are included as additives to bind chlorine, sulphur and phosphorus, so that the most important reaction products are carbon monoxide and hydrogen. They can be utilized subsequently for steam production.

Some completely new methods have been investigated. A process has been developed in the USA, for example, for the decomposition of toxic organic wastes by means of a microwave plasma [63]. A plasma is a partially ionized mixture of gases, consisting of free electrons, ions and various neutral species. The initiators of the plasma reaction are the free electrons. If these encounter molecules of the compounds to be destroyed, they either cause ionization with the production of further electrons and ions, or dissociation into free radicals, which rapidly react further in consequence of the unpaired electrons they posses. The plasma is maintained by means of an oscillating electrical field, produced by microwave energy, in general reduced pressures of from 1–100 Torr are employed. If oxygen is employed as the reaction gas, then atomic oxygen also comes into action as an oxidizing agent for the organic wastes. Even compounds as stable as PCBs can be completely decomposed to CO, CO_2, HCl, Cl_2 and ClO_2 by plasma processes. Dangerous products such as phosgene or vinyl chloride could not be detected by GC-MS analysis. The mechanism of decomposition is thought to include both direct decomposition by energetic electrons and the effect of atomic oxygen. The employment of the process on an industrial scale is envisaged [64]. It is also possible to destroy the mixture used by the US navy for flares and smoke signals, which consists of 55.4% xylene-azo-β-naphthol, 18.8% 1-methylaminoanthraquinone, 18% sucrose, 5.9% silica binder and 1.8% graphite, using plasma with oxygen as carrier gas yielding 99.99% H_2O, CO, CO_2, NO and SiO_2 [65].

Superciritical liquids are reported to be very efficient media for thermal decomposition. Thus, under supercritical conditions (above 374 °C and 218 atm) water is a very good solvent for organic material; high molecular weight substances are cleaved to smaller fragments; in the presence of oxygen rapid oxidation of carbon and hydrogen occur; heteroatoms such as halogens, sulphur, phosphorus and metals form insoluble salts that can be filtered off. Industrial application is not yet feasible [70].

The heavy oil residues produced in oil refining can be utilized and thereby disposed of without pollution by the employment of modern, high pressure gasification. Their utilization in power stations for example is scarcely possible or results in high flue gas scrubbing costs, because of their very high sulphur content, which can amount to more than 5%. In contrast high pressure gasification in Claus plants yields large quantities of hydrogen and carbon monoxide (together with products such as ammonia, methanol and urea) with simultaneous sulphur recovery and leads to a very large reduction in pollution. The processes are based on partial oxidation at pressures of 60–80 bar. The water vapour used as moderator can be generated in the reactor itself. After the gasification comes a cooling step (e.g. steam generation from waste heat) and soot removal. The recovered soot is returned to the process in the oil feedstock. Organic absorbents such as methanol or polyethyleneglycol ether are employed for the removal of hydrogen sulphide. The CO_2 which is also obtained on desorption is removed in a second scrubbing process. If the process is laid out for the recovery of H_2 or NH_3 the carbon monoxide must also be removed (otherwise methane formation and catalyst poisoning occur at the NH_3 synthesis step); this is accomplished either by complex formation with copper (copper aluminate in organic solutions or copper acetate in ammoniacal solutions) or by scrubbing with liquid nitrogen.

2.2 The Chemical Treatment of Wastes

Solid and liquid waste chemicals can, under certain conditions and particularly in small quantities and if other methods are inapplicable, be rendered nontoxic and suitable for disposal by chemical reaction. In general it is necessary to dissolve them in a solvent. Water is to be preferred if this is possible; emulsifiers and dispersants may be employed. Detoxification reactions in organic solvents constitute the exception, since the organic solvent — particularly if it is halogenated — creates new disposal and destruction problems. They are, however, indespensable for some problems encountered in the laboratory or in accident situations.

The basic methodological possibilities for the detoxification of aqueous solutions are discussed in the section entitled "The detoxification of industrial waste waters" so that here the discussion can be limited to a few specific examples of chemical disposal not discussed elsewhere.

2.2.1 Waste Neutralization

During the reprocessing of used oils by the sulphuric acid method these are cleansed of all their impurities and dewatered by the use of sulphuric acid. This results in the production of large quantities of toxic, tar-containing and evil-smelling acid resins, which, although they have a high calorific value, nevertheless emit too much SO_2 (from sulphuric acid, sulphonic acids etc.) on combustion. One possibility of converting such resins to a form suitable for landfill disposal is neutralization. This, however, is not technically satisfactory if quick or slaked lime is used, since, on the one hand, the large amount of heat evolved (exothermic reaction) causes problems and, on the other, large quantities of SO_2 are also evolved. Additionally during the admixture of resin and neutralizing agent a hardening process takes place, so that large amounts of energy are required for mixing [21]. Better results are obtained if the acid resins are dispersed as particles with sizes of from 20–300 μm with the aid of additives (particularly clays). Simultaneously the neutralizing agent is added (lime, milk of lime or waste caustic lyes) [21]. The exothermic reaction is so controlled by the addition of water that the temperature does not rise above 60 °C. The evolution of SO_2 is avoided and it is possible to adjust the pH precisely. The crumbly neutralization product can be tipped alongside household waste. It consists principally of chalk, calcium sulphate and finely divided hydrocarbons. Its utilization in the building industry is under investigation [21].

2.2.2 Hydrolytic Processes

A whole range of liquid and solid industrial and laboratory wastes and some gases and vapours can be detoxified by hydrolysis; however, even at elevated temperatures the reaction times can be extended so that the use of catalysts is necessary; in the simplest instance of the laboratory detoxification of many materials these can be hydrogen or hydroxyl ions (see also p. 256). Individual hydrolytic techniques or combinations of techniques are also being increasingly investigated for the disposal of industrial wastes and poisons.

Thus investigations are under way of the pyrolysis of plastics, old tyres and used oils with superheated steam at 500°–650 °C from waste incineration plants (see Fig. 2.3) [9, 22]. Investigations have also revealed that, for instance, polyurethanes can be cleaved by steam to toluene diamine which can be reacted with phosgene to yield toluene diisocyanate again. The polyether that is also produced is the same as that produced from glycol and alkylene oxide.

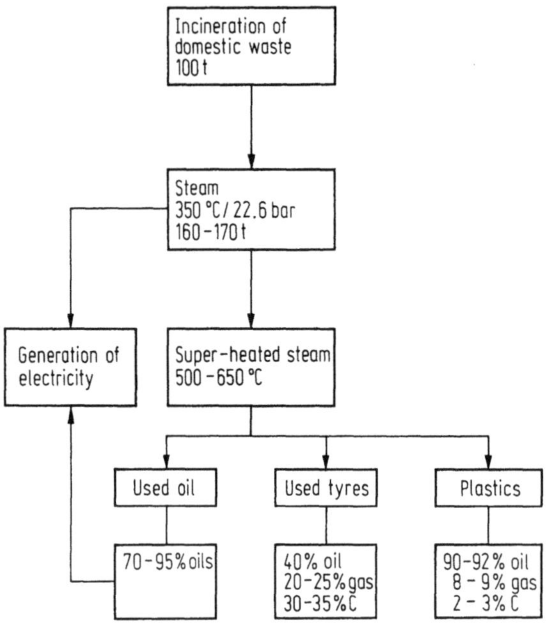

Fig. 2.3. Suggested steam pyrolysis process for industrial wastes

As has been repeatedly reported, processes for the disposal of solid cyanides and of cyanide solutions of various concentrations play an important role in industry. According to Schindewolf et al. [24] waste cyanides can also be disposed of hydrolytically. It has been known since 1832 that cyanides can be rendered harmless in this way [25]. Ammonia and sodium formate are obtained.

$$NaCN + 2\,H_2O \rightarrow NH_3 + HCOONa$$

The advantages of this hydrolytic process are [24]:

— Only water and heat are required and relatively harmless products are obtained that are easily disposed of.

— Impurities in the cyanide such as $BaCl_2$, NaCl, $NaNO_2$, NaOH (i.e. case-hardening salts for example) do not affect the reaction even when they are in great excess and some of them can be recovered.

— Concentrated or dilute solutions can be treated.
— The simplicity of the reaction simplifies the measuring techniques.

Against this there are the disadvantages that when operating at high temperatures and pressures special steels are necessary because of the corrosiveness of the cyanides; furthermore, pressure of up to 100 atm have to be employed, which cannot be achieved in every plant. Nevertheless, the costs of such a process are estimated to be less for concentrated cyanide solutions than those of chloroxidation [26]. The exothermic reaction takes place very rapidly above 150 °C, the equilibrium of the reaction is completely on the side of the reaction products. The NH_3 produced causes the pressure in the system to increase. Thus, as a result of the exothermic reaction of a 10 M cyanide solution, the temperature was observed to increase from 120° to 240 °C and the pressure to 120 atm within 5 minutes. At 190 °C the reaction is practically complete within an hour and the residual cyanide concentration is less than 1 mg/l. In the concentration range 1 mg cyanide/l to 100 g cyanide/l the half-life is reported to vary from 1–500 min within the temperature range 100° to 190 °C. The upper limit to the cyanide concentration that can be processed is determined by the pressure increase as a result of ammonia formation. It is an advantage that even complex cyanides are also decomposed, even if a great deal more slowly because of the very low concentrations of free cyanide present, thus the stable nickel complex is destroyed to the extent of about 50% after 12 minutes at 230 °C. The same reaction of solid cyanides can be achieved in the melt with steam at temperatures above 700 °C without the use of increased pressure, in order to avoid solution in large volumes of water [26]. The formate formed is decomposed to carbonate, carbon monoxide and hydrogen under these conditions. These gases, together with the ammonia produced in the first part of the process, burn off above the melt. In practice the steam hydrolysis of solid cyanides is performed at 850°–990 °C. Below 850 °C the melt solidifies because of the build up of sodium carbonate. Residual concentrations of less than 0.5 mg CN^-/l are reported. Nitrogen can also be produced alongside the gaseous products mentioned above and up to 2% of the cyanide can escape as HCN. The toxic products are completely destroyed during the burning off of the gases. Any cyanates present in the waste salts are converted during the reaction to carbonates and cyanides, the latter is then hydrolyzed to formate and ammonia:

$$4\,NaCNO \xrightarrow{H_2O} Na_2CO_3 + NaCN + CO + N_2$$

The salts must, however, be free from nitrites, since nitrites and cyanides can react explosively with each other at high temperatures. Even though the techniques for the hydrolytic detoxification of cyanides have not yet been optimized for use on an industrial scale, they are certain to find a place in the centralized detoxification plants that will be built in the future [27]. For example 10–15 g steam/min can destroy 11 kg of cyanide salt in 7 hours.

2.2.3 Molten Salt Processes

The mutual reaction of the cyanide and cyanate components with the nitrite and nitrate components is an interesting method of detoxifying case-hardening salt wastes [27–29]; these generally contain, on a weight basis, 20% Ba, 20% nitrite and nitrate, with 5% cyanide and cyanate as toxic constituents [28]. Amongst the older suggestions for the treatment of case-hardening salts are the reaction with ferric oxide or hydroxide [30] at 900 °C in the molten state to yield cyanate or NH_3 and CO_2 and reaction with added sand or clay at temperatures of more than 900 °C to produce glass-like products suitable for landfill disposal [31, 32] and processes involving partial reutilization like precipitation of the complex iron salts from aqueous solution [33], the recovery of the barium carbonate after treatment with hydrogen peroxide [34], the treatment of the melt with water vapour as has already been discussed for the disposal of cyanides [26] or hydrolysis under pressure at elevated temperatures [24]. These, however, do not take advantage of the very high nitrite/nitrate contents. Mutual destruction occurs with adequate speed in the homogeneous state i.e. in the melt [28, 29].

$$CN^- + NO_3^- \rightarrow CO_3^{2-} + N_2$$

$$CNO^- + NO_2^- \rightarrow CO_3^{2-} + N_2$$

$$5\,CNO^- + 3\,NO_3^- \rightarrow 4\,CO_3^{2-} + CO_2 + 4\,N_2$$

$$3\,CN^- + 5\,NO_2^- \rightarrow 3\,CO_3^{2-} + O^{2-} + 4\,N_2$$

This state of homogeneity is reached at 380 °C. The reactions are strongly exothermic and proceed explosively above 600 °C, so that care must be taken to ensure that the temperature does not rise above the 450°–550 °C range. It should be arranged that the reaction takes place either in a thin layer of the mixture, whose cyanide/cyanate content is lower than 5% by weight, or continuously whereby the mixture to be reacted is added at a constant rate to the already reacted melt [28]. The excess nitrite/nitrate does not react any further. The solidified melt is leached with water, the insolubles contain 95% barium carbonate, which can be used for the production of utilizable barium compounds such as barium chloride. The major portion of the soluble matter consists of alkali metal nitrates and nitrites, carbonates and chlorides.
The carbonates can either be eliminated by reaction with nitric acid

$$CO_3^{2-} + HNO_3 \rightarrow HCO_3^- + NO_3^-$$

$$CO_3^{2-} + 2\,HNO_3 \rightarrow CO_2 + 2\,NO_3^-$$

or by treatment with alkaline earth nitrates

$$CO_3^{2-} + Ba(NO_3)_2 \rightarrow BaCO_3 + NO_3^- .$$

The chloride is removed in a first concentration step. A second such step yields at 160 °C a salt melt that can be directly applied in hardening. A process suitable for industrial application has been developed and tested on the pilot plant scale [28].

The first thermal-catalytic process to find commercial application was the molten salt oxidation process, employing molten sodium carbonate at between 750° and

1000 °C. In this temperature range the molten salt possesses catalytic activity and effects a large acceleration of the oxidation of organic substances, the heat of combustion released being generally employed to maintain the necessary temperatures. The advantages of sodium carbonate are its relatively low price, its nonvolatility and its nontoxicity. The only gases evolved from the melt are CO_2 and steam. The HCl or chlorine evolved by chloride-containing wastes and halogenated organic compounds are bound as sodium chloride and sulphur, phosphorus and arsenic also form salts, which remain in the melt. Because of the relatively low reaction temperature of the process no oxides of nitrogen are detectable in the exhaust gases, in spite of this carbon is completely oxidized to CO_2. Entrained salt particles are removable by a simple filtration plant. According to the literature such problematical substances as halogenated organics and polychlorobiphenyls can be destroyed. The efficiency of the process amounts to 99.99% [70].

2.2.4 Specific Chemical Methods

"Perhydrolysis" plays a particular role in the detoxification of the esters of phosphoric and phosphonic acids. On the basis of investigations of the Schönemann reaction [23] it was discovered that the rate of hydrolysis of phosphate and phosphonate esters in alkali was considerably accelerated by the addition of H_2O_2. It is thought that the reaction proceeds via intermediate peroxyphosphate or peroxyphosphonate esters, which either react with a further molecule of hydrogen peroxide or with a molecule of the starting ester to form the corresponding phosphate. Epstein et al. [35] using the examples of sarin and paraoxon were the first to establish a significant acceleration of basic hydrolysis by means of H_2O_2. In the case of paraoxon the course of the reaction can easily be followed optically because of the formation of the highly coloured p-nitrophenate anion. The probable mechanism is as follows: —

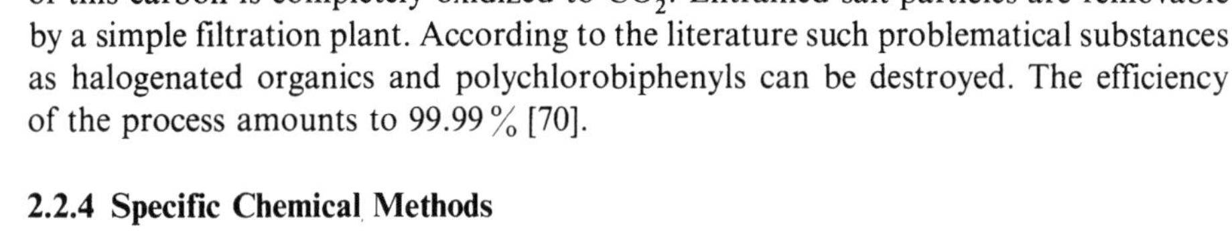

The reaction velocity of the "perhydrolysis" is ca. 10^2 times greater than that of the alkaline hydrolysis, even though the OH^- is more basic than the perhydroxyl amion. This has led to the postulation of a push-pull mechanism.

The reactivity of hydrogen peroxide is also utilized in a new method of detoxifying cyanide [36, 37]. The inorganic cyanide is first reacted with formaldehyde to yield formaldehyde cyanhydrin, which is then transformed into glycolic acid amide by the catalytic action of formaldehyde and this is hydrolysed to glycolic acid which is biologically degradable. However, a small proportion of the cyanide is oxidized to cyanate in a side reaction.

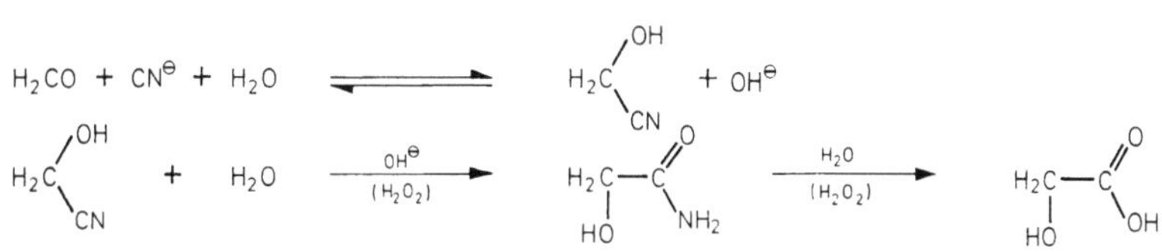

This method is particularly suitable for the detoxification of low concentrations of free alkali metal cyanides, but the cyanide complexes of metals can also be disposed of in the same way. The metal is quantitatively deposited as the hydroxide or the carbonate (this process is largely completed during the initial reaction with formalin):

$$[Zn(CN)_4]^{2\ominus} + 4\,H_2O + 4\,H_2CO \rightleftharpoons Zn(OH)_2 + 4\,H_2C\begin{smallmatrix}OH\\CN\end{smallmatrix} + 2\,OH^{\ominus}$$

The process is particularly suitable for the cleansing of laboratory ware and electoplating or hardening apparatus contaminated with cyanide by immersing them in the reaction solutions; one of its industrial applications is the detoxification of cyanide-containing blast furnace waste waters [37]. H_2O_2 also reacts with nitriles in neutral or slightly alkaline media (pH $<$ 8) at elevated temperatures to form intermediate percarboxylic acid imides [39], which decompose at more alkaline pHs to yield amides of carboxylic acids, water and oxygen.

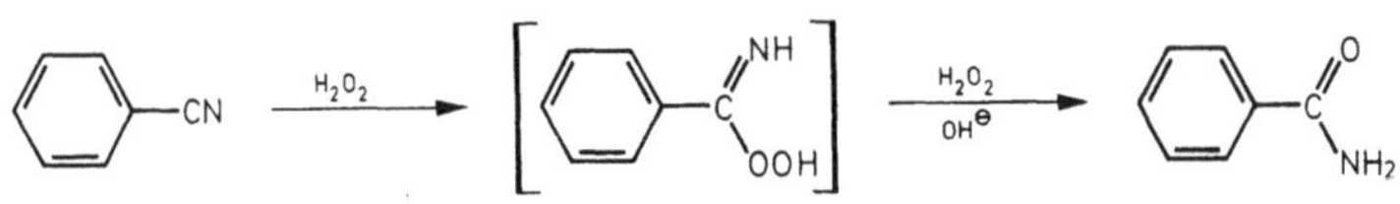

Alcohols are added to aid dissolution. Aliphatic nitriles such as butyronitrile or valeronitrile in aqueous alcoholic solution give the corresponding amides in 60% yields after treatment with 6% H_2O_2, followed by addition of caustic soda. Benzonitrile reacts very readily in slightly alkaline solution to yield benzamide, the reaction with HO^- being 10^4 times faster than that with OH^- ions. The same applies to benzonitriles with a second substituent. With alkaline 6% H_2O_2 solution p-nitrobenzonitrile yields 90% of the corresponding amide in 4 hours. If unstaturated amides are produced there is a possibility of further reaction at the C=C double bond to form glycide amides. This method of decomposition is particularly suitable for aromatic nitriles. The treatment of nitriles with NaOH and H_2O_2 in aqueous alcohol is one of the standard methods of synthesising amides [40].

Apart from the chemical disposal methods already described and the normal methods of treatment of aqueous waste waters and concentrates there is a whole range of organic reaction types and individual chemicals suitable for particular cases of detoxification processes and especially for detoxification or immobilization in the laboratory; for instance, substitutions, additions and eliminations. The best-known substitution reaction used for detoxification purposes is probably the decomposition of organic halogen compounds with inorganic sulphides. Aqueous sodium sulphide in the presence of dissolution promoters is well-known as a detoxifying reagent for N and S mustard gases, bromobenzyl cyanide, haloketones, haloethers, halocarboxylic acid derivatives haloalkylbenzene compounds, when aliphatic or cyclic thiocompounds are produced [41]. In general a 20% solution of alkali metal sulphide containing alcohol or 1% of emulsifying agent is employed for detoxification. The use of aqueous two-phase systems in the presence of phase transfer catalysts is also described in the literature [38, 42–44]. Organic phosphonium or ammonium salts are employed as catalysts.

A patent [45] describes the removal of anthraquinonesulphonic acid from waste waters from the sulphonation of anthraquinone, by conversion into the insoluble haloanthraquinone, which is an intermediate in the manufacture of anthraquinone dyes. The technical sulphuric acid produced is also a welcome by-product.

Some organic materials can be "precipitated" by addition reactions with inorganic substances and thus immobilized. The best-know "precipitating agent" is probably sodium bisulphite, which yields solid addition compounds with, for example, aldehydes and unsaturated compounds. Thus, the addition of sodium bisulphite to acrolein in aqueous solution occurs spontaneously in two stages; the addition to the aldehyde group which can be reversed by adding acid and heating and the irreversible addition to the C=C double bond:

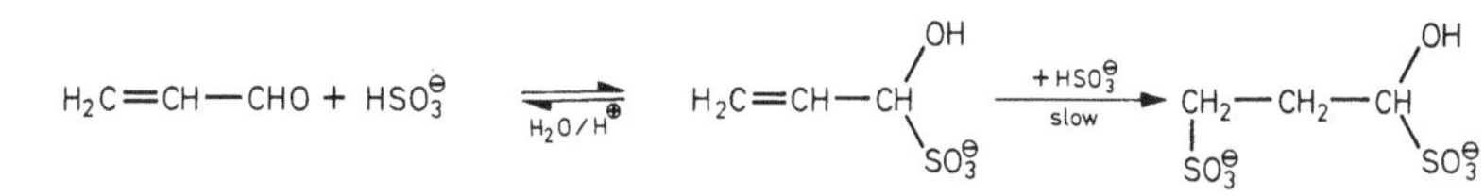

Liquid amine bases can also be immobilized as solids, aliphatic amines by addition of sodium bisulphite, primary aromatic amines by agitating with a strong acid (the formation of hydrochloride for example). Elimination reactions such as dehydrohalogenation, dehalogenation, demethylation, decarboxylation, desamination can also be associated with a reduction in toxicity, so that in particular cases they can be employed as detoxification reactions. This is particularly the case for valuable chemicals that can be utilized for other purposes. Newer reaction principles can also be employed; for instance the utilization of polymeric reagents. Thus the use of cross-linked polyvinylpyridine as an HCl acceptor instead of the usually employed tertiary amine has been described [46]. It is unlikely that such methods will find immediate application in practical detoxification and in the future they will probably remain limited to specific problems. However, elimination reactions — like all other types of reactions — can also be associated with an increase in toxicity. An example is the elimination of HCl from the phosphonate ester trichlorphon, which leads with rearrangement to the phosphate ester DDVP (dichlorvos) which possesses a considerably higher acute toxicity.

New very specific methods for particular chemical structures are being continually developed in the field of the chemical treatment of dangerous wastes. For instance, recently investigations have been reported concerning the possibility of destroying PCBs using sodium naphthalenide (prepared in situ from sodium and naphthalene) in THF within 15–30 minutes [66]. The effect of polyethylene glycol and metallic sodium (heated to the boiling point of sodium, when a uniform suspension is produced) also leads to virtually complete dechlorination in an exothermic reaction. The products are polyhydroxybiphenyls and common salt. Halogenated pesticides such as hexachlorocyclohexane and 1,3,5-trichlorobenzene can also be transformed into hydroxylated compounds in this manner [67]. Reference [68] contains a review of the treatment of PCBs.

A recent chemical method of the destruction of 2,3,7,8-tetrachlorodibenzodioxin (TCDD) is based on the cleavage of the ether linkage of the aromatic system by the use of quaternary ammonium chloroiodides in 0.1 M aqueous ammonium chloride solution. Thus TCDD at a concentration of 10 µg/ml in 0.1 M cetylpyridinium chloride was 92 % destroyed in the presence of 0.5 % cetylpyridinium chloroiodide. The decomposition products (such as chlorophenols, phenols and 2-phenoxychlorophenol) have not been individually investigated as yet. The mechanism is thought to be similar to that of the direct action of iodic acid. The advantages of this reaction are that it takes place at room temperature, in aqueous solution, in the absence of UV irradiation [69].

2.3 The Detoxification and Destruction of Important Laboratory Wastes

It is important not to underestimate the chemical wastes which have to be disposed of by the laboratories and pilot plants of universities, research institutions, industry and public bodies, neither quantitatively nor in their qualitative variability and continually changing composition. Thus in 1975 the laboratory wastes of the chemical industry alone were estimated to amount to some 20000–40000 t in the Federal Republic of Germany [52]. The structurally very different inorganic and organic laboratory wastes should be divided according to their properties and the possibilities available for their disposal.

A. Refuse

This includes harmless solid inorganic and organic wastes insofar as they do not contain dangerous or radiocactive materials from groups C and D and which can be disposed of without danger along with domestic waste or be incinerated. They incinerated. They can be prepared for disposal in closed sacks or containers.

B. Chlorinated solvents and other organic solvents and liquid chemicals, insofar as they do not contain dangerous or radioactive materials from groups C and D, including organic substances dissolved in them, insofar as they do not react with water or with one another.
The total chlorine content must lie below 5 %. As a matter of course the attempt is made to utilize solvents with a low degree of chlorination as much as possible (e.g. methylene chloride instead of carbon tetrachloride) — and to reuse organic solvents if possible (e.g. distilled-off solvents) or to redistil them (e.g. wash acetone). Non-reusable waste solvents and liquid chemicals are collected and sent to a central incineration plant. If this is not possible such wastes must be burned off carefully in small quantities, away from built-up areas and taking all necessary precautions.

C. Dangerous materials

These include toxic, corrosive, oxidizing, pyrophoric, easily inflammable, potentially explosive or obnoxious smelling chemicals or solutions. Radioactive substances may not be present. Dangerous chemical wastes must be detoxified or immobilized without delay as they are produced. If this is impossible they must be passed on in separate, closed, labelled containers, with precise detoxification instructions. Storage times of more than 2 months should be strictly avoided. Great care is essential during all detoxification operations; protective goggles, protective clothing and gloves (possibly rubber aprons) should be worn. Overhasty manipulation and working alone should be strictly avoided. The work should be done in an area specifically set aside for the purpose with adequate forced ventilation. The correct order of mixing, not too large a quantity and slow addition should be the order of the day. The detoxifying agent should usually be in excess; after the completeness of the detoxification process has been verified the excess should be destroyed. Particularly in the laboratory, where the most widely varying classes of substance occur as wastes — the choice of materials and apparatus, the size of the batch, the means of recognizing completeness and necessary safety precautions should be carefully considered before commencing detoxification. It must be carefully considered whether the chemicals to be destroyed and the proposed detoxifying agent might react with each other violently or explosively (see Table 2.6) or could form dangerous products. Examples of the latter are the formation of chloropicrin during the oxidation of sodium picrate with chloride of lime, the production of phenylcarbylamine chloride by the action of chlorine on mustard oil or the formation of cyanogen chloride when chloramines and sulphuric acid come into contact.

D. Radioactive wastes [see special reviews]

In principle detoxification reactions in aqueous or nonaqueous solvents or, exceptionally, in the absence of solvent can be carried out. The use of aqueous solutions often has the disadvantage that the poisons to be detoxified are insoluble or difficultly soluble. In a number of cases it is therefore necessary to utilize nonaqueous solvents, which are good solvents for the poison and the detoxifying agent and thus guarantee a rapid reaction, usually at a low temperature. This, however, is only advantageous for the destruction of toxic materials in the laboratory and in cases of accident, but not for large quantities of industrial chemical waste.

Nonaqueous detoxification agents have been expecially developed for military decontamination [55], for example chloramine in dichloro or tetrachloroethane as aprotic solvents (indifferent, dissolve only nonelectrolytes). On treating coated metallic surfaces solvents can be chosen that are less corrosive than water. However, some organic solvents react aggressively with metals and other constructional materials (plastics, rubber, leather, textiles). It should be noted that many solvents, especially the chlorinated hydrocarbons, that are particularly useful in so many situations, are themselves toxic or produce toxic vapours and are thus the cause of disposal problems. Many organic solvents catch fire easily. The fact that detoxifying agents and solvents react with each other solowly can constitute another problem, so that such detoxification mixtures must be freshly made up. Water-miscible organic solvents (lower alcohols, DMSO, DMF) are often added to aqueous solutions, where they act to improve solubility. The so-called universal detoxifying agents contain neutral and basic (nucleophilic) solvents possessing autoprotolytic or amphiprotolytic properties [55] such as amines or amine alcohols, which, like water, are autoprotolytic and generate lyonium and lyate ions

$$2\,RNH_2 \rightleftarrows RN^+H_3 + RNH^-$$

The most reactive component is a powerful nucleophile, usually an alcoholate; the reactivity depends on the low solvation.

Table 2.6. Incompatible chemicals (ref. [57] with additions)

Substance	Must not come into contact with
Acetylene	halogens, copper, mercury, silver
Active carbon	calcium hypochlorite, all oxidizing reagents
Alkali metals plus aluminium and magnesium powders	water, halogens, chlorinated hydrocarbons, HCN, CO_2 (fire extinguishers)
Ammonium nitrate	metal powders, acids, chlorates, nitrites, sulphur, flammable solvents, finely divided flammable solids
Anhydrous ammonia	Hg (manometer), halogens, calcium hypochlorite, HF
Aniline	nitric acid, conc. hydrogen peroxide
Bromine and chlorine	metal powders, hydrogen, ammonia, acetylene, butadiene, methane, propane, butene, turpentine, benzene
Carbon disulphide	sodium peroxide
Chloramines	NH_3, NH_4^+, urea, NH compounds
Chlorates	metal powders, ammonium salts, acids, sulphur, finely divided, flammable solids
Chlorides	peracetic acid [54]
Chlorinated lime (dry)	oxidizable substances
Chlorine dioxide	ammonia, H_2S, phosphine, methane, sulphur
Chlorites	Na_2SO_3, $NaHSO_3$
Chromic acid	acetic acid, naphthalene, camphor, glycerol, turpentine, alcohol, flammable solvents
Cumene hydroperoxide	inorganic and organic acids
Flammable solvents	ammonium nitrate, chromic acid, halogens, conc. hydrogen peroxide, sodium peroxide, nitric acid
Fluorine	do not bring into contact with anything!
Glacial acetic acid	chromic acid, nitric acid, perchloric acid, peroxides, permanganate
Hydrocarbons aliphatic and aromatic (e.g. propane, butane, kerosine, turpentine, benzene)	halogens, chromic acid, sodium peroxide
Hydrocyanic acid	nitric acid, alkali metals
Hydrogen fluoride (anhydrous)	ammonia (aqueous or anhydrous)
Hydrogen peroxide (conc.)	alcohols, aldehydes, ketones, carbohydrates
Hydrogen sulphide	fuming nitric acid, oxidizing gases
Iodine	ammonia (aqueous or anhydrous), hydrogen, acetylene
Oxalic acid	mercury, silver
Oxygen (liquid)	all organic flammable substances (asphalt!)
Perchloric acid	bismuth and its alloys, acetic anhydride, alcohol, paper, wood, sugar
Potassium perchlorate	particularly sulphuric acid
Potassium chlorate	inorganic acids
Potassium permanganate	sulphuric acid, ethylene glycol, benzaldehyde, glycerol
Silver	ammonium compounds, acetylene, oxalic acid, tartaric acid
Sodium peroxide	ethyl acetate, ethyl alcohol, ethylene glycol, benzaldehyde, glacial acetic acid, acetic anhydride, glycerol
Sulphuric acid	potassium chlorate and perchlorate, potassium permanganate

Table 2.7. Important laboratory detoxifying agents and ancillaries

Neutralizing agents	
Bases Caustic soda and caustic potash and their conc. solutions (30%); Aq. ammonia (NH_4OH), ammonia gas (NH_3), washing soda (Na_2CO_3), potassium carbonate (K_2CO_3); Slaked lime $Ca(OH)_2$	Acids hydrochloric acid (32–36%) and dilute (2 N), sulphuric acid conc. (95–97%) and dilute (4 N); carbonic acid (CO_2); sulphamic acid (NH_2SO_3H)
Redox reagents	
Reducing agents Ferrous sulphate ($FeSO_4$); Sodium bisulphite ($NaHSO_3$); Sodium sulphite (Na_2SO_3); Sodium thiosulphate ($Na_2S_2O_3$) Sulphur dioxide (SO_2)	Oxidizing agents sodium hypochlorite (12–15% active chlorine NaOCl), potassium hypochlorite (Javelle water, KOCl), chloride of lime and calcium hypochlorite [$(Ca(OCl)_2)$]; hydrogen peroxide 3%, 15% and 30% (H_2O_2); Caro's acid (peroxomonosulphuric acid, H_2SO_5); potassium permanganate ($KMnO_4$); conc. nitric acid (HNO_3); chromic acid (CrO_3/H_2O) and chromic-sulphuric acid (CrO_3/H_2SO_4); chloramines (R^1R^2NCl); chlorine gas (Cl_2); cerium ammonium nitrate $(NH_4)_2[Ce(NO_3)_6]$; hopcalite (oxidizing mixture, major component MnO_2)
Precipitating agents	
Milk of lime [$Ca(OH)_2$] and caustic soda solution; Washing soda [(Na_2CO_3)]; Ammonium sulphide $(NH_2)_2S$; Sodium sulphide (Na_2S); hydrogen sulphide (H_2S); Ferric chloride ($FeCl_3$); aluminium sulphate [$Al_2(SO_4)_3$]	
Miscellaneous substances	
Powdered limestone ($CaCO_3$), kieselguhr (diatomaceous earth), vermiculite (mica type); soap powder; solvents (C_1—C_4 alcohols, dioxane, acetone, toluene etc.), active carbon, calcium polysulphide, sulphur, zinc or brass powder	
Ancillary materials, apparatus	
pH papers, nitrite paper (o-aminobenzalphenylhydrazone); starch iodide paper; KI/ammonium molybdate or titanyl chloride (peroxide); plastic and paper sacks; cellulose, sawdust, sand; wood wool; as well as normal laboratory apparatus, water sprays, buckets and iron baths, for open-air incineration, and suitable fuses should be available. Cooling agents or cooling mixture; Personal protective clothing and fire extinguishers	

Table 2.8. The destruction of oxidizing and reducing substances

Type of substance	Spilled material	Waste
Oxidizing agents	mix with a reducing agent (Na_2SO_3, $NaHSO_3$ or $FeSO_4$) and spray with water (with sulfite and ferrous salts use ca. 3 M H_2SO_4); neutralize the suspension with anhyd. sodium carbonate in a large beaker and wash down the drain with ample water; wash the contaminated areas with soap solution containing a little reducing agent	add an excess of a conc. reducing agent and acidify with 3 M H_2SO_4; neutralize after reduction and wash down the drain using ample water
Reducing agents	mix with anhyd. sodium carbonate or $NaHCO_3$ and spray with water; place in a large beaker and carefully add an equal volume of calcium hypochlorite solution; dilute with water and allow to stand for 2 hours; if necessary neutralize (pH paper) and wash down the drain with ample water	

Table 2.9. Simple, rapid methods of detoxifying important classes of organic substances [47–51]

Compound	Spilled material	Laboratory waste
Aflatoxins	treatment with 3% NaOCl solution; see review [58]; A. in fodder is destroyed by gassing with NH_3 [59]	
Alcohols	soak up with cellulose or paper and burn	burn alone or in a solvent, small amounts of lower alcohols may be washed down the drain with plenty of water
Aldehydes	soak up small quantities with cellulose, allow to evaporate and burn the cellulose; cover larger quantities with $NaHSO_3$, mix with a little water and allow to stand for an hour in a large beaker, wash down the darin with plenty of water	soak up in vermiculite or dilute with a solvent (ethanol, acetone, benzene[1]) or burn; formaldehyde can also be oxidized to formic acid with H_2O_2, which can be diluted with plenty of water after neutralization
Alkyl nitrates or nitrites	can be hydrolysed by refluxing with 2 N NaOH which can be reduced to N_2 with sulphamic acid or oxidized to nitrate with hypochlorite; or incinerate in a solvent (NO_x formation!)	

Table 2.9 (cont.)

Compound	Spilled material	Laboratory waste
Amines	cover aliphatic amines with $NaHCO_3$, spray with a little water and then wash down the drain with plenty of water; an alternative for liquid amines is neutralization with $NaHSO_4$, the solid material so formed is incinerated; adsorb or cover aromatic amines with sand/anhyd. Na_2CO_3 (9:1) and burn with paper/wood in an open pit; cleanse contaminated areas with soap solution	admix with sand/Na_2CO_3 (9:1) and burn with paper/wood or in a solvent (alcohol, benzene[1]), (NO_x formation!); aliphatic amines can also be mixed with $NaHSO_4$, moistened and poured down the drain with plenty of water after neutralization
Azo compounds (only small quantities as they are very toxic and explosive)	adsorb in vermiculite or cellulose (moisten solids and take up with paper) and burn; or take up on a wet sponge, then detoxify with 10% cerium ammonium nitrate or 20% sodium nitrite solution	add an excess of cerium ammonium nitrate solution while stirring and cooling, dilute with ample water; diazoalkyl compounds and diazo esters can be decomposed in suitable laboratory apparatus with 2 N HCl in such a manner that the corresponding organic halide and N_2 are formed (follow gasometrically), the halide can be burned, neutralize the aqueous phase with NaOH and dilute with water; diazomethane can also be detoxified by esterification, by adding a carboxylic acid in CH_2Cl_2 with cooling and following the N_2 formation gasometrically
Carbon disulphide	absorb in cellulose, allow to evaporate in the open air, burn the cellulose	absorb in vermiculite or cover with water and bring into the open air and ignite from a safe distance by means of a fuse
Carboxylate esters	see alcohols	
Carboxylic acid amides	adsorb into cellulose, mix with alcohol and burn	

Table 2.9 (cont.)

Compound	Spilled material	Laboratory waste
Carboxylic acid chlorides and anhydrides	absorb with Na_2CO_3 or $NaHCO_3$ and dilute with water, allow to stand for 1 hr	add dropwise to NaOH solution with efficient stirring, evaporate and burn the residue
Carboxylic acids unsubstituted	cover with anhyd. Na_2CO_3 or $NaHCO_3$, spray with a little water and wash the paste into a beaker with water, neutralize and dilute; cleanse the contaminated areas with washing soda solution	mix with solvents (solid acids also with paper/wood) burn; neutralize small amounts of lower aliphatic carboxylic acids (formic and acetic acids) with Na_2CO_3 and dilute
substituted	mix with large quantities of $NaHCO_3$ and allow to stand in an open container (a large beaker for small quantities), with a large excess of water, for 24 hours; an alternative is to burn them in a solvent	
Carcinogens	see review [62]	
Chlorohydrins	adsorb on anhyd. Na_2CO_3, spray with water and allow to stand with ample water in a large beaker for 2 h, then neutralize and cleanse contaminated areas with soap solution	mix with anhyd. Na_2CO_3 add to ample water and neutralize after several hours and wash down the drain with plenty of water; hydrolyse ethylene chlorohydrin with $NaHCO_3$ to glycol, epichlorohydrin with dil. NaOH to glycerol; burning after mixing with vermiculite or solvent is another alternative
Epoxides, ether, (aliphatic and cyclic)	absorb in cellulose, allow to evaporate in the fume cupboard and burn the cellulose	dissolve in butanol or benzene[1] and burn; if the presence of peroxides is suspected in ethers surround the container with wood wool for protection and break it open from a safe distance, ignite the contents with a fuse.
Halogenated aromatic amines	absorb in cellulose, vermiculite or $NaHCO_3$ and burn with paper/wood in an open pit; cleanse contaminated areas with soap solution	mix with $NaHCO_3$ or sand/ $NaCO_3$ (9:1) or take up in a solvent (alcohol, benzene[1]) and burn
Halogenated hydrocarbons	take up with cellulose or vermiculite (or kieselguhr	mix with sand/anhyd. Na_2CO_3 (9:1) (in the

Table 2.9 (cont.)

Compound	Spilled material	Laboratory waste
	etc.) and burn with paper/ wood or in a solvent	case of fluorinated compounds add slaked lime too) and burn with paper/ wood or in a solvent (not more than a 25% solution) (take into consideration possible phosgene formation e.g. from dichloroethane, tetrachloroethane, tetra-chloroethylene!); larger amounts should be recovered by distillation; simple compounds such as CH_3Br can be hydrolysed at elevated temperatures with alcoholic KOH; carefully add highly reactive substances such as benzotrichloride to water, neutralize and evaporate, burn the residue in solvent; neutralize trichloroethylene with Na_2CO_3 or $Ca(OH)_2$ (not with KOH or quick lime (danger of explosion because of chloroacetylene formation!) for recovery and then steam distil (injection temperature less than 105 °C). Bury small quantities of pesticides such as DDT and lindane observing the relevant regulations
Hydrocarbons	absorb in cellulose or take up on paper and burn	burn in admixture with a flammable solvent (not usually necessary for liquid HCs) or vermiculite
Isocyanates, diisocyanates (polyurethane materials)	burning in vermiculite is a suitable method, reaction with primary and secondary amines leads to urea formation; reaction for 2 h with a 20-fold excess of a mixture of 50 parts ethanol, 42.5 parts water and 7.5 parts conc. ammonia followed by burning is recommended for diisocyanates; polymerization with polyols and dumping is also a possibility	
Ketones	see alcohols	
Medicaments	see review [60, 61].	
Mercaptans (thiols)	add aqueous up to 15% $Ca(OCl)_2$ solution (or NaOCl solution), establish the presence of excess with nitrite paper and allow to stand in the fume cupboard for 12 h in	

Table 2.9 (cont.)

Compound	Spilled material	Laboratory waste
	a large beaker, if necessary neutralize and wash down the drain with ample water; cleanse contaminated areas with soap solution containing hypochlorite	
Nitriles	add ample NaOH and $Ca(OCl)_2$ solution and allow to stand 1 h in a beaker, dilute with ample water; cleanse contaminated areas with soap solution containing a little hypochlorite	add NaOH in a great deal of alcohol, after 1 h evaporate the alcohol off and add an excess of $Ca(OCl)_2$ solution and wash down the drain with ample water after 24 h; alternatively burn in a solvent; after being properly polymerized acrylonitrile can be dumped
Nitro and nitroso compounds	absorb in or mix with $NaHCO_3$ and burn with a solvent (beware, highly nitrated compounds have a tendency to deflagration and explosion); destruction via the hydroxamic acids (addition of conc. HCl), which are further decomposed to carboxylic acids, is relatively laborious; nitrocellulose and collodion are spread out in a thin film in an iron bath (max. 5 cm) and burned with paper/wood or alcohol (ignite with a fuse); smaller quantities (up to 250 cm^3) can be put in a large stainless steel container (2–3 l), placed in a fireproof bath and treated with an equal volume of 10% caustic alkali (exothermic) most of the nitrocellulose is destroyed after 20 min, after 1 h dilute with plenty of water	
Nitrogen heterocyclics	take up solvent (alcohol, benzene[1]) and burn	
Nitroparaffins	cover with anhyd. Na_2CO_3, spray with a little water, allow to stand in abuchet for 2 hours, neutralize and dilute with ample water; or mix with $NaHCO_3$ and burn with paper/wood in an open pit	mix with a large excess of anhyd. Na_2CO_3 and neutralize after several hours; alternatively burn in sand/anhyd. Na_2CO_3 (9:1), with wood/paper or in a solvent; nitroglycerol can be decomposed with aqueous Na_2S
Nitrosamines, nitrosamides [71]	treat with HBr/acetic acid (denitrosation); HCl gas, hydrochloric acid (not sulphuric acid!). The nitrosamine content of a dinitroaniline can be reduced from 68 to 1 ppm by reaction with Br_2, Cl_2 or N-bromosuccinimide [56]. Dimethylnitrosamine can also be quantitatively reduced to dimethylamine using a Ni-Al alkali reagent (solvent H_2O or CH_2Cl_2) [57]	
Organometallics and metal alkoxides	cover small quantities with dried anhydrous Na_2CO_3,	mix with dried anhydrous Na_2CO_3 and burn in an

Table 2.9 (cont.)

Compound	Spilled material	Laboratory waste
	mix and gradually add to butanol, after 24 h dilute with a great deal of water	iron bath with paper/wood or blow dry steam onto the waste which is spread out in an iron bath (danger of spluttering)
	organolithium compounds can be taken up in dioxane and slowly decomposed by the addition of ethanol until no more hydrogen is evolved, then diluted with water; dilute acid is added until the solution is clear; organo Hg compounds can be converted to the nitrate with conc. nitric acid, from which the Hg can be precipitated after dilution as the sulphide; Al alkyls are best dissolved in paraffin (up to 10% solutions) and burned Pb tetraethyl is decomposed by conc. H_2SO_4; contaminated apparatus is cleansed by treating with a mixture of petrol and suphuryl chloride, chloride of lime and water (1:1) and aqueous $KMnO_4$ (20:1) are also suitable	
Peroxides and per acids	absorb on or mix with vermiculite, spread out in an iron bath or a pit in the ground with a spatula and ignite with a fuse; wash the apparatus used in a 30% solution of $FeSO_4$; possible alternative (esp. for per acids) is reduction in an acidic solution (eg. with $FeSO_4$, $NaHSO_3$ or Na_2SO_3)	after admixture with vermiculite moisten with 10% NaOH and burn the mixture; reduction in acidic solution is also practicable
Phenols	see alcohols	
Phosphoric and phosphonic acid derivatives	absorb with cellulose or take up on paper and burn with a solvent	mix with equal volumes of sand and powdered chalk and burn after wetting with alcohol or benzene[1]; an alternative is basic hydrolysis (reflux 30 min) (exception diazinon acidic) then evaporate and burn the residues (take care with halogen and cyanide-containing pesticides — toxic vapours can be evolved)
Quinones	see alcohols	
Sulphate esters (e.g. dimethyl sulphate)	add dropwise to dilute caustic, neutralize with HCl and wash down the drain with ample water	

Table 2.9 (cont.)

Compound	Spilled material	Laboratory waste
Sulphonic acids	take up in vermiculite, burn with paper/wood or in solvent	
Thioethers (sulphides)	pour on aqueous up to 15% $Ca(OCl)_2$ solution (or NaOCl solution), establish the presence of excess with nitrite paper, leave for 12 hours in a beaker in the fume cupboard, if necessary neutralize and wash down the drain with plenty of water, cleanse contaminated areas with soap solution containing hypochlorite	

If it is not possible to burn the materials in a technical incinerator with flue gas purification then the wastes should be burned in small portions in the open air in an iron bath or a pit in the ground, taking all the legally prescribed precautions, in isolated areas; ignition must always be performed using a fuse (e.g. a wood-wool cord)

[1] If the hepatotoxic and cancerogenic benzene does not result as an immediate waste solvent, then the very much less toxic solvent toluene should be substituted.

Table 2.10. Simple, rapid methods of detoxifying important inorganic elements and compounds [47–51]

Compound	Spilled material	Laboratory waste
Acids	cover with anhyd. sodium carbonate/slaked lime (1:1), mix with water and dilute; $NaHCO_3$ can also be exmployed to advantage; in the case of HF transform to CaF_2 by reaction with milk of lime at pH 12 (use polyethylene containers for the reaction!); large quantities of phosphoric acid should be precipitated as phosphates using aluminium or iron salts and incinerated or dumped	
Active chlorine compounds	are treated with $Na_2S_2O_3$ solution (1 kg of chlorinated lime requires 0.3 kg $Na_2S_2O_3$)	
Alkalis and aq. ammonia	take up with cellulose or paper or dissolve in water, dilute and neutralize with 6 M HCl	
Amides	take up with paper, stir into cold water and neutralize; alkali amides can also be suspended in dioxane and decomposed by the dropwise addition of ethanol (for large quantities use isopropyl alcohol), after the production of NH_3 has ceased the alcoholate formed is carefully decomposed with water and after neutralization the solution is washed down the drain with plenty of water	
Azides	take up with cellulose or paper and detoxify with 10% cerium ammonium nitrate solution	add the azide dropwise in water to a solution of cerium ammonium nitrate (follow the formation of N_2

Table 2.10 (cont.)

Compound	Spilled material	Laboratory waste
		gasometrically), carefully evaporate the reaction mixture and burn the residue
Bromine	add with stirring to dilute caustic, destroy the hypobromite formed using $Na_2S_2O_3$ solution, dilute a great deal and pour down the drain	
Carbides	cover with vermiculite and transport to a safe place in the open air in a dry brucket, mix with plenty of water and ignite the hydrocarbons formed with a flame. Allow to stand for 24 h and convey the solid residues to a dump	
Chlorinated sulphur compounds	chlorosulphonic acid, thionyl chloride, sulphuryl chloride and sulphur chlorides are added dropwise to ice water with stirring, neutralized with NaOH and washed down the drain with plenty of water (caution skin-irritating vapours!)	
Hydrazine	take up and dilute to a maximally 5% solution, decompose with either hydrogen peroxide (1 mol per mol hydrazine) or $Ca(OCl)_2$ (7–10 parts by weight for every part by weight of hydrazine)	
Hydrides	place in a dry plastic bag inflated with nitrogen and burn in the open air, pour water on the combustion residues	burn in an iron bath or open pit; or mix with dry sand and slowly decompose in the open air with anhydrous butanol, then spray with water until the last traces of hydride are decomposed, dilute with water, if necessary neutralize, convey solid residues to a dump
	alkali metal and alkali earth hydrides can be suspended in anhydrous dioxane and decomposed by the dropwise addition of isopropyl alcohol whilst stirring until no more hydrogen is evolved, the alcoholate is then carefully decomposed with water and washed down the drain after neutralization with plenty of water (first separate out the metal hydroxide that is formed when treating large quantities)	
Specifically: Alkali metal borohydrides	alkali metal borohydrides are dissolved in methanol and plenty of water and then decomposed dropwise with HCl, the toxic diborane that is formed initially (fume cupboard!) decomposes to boric acid and hydrogen (!) and the mixture can be washed down the drain with plenty of water after 24 hours	

Table 2.10 (cont.)

Compound	Spilled material	Laboratory waste
Lithium aluminium hydride	suspend lithium aluminium hydroxide in anhyd. dioxane and decompose dropwise with ethyl acetate while stirring and cooling, the alcoholate formed is carefully decomposed with water, neutralized and diluted with plenty of water	
Inorganic anions	cover with anhyd. Na_2CO_3 and mix, allow to stand in a large beaker for 24 h with plenty of water and neutralize (in the case of fluorides also add burnt lime)	
Specifically: Chlorides	thoroughly mix with $NaHCO_3$, spread on a porcelain tray (fume cupboard) and spray with 6 M ammonia solution while stirring, after the evaporation of NH_4Cl vapour add ice water, dilute in a large container and neutralize	
Chromates (and CrO_3)	adjust to pH 2–3 and reduce to Cr^{3+} by the addition of $NaHSO_3$ (ca. 40 g for 20 g CrO_3) (90 min), bring to pH 8.5 with NaOH and separate the precipitated $Cr(OH)_3$ (aim for recycling). Neutralize the filtrate and wash down the drain with plenty of water; chromium-containing wastes must not be burned, since Cr^{VI} is formed at high temperatures	
Cyanides	mix with vermiculite and take up, make alkaline with NaOH in a large beaker then react with excess $FeSO_4$ solution, after 1 h adequately dilute and wash down the drain; or mix the cyanide-containing suspension with excess sulphur and immobilize as thiocyanate by heating, dilute this with ample water	add to an excess of strongly alkaline $Ca(OCl)_2$ or NaOCl solution) and dilute with plenty of water, after 24 h (if necessary neutralize) check on the presence of excess hypochlorite with nitrite paper and destroy this with $Na_2S_2O_3$
Nitrite	reduce to nitrogen with sulphamic acid at pH 3 (do not work at too acid a pH because of the danger of formation of nitric oxide); or oxidize to nitrate at pH 4 (hypochlorite) and neutralize	
Sulphides	take up in vermiculite or dissolve in water, stir in $FeCl_3$ solution in the open air, add a slight excess of anhydrous sodium carbonate and wash down the drain with ample water; cleanse contaminated areas with soap solution	
Thiocyanates	decompose with half conc. H_2SO_4 to COS and bisulphate, in water COS slowly decomposes to CO_2 and H_2S (fume cupboard!)	

Table 2.10 (cont.)

Compound	Spilled material	Laboratory waste
Intermetalloid compounds	cover with $NaHCO_3$ or anhyd. Na_2CO_3/slaked lime (1.1), mix and spray with water (beware flames!), if no further reaction occurs place in a large container with plenty of water and neutralize	
Metal carbonyls	can be decomposed with H_2SO_4; but treatment with a mixture of hopcalite and active carbon is better, the carbon causes catalytic decomposition, the hopcalite oxidizes the CO	
Metals[1]	cover alkali and alkaline earth metals with anhyd. sodium carbonate, take up and add to isopropyl or tert-butyl alcohol and wash down drain with plenty of water after 24 h (should no reaction occur then add a little methanol); amalgamate scattered drops of mercury (tin foil, Zn or brass powder), destroy in cracks with Ca polysulphide and plenty of sulphur	Na or K can be added slowly in small pieces to an excess of isopropyl alcohol (tert-butyl alcohol for K) possibly adding methanol the (ethanol for K) to accelerate the process (care H_2 gas!) decompose the alcoholate so formed with water and after neutralization wash down the drain with plenty of water; because of its low reactivity Ca can be added directly to water, which it is then necessary to heat to 60 °C; another possibility is to destroy by burning with paper/wood in the open air
Metal ions[1] (specifically heavy metals and transition elements)	precipitate as insoluble compounds (carbonates, hydroxides, or oxides, sulphides), recycle or send to a chemical waste disposal site	
Peroxides	cover with plenty of sand/anhyd. sodium carbonate (9:1), mix throughly and transfer to a large beaker containing plenty of $NaSO_3$ solution with a plastic spatula, neutralize with dilute H_2SO_4 and dilute with plenty of water	

[1] *Recovery of metal compounds and metals*
Many elements and compounds should be collected in the laboratory and processed for recovery, for example, compounds (and the elements themselves) of antimony, arsenic, lead, cadmium and mercury (mercurous) compounds must first be oxidized to the mercuric state) should be precipitated as the sulphides and worked up; beryllium ions precipitated as the hydroxide (with NH_4OH) and then ignited to BeO; barium ions by precipitation as the sulphate; strontium ions by precipitation as the carbonate; silver ions by precipitation as the chloride; stannous compounds — after oxidation to stannic compounds — by precipitation as SnO_2; thallium salts — after dissolution in HNO_3 — by precipitation as Tl_2O_3 (with H_2O_2) or Tl_2S (with $(NH_4)_2S$); selenium and tellurium compounds by reduction to the element (e.g. with Na_2SO_3 in acid solution).

Table 2.10 (cont.)

Compound	Spilled material	waboratory waste
Phosphorus	cover yellow phosphorus with sand, moisten and spread out in an iron bath in the open air, the phosphorus ignites spontaneously in the air as it dries out; red phosphorus is also burned	
Phosphorus pentoxide, oxychloride and pentachloride	stir into plenty of ice water (caution skin-irritating vapours), neutralize with NaOH and dilute with plenty of water (in large quantities the phosphates formed should be precipitated by the addition of $FeCl_3$ before washing down the drain)	
Phosphorus pentasulfide	stir into a suspension of iron salts, dump the sulphides and phosphates formed	

Let us now consider detoxification in aqueous solution, which aside from economy also has the advatage that no additional problematical substances need to be introduced. If the poisons are easily hydrolysed, pure water and the application of heat often effect a sufficient degree of detoxification, for other substances aids to hydrolysis (acids, bases, catalysts) must also be added, since water alone is not sufficiently nucleophilic for the attack. Finally, for stable poisons strongly nucleophilic, chlorinating, oxidizing or other agents are necessary. The most various dissociation and hydrolytic equilibria are then set up in the resulting aqueous solutions so that the most complicated interactions can occur between the components. The equilibria are often affected by impurities in the technical grade detoxifying agents. The hardness of the water also has an effect on certain detoxification reactions. When calcium hypochlorite is employed water containing a great deal of lime can displace dissociation of this in favour of the undissociated compound. The dissociation and hydrolysis constants of a detoxifiying agent are decisive for its effectivity. With poorly soluble poisons and detoxifying agents the addition of solubilizers or detergents can be helpful as can raising the temperature. Table 2.7 reviews the most important detoxifying agents used in the laboratory. Tables 2.8–2.10 summarize the simplest and most rapid methods for the disposal of important laboratory chemicals.

References

1. Heitmann, A.: Chemie-Ing.-Tech. 50 (1978) 354.
2. Informationsschrift Sonderabfälle. Hrsg. von der Länderarbeitsgemeinschaft Abfallbeseitigung. Berlin/West: Erich Schmidt Verlag 1975.
3. Matthes, B.: Müll u. Abfall 7 (1975) 225; Leib, H.: Chemie-Ing.-Tech. 46 (1974) 319.
4. Rasch, R.: Chemiker-Ztg. 99 (1975) 226; ISWA Information 1978 (24), 11.
5. Martinetz, D.: Wissenschaft u. Fortschritt 31 (1981) 333.
6. Fichtel, K.: Das techn. Umweltmagazin. Februar 1976, 32; Tabasaran, O. et al.: Müll u. Abfall 9 (1977) 293.
7. Mosch, H.: Müll u. Abfall 8 (1976) 87.
8. Barniske, L.: Müll u. Abfall 7 (1975) 125.

9. Sinn, H. et al.: Angew. Chem. 88 (1976) 737.
10. Kaminski, W. et al.: Chem.-Ing.-Tech. 51 (1979) 419; Kunststoffe 68 (1978) 284.
11. Collin, G. et al.: Chem.-Ing.-Tech. 50 (1978) 836.
12. Fitzer, E.: Angew. Chem. 92 (1980) 375.
13. Schnecko, W.: Chemie-Ing.-Tech. 48 (1976) 443.
14. Sinn, H.: Chem.-Ing.-Tech. 46 (1974) 579.
15. Das techn. Umweltmagazin. September 1977, 31.
16. Nowack, F.: Das techn. Umweltmagazin. August 1976, 28.
17. Schüer, U.: Wasser, Luft u. Betrieb 21 (1977) 362.
18. Müller, H., Denne, A.: Wasser, Luft u. Betrieb 21 (1977) 50.
19. Schmidt, P.: Umwelt 1974 (4), 39; Eisenburger, J. P.: Wasser, Luft u. Betrieb 20 (1976) 186; Beitz, H. et al.: Zum Stand der Forschung bei der Beseitigung von pflanzenschutzmittelhaltigen Abwässern und Restbeständen in der Landwirtschaft. Akademie der Landwirtschaftswissenschaften der DDR 1978.
20. Thomé-Kozmiensky, K. J.: Umwelt 1977 (1), 15.
21. Das techn. Umweltmagazin. Februar 1975, 40.
22. Tsutsumi, S.: Cônversion of Refuse to Energy. Conference Papers IEEE. Montreux 1975.
23. Lohs, Kh.: Synthetische Gifte. Berlin: Militärverlag der DDR 1974.
24. Schindewolf, U.: Chemie-Ing.-Tech. 44 (1972) 682.
25. Geiger, Ph. L.: Ann. Pharm. 1 (1832) 44; Pelouze, J. P.: Ann. Pharm. 2 (1832) 84.
26. Hoerth, J. et al.: Chem.-Ing.-Tech. 45 (1973) 641.
27. Wolfbeiß, E., Schindewolf, U.: Chem.-Ing.-Tech. 48 (1976) 63.
28. Müller, W., Witzke, L.: Chem.-Ing.-Tech. 47 (1975) 435.
29. Müller, W., Witzke, L.: Chem.-Ing.-Tech. 45 (1973) 1285.
30. DDR-Pat. 11840 (1955).
31. BRD-Pat. 2146257 (1971), Mohr, E.
32. BRD-Pat. 2150679 (1971), Mohr, E.
33. Gmelins Handbuch der Anorganischen Chemie. Bd. Eisen, Teil B, Weinheim: Verlag Chemie 1932, S. 562.
34. Oehme, F., Disman, J.: Galvanotechnik 58 (1967) 236.
35. Epstein, J.: J. Org. Chem. 21 (1956) 796.
36. BRD-Pat. 2109939 (1971) Lawes, B. C., Mathre, O. B., Du Pont de Nemours and Co.
37. Fischer, B. et al.: Z. Wasser Abwasser Forsch. 14 (1981) 210.
38. Martinetz, D., Hiller, A.: Z. Chem. 18 (1978) 61.
39. Weigert, W. M.: Chemiker-Ztg. 99 (1975) 106.
40. Fieser, L. F., Fieser, M.: Reagents for Organic Synthesis. New York: S. Wiley 1967, S. 469.
41. Martinetz, D.: Z. Chem. 16 (1976) 1.
42. Landini, D., Rolla, F.: Synthesis 1974, 565.
43. Kim, I. K. et al.: Taehan Hwahak Hoechi 1974, 421.
44. Martinetz, D., Hiller, A.: Z. Chem. 16 (1976) 320.
45. BRD-Pat. 2603591 (1977), Schmitz, R. et al., Bayer AG.
46. Hallensleben, M. L., Wurm, H.: Angew. Chem. 88 (1976) 192.
47. Allisson, S.: Chem. Rundsch. 27 (1974) 28, 17.
48. Roth, L.: Giftliste. München: Verlag Moderne Industrie 1976.
49. Laboratory Waste Diposal Manual. Manufacturing Chemists Association (USA) 1969 (revised 1975).
50. Fa. J. T. Baker: Chemikalienkatalog 780, S. 24ff.
51. Ivić v. Rechenberg, H.: Chem. Rundsch. 1981, Nr. 33.
52. Abfallbewirtschaftungsprogramm der BRD 1975, Umweltbrief 1976, Nr. 13, S. 61.
53. Müller, R. K., Keese, R.: Grundoperationen der präparativen organischen Chemie. Zürich: Juris-Verlag 1975, S. 77.
54. Wienhöfer, E.: Chemiker-Ztg. 104 (1980) 146.
55. Franke, S.: Lehrbuch der Militärchemie. Bd. 1 und 2. Berlin: Militärverlag der DDR 1977.
56. Eizember, R. F. et al.: J. Org. Chem. 44 (1979) 784.
57. Lunn, G. et al.: Fd. Cosmet. Toxicol. 19 (1982) 493.

58. IAR Scientific Publications Nr. 37, Lyon 1980: Laboratory Decontamination and Destruction of Alfatoxins B_1, B_2, G_1, G_2 in Laboratory Wastes.
59. Lüthy, J.: Chem. Rundsch. 31 (1981) 22, 3; Tagungsbericht der Arbeitstag. „Gesundheitsgefährdung durch Aflatoxine" der ETH und der Universität Zürich 20./21. 3. 1978.
60. Der Städtetag 1977 (6), 347.
61. Hofmann, H.: Pharmaz. Zeitung 126 (1981) 2129.
62. Castegnaro, M., Michelon, J.: Methods for Destruction and Disposal of Laboratory Wastes contamined with some Chemical Carcinogens. International Cancer Agency for Research on Cancer, Lyon.
    Lansone, E. B., Lunn, G.: American Laboratory, September 1983, 82.
63. Bailin, L. J., Hertzler, B. L.: Environm. Sci. Technol. 12 (1978) 673.
64. Hiraoka, K. et al.: Chemistry Letters 1979, 739.
65. Hardt, A. P. et al.: Detoxification of Pyrotechnic Agents Using a Microwave Plasma. (private communication).
66. Smith, J. G. et al.: J. Chem. Technol. Biotechnol. 30 (1980) 620.
67. Pytlewskii, L. L. et al.: Mid-Atl. Ind. Waste Conference [Proc.] 1979, 11, 97.
68. Craddock, J. H.: Environm. Symp. [Proc.] March 8–10, 1982, S. 161.
69. Botré, C. et al.: Environm. Sci. Technol. 13 (1979) 228.
70. Chemical and Engineering News 60/10, 10 (1982).
71. Lunn, G. et al.: Carcinogenesis 4 (1983) 3, 315.

3 Detoxifying Industrial Waste Waters

3.1 Processes for the Removal of Toxic Waste Water Constituents

Just as with the other areas, exhaust air and chemical waste, the possibility should be checked of utilizing other raw materials or newer processes or technologies to minimize the amounts of pollutants in the water. At the moment, however, industrial waste waters must be treated mechanically, biologically and chemically in order to allow them to flow harmlessly into the drains. Combinations of processes are necessary for the purification of waste waters containing complex mixtures of pollutants. In order to decide on the combination necessary a whole range of boundary conditions, as well as the sort, amount and concentration of the waste water and not least economic cost benefit considerations, have to be taken into account. It is not possible to adequately purify most industrial waste waters in mechanical-biological water treatment plants; this is particularly true for specific constituents and high concentrations [1].

The mechanical processes employed as the first stages of purification for the separation of solids from the waste water (for example by raking, filtration or in sedimentation tanks), the separation of suspended materials (e.g. by flotation), the separation of suspensions (e.g. with centrifugal separators) and the separation of oils can only be mentioned briefly here.

The physico-chemical processes or subprocesses for the separation of dissolved, suspended or colloidal pollutants and the chemical subprocesses for transformation to less toxic or biologically degradable substances will be reviewed in the following section (see also Table 3.1).

Chemical subprocesses bring about structural alterations to the substances contained in the water that lead to forms, which either allow a phase separation and thus removal of the pollutant (e.g. precipitation), are more easily removed by subsequent processes (e.g. ion exchangers) or can be discharged harmlessly in the waste water (e.g. neutralization of acids and bases). Physico-chemical processes only achieve a separation from the waste water flow and do not bring about actual detoxification. This has to flow. Whether the physico-chemical or chemical treat-

Table 3.1. Main physicochemical and chemical processes for the purification of waste waters

Physicochemical process mainly for		Chemical process
Dispersed substances	substances in true solution	
Flocculation Ultrafiltration Electrophoresis	stripping evaporation reverse osmosis electrodialysis adsorption	ion exchange precipitation hydrolysis neutralization oxidation
Extraction	extraction	waste water incineration electrolytic reduction

ment should follow or precede a biological treatment depends on the specific instance. In the treatment of local authority and slightly polluted waters the biological purification is generally the second step and a further chemical treatment can follow. Figure 3.1 illustrates, as an example, the flow diagram of a waste treatment plant with mechanical, biological and chemical steps [2].

Highly polluted industrial wastes must be purified of the major portion of their pollutants by physico-chemical means or by chemical detoxification before the biological treatment, whereby each individual industrial waste water flow must be investigated for "problem substances". Table 3.2 gives a costing comparison for the individual treatment processes.

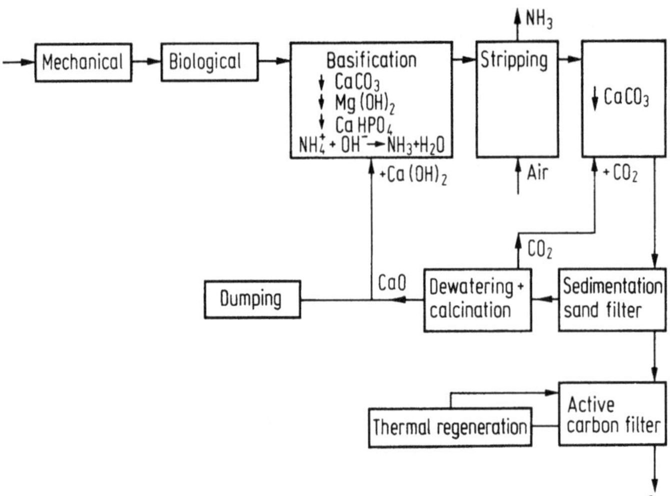

Fig. 3.1. Flow chart of waste water treatment plant for communal waste waters [2]

Table 3.2. The cost of waste water purification [2]

	DM/t organic substance	Suitable for
Biological water purification	1000–3000	dilute waste waters
Adsorption on active carbon	5000–50000	dilute waste waters
Oxidation (total) in aq. phase with ozone	ca. 25000	dilute waste waters
Waste water incineration	3000–6000	highly concentrated waste waters

3.1.1 Physical-Mechanical Processes

The unit physical-mechanical processes for the removal of solid, dispersed and non-water-miscible liquid, unwanted waste water components, such as sedimentation, decantation, centrifugation, flotation [14, 15] etc. will not be discussed in any detail.

Flotation, for instance, has been applied on a large scale industrially for more than 80 years for the separation of the solid phases from solid/liquid phase systems, particularly in ore dressing. Nowadays flotation is also utilized as a method of recovery for recycling purposes in the paper, cellulose, textile, fat and oil-processing industries. Its utilization for the removal of suspended impurities from communal and industrial waste waters is, however, new. The mechanism of flotation depends on the chemical, physical and physico-chemical processes occurring in three-phase systems. The basic processes occur at the interfacial surfaces between the phases. The solid/liquid separation occurs because the solid phase (in spite of its higher specific gravity) is made to float by adsorption on air bubbles. After separation the air bubbles are stabilized by "foaming agents". Various froth flotation processes are in technical use; e.g. pressure flotation, vacuum flotation or electroflotation (here electrolytically produced bubbles of hydrogen and/or oxygen are employed) [14, 15].

Neither do we wish to treat the emulsion-splitting processes in detail because of their very heterogeneous nature, ranging from the physico-chemical processes (adsorption, electrophoresis, ultrafiltration, flotation) and thermal processes (heating, evaporation, burning) to chemical processes, which destroy emulsions by the addition of acids, salts, de-emulsifiers and flocculating agents. Often magnesium salts are employed for splitting in the presence of adsorbents (silicic acid preparations) or de-emulsifiers. Combinations of processes are employed in the majority of cases. We will refer to the possibilities of emulsion splitting during the discussion of individual waste water purification processes.

A further group of processes employed in the removal of pollutants from industrial waste waters are the physico-chemical processes, which are reviewed in the next section. They are also suitable for the treatment of aqueous laboratory wastes, if under somewhat different conditions. The physicochemical methods are not of themselves direct detoxification methods, but methods for the separation of toxic components from waste waters and concentrates, so that, in general, the separated materials must be subjected to chemical detoxification or incineration [16].

3.1.2 Evaporation and Concentration

Volatile components of waste waters, such as hydrocarbons, can be removed by evaporation (stripping), when the process should be performed on the undiluted waste water at the site of its production. In practice submersed burners, carrier gases or steam are employed and the stripped vapours are either condensed for recovery or catalytically incinerated. Stripping with air or with N_2 and CO_2 at 20–40 °C with an adequate gas flow can be used successfully for the removal of volatile hydrocarbons [6, 17].

Combinations of chemical reaction and stripping are employed in the utilization of CO_2 or CO_2-containing flue gases for the elimination of sulphide. Hydrogen sulphide is released first and this can be bled off at normal pressure at about 95 °C from the head of the column [6, 17]. Soviet processes for the removal of sulphides from waste waters employ a combination of oxidation and stripping. The sulphide-containing wastes are treated with a countercurrent of air at pH 6 in a packed (iron turnings) column, when a portion of the sulphide is oxidized and the majority driven off as hydrogen sulphide [6]. The CHEVRON waste water treating process [18] developed in the USA may be mentioned as an example of the treatment of waste water from oil refining, which contains H_2S and ammonia. The waste water is first pumped from collection tanks to a H_2S stripper column, where H_2S is driven off at the column head and processed to elementary sulphur. The aqueous ammonia sinks to the foot of the column where it can be drawn off. Gaseous ammonia is distilled out in a second column for reutilization.

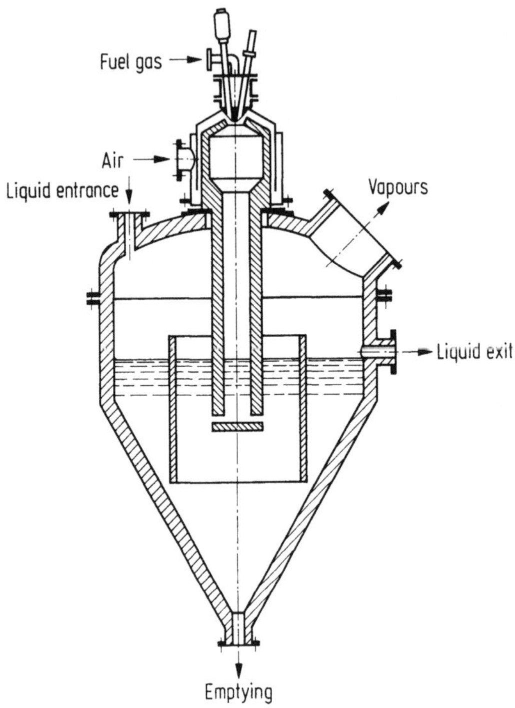

Fig. 3.2. Section through an immersion burner [21]

Concentration processes [19] serve to separate out non-volatile waste water components; both direct and indirect evaporation techniques are employed [20]. A submersed burner with direct heat exchange between the flue gases and waste water is a frequently utilized process [21] (see Fig. 3.2), which amongst other uses is employed for the evaporation of oil-containing waste waters. Depending on the porperties of the pollutants other types of evaporators are also employed, such as stirring, recycling or film evaporators.

Evaporation processes always have a stripping effect too, i.e. volatile components are driven out with the steam; however, this can be desirable. How far a particular waste water is concentrated depends on the properties of its constituents (e.g. tendency towards encrustation, sublimation, etc.) and the further treatment or disposal procedure envisaged for the concentrate or residue (e.g. reutilization, incineration or disposal). One of the main aims of evaporation processes is the recovery of raw materials, when very great care must be taken to avoid thermal damage to the residue.

3.1.3 Extraction

Toxic waste water components in high concentrations can be extracted from the waste water flow or from aqueous laboratory wastes with non-aqueous solvents, by exploiting the superior solubility of the material to be extracted in the solvent used. The extracting agent residues remaining in the water after the extraction processes need to be stripped out, particularly in large-scale industrial applications. The requirement for extraction is intimate mixing of the liquid being extracted with the extractant until the distribution equilibrium is set up between the extract and the raffinate phases. The phases are then separated in an appropriate installation. Extraction and separation constitute one theoretical extraction step and in practice are not very effective. Better results are obtained using multistage concurrent extraction. The most effective method employed industrially is continuous countercurrent extraction for continuously produced waste waters of approximately constant composition.

The choice of the extractant must also be considered for each application. Alongside the requirement of better solubility of the substance to be extracted in the extractant and the required selectivity, the extracting agent must only have a very limited miscibility with the liquid being extracted, the phases should separate as quickly as possible after the extraction process (e.g. because of a large density difference) and no irreversible reactions must occur with the substance being extracted; the reversible reactions such as complex formation, the formation of addition compounds and associates is often desirable in specific instances [22]. For example, it is possible to utilize liquid ion exchange processes for the removal of certain metals (e.g. Cu, V) or acids from waste waters, using tertiary amines dissolved in kerosine [23].

In order to achieve all the required properties it is often necessary to employ solvent mixtures. In waste water purification it is always an aqueous phase that is to be extrated. Marvel and Richards [24] have set up a series of solvents for the partition of mono and dicarboxylic acids between the aqueous and organic phases, that can be used in general for extractions from the aqueous phase [22]. The solvents are ranged in order of increasing partition coefficients in favour of the organic phase: heptane, carbon tetrachloride, benzene, chloroform, diisopropyl ketone, butyl acetate, diethyl ether, methyl isobutyl ketone, ethyl acetate, methyl propyl ketone, methyl ethyl ketone, cyclohexanone, n-butanol. This series which begins with substances incapable of forming H bonds, proceeds to H donor solvents then to solvents with both H donor and acceptor properties — such as n-butanol. This endows the latter with a high extraction ability from aqueous solutions, — at the expense of a low selectivity, however. For large-scale use preliminary experiments should be performed to determine which extractant is to be employed, whereby the pH of the solution to be extracted must also be taken into account.

The following general principles apply for the application of extraction techniques [22]:

- Extraction is a preliminary purification step, which has to be followed by chemical and/or biological steps.
- When possible extraction should be performed at the site of production of the waste water, since the concentrations of the pollutants concerned are highest there.

- Multistage countercurrent extraction is the most economical process; however, it can only be employed for large quantities of waste water produced continuously and at a relatively constant rate.
- Extraction processes usually extract not only the inhibitory substances or those which difficult to decompose, but also other biologically degradadable substances (e.g. alcohols, acids, etc.) and thus lower the BOD_5 and COD.
- Solvents should be employed which posses a high partition coefficient for the substances to be removed, if possible they should be waste solvents.
- Extracts are recovered and the extracting agents recycled. If this is not economic, the extract can be incinerated.

Extraction processes acquire particular importance in waste water purification, when valuable materials can thereby be recovered.

The recovery of phenol from coking and carbonization plant effluent is an important practical example of waste water purification [26]. An economic extraction is only possible, however, if the concentrations present lie above 1 g/l and at least 10 m^3/h of waste water is produced. The important processes are the benzene-caustic process, the phenosolvan process and the ifawol process [22].

The benzene-caustic process is the oldest process for phenol recovery from prefiltered coking oven waste waters. Here the phenol is extracted from the aqueous phase at ca. 50 °C with benzene in a Raschig ring-filled extraction tower. The phenol is then extracted from the benzene as phenate by washing several times with caustic soda solution. Pure phenol is prepared from this caustic soda solution, which contains 25–32% phenol. The benzene is recovered and recycled. The efficiency of a modern phenol recovery plant working on these principles is 94%, i.e. a biological purification must follow. The most modern processes, employing multistage, rotating countercurrent extractors, achieve efficiencies of up to 98% [26].

The phenosolvan process works at ca. 35 °C with diisopropyl ether (formerly butyl acetate) in a multistage continuous countercurrent process employing the mixer settler principle (mixing and separation processes in different chambers). Diisopropyl ether with a distribution coefficient for phenol of 20 is a better extractant than benzene with one of 2.2, so that the amount of recycling extractant can be kept relatively small. The extract is processed by distillation, the solvent recycled and the raw liquid phenol processed to give pure phenols. The efficiency of the process can be up to 99%. In practice this means that residual phenol concentrations of less than 10 mg/l can be attained. The other organic constituents of the coking plant effluent (ca. 50% with 50% phenol) are simultaneously extracted to the extent of 66%–77% and the BOD_5 is lowered [26].

A third process is the ifawol process [22], that employs a high boiling solvent (hydrated primary and secondary oxo process products) with a distribution coefficient for phenol of 12 for countercurrent extraction in packed columns. The efficiency is reported to be up to 98%. One of the advantages claimed is that raw phenol is obtained with a 80%–85% purity.

Extraction processes can also be employed for treating oilwater emulsions and for the extraction of hydrocarbons from cracking processes, for the recovery of acetic acid from the aqueous effluents from the manufacture of cellulose acetate and for the removal of chlorobenzenes and nitrocompounds from waste waters [27].

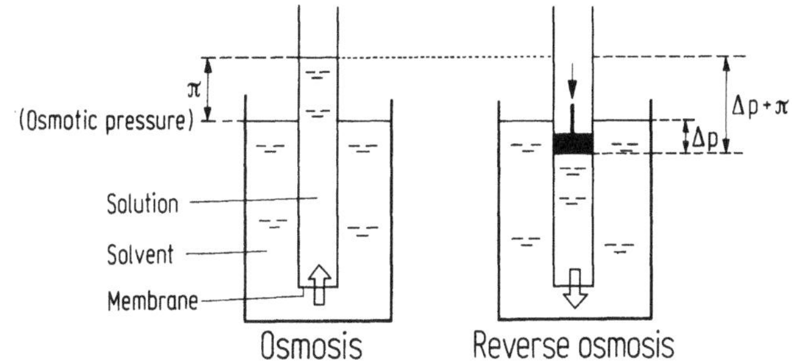

Fig. 3.3. Schematic representation of osmosis and reverse osmosis

3.1.4 Reverse Osmosis and Ultrafiltration

Membrane separation consists of waste water treatment processes where liquid phases flow through membranes under the influence of a pressure gradient [28], i.e. reverse osmosis and ultrafiltration [29], or without the application of pressure as in electrodialysis. Membrane processes are reviewed in detail in [30]. The term reverse osmosis was chosen because the normal osmotic flow is reversed by the application to the system of an opposing pressure higher than the osmotic pressure (see Fig. 3.3). In contrast to normal filtration, in reverse osmosis and ultrafiltration the flow is maintained parallel to the separating membrane, so that separation occurs into two flows of differing concentration. In the ideal case all the dissolved particles are in one stream and the permeate stream consists of pure solvent or water. Ions, molecules, colloids or emulsified particles are held back. The concentrate can be recycled or burned; depending on the remaining contaminants the permeate can be used for washing purposes, subjected to further processing or led down the drain [31]. Both processes are theoretically reversible, constant temperature processes for the concentration of dilute solutions with a minimum expenditure of energy.

Reverse osmosis achieves the enrichment of electrolytes and/or low molecular weight compounds at pressures of ca. 100 bar and particle sizes of ca. $5 \times 10^{-7} - 10^{-6}$ mm, ultrafiltration employs lower pressures of <10 bar, preferably about 3 bar, to separate high molecular weight particles from 10^{-6} to 10^{-2} mm [33] (cf. Fig. 3.4).

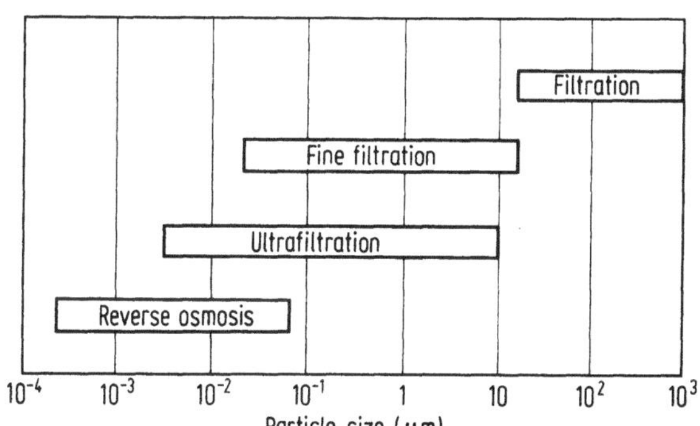

Fig. 3.4. Areas of application of the individual filtration techniques [32]

Separation by reverse osmosis is not quantitative and is described in terms of degree of retention; ultrafiltration, particularly of high molecular weight substances, can be quantitative.

The pollutants are separated out of the prefiltered waste water by the membranes themselves, which are thin films manufactured from many different types of polymer materials. In order to achieve high permeate flow rates with low residual contamination, membranes with good separating properties must be employed that are as thin as possible, working at as low a concentration difference as possible. In practice, all these requirements cannot be realized simultaneously. Up to now asymmetrical cellulose acetate membranes with their good separating properties and high water permeability (between 10° and 30 °C the water permeability increases markedly with the temperature while the salt retention capacity remains relatively constant [33]) have proved to be superior to most polymeric membranes. The assymetry is achieved by a particular casting process, which produces a membrane about 100 μm thick, supporting a small-pored layer ca. 500 to 5000 Å thick [28]. Their disadvantages are a limited temperature range, sensitivity to hydrolysis at high pH and susceptibility to certain micro-organisms. Various other polymers have been investigated [34], e.g. polysulphones, aliphatic and aromatic polyamides, polyethers and polyamidehydrazides, many of which can be employed in the pH range 1–13 and at temperatures of up to 80 °C.

The units fabricated from semipermeable membranes are known as modules. Commercially available modules include tubular modules, plate modules, hollow fibre modules and spiral modules.

Reverse osmosis can be usefully employed for electrolyte solutions with salt concentrations of between 0.5 and 50 g/l [33]. Sorption and ion exchange processes are more effective at lower concentrations, the upper concentration limit for reverse osmosis is set by the maximum working pressure of the module employed which is about 50–80 bar [33]. In such cases evaporation or incineration techniques are to be preferred. Let us take as an example the treatment of 1 M = 58 g NaCl in one litre water. The solution containing 2 mol ions exerts an osmotic pressure of 44.8 bar, which must be overcome. The difference between this and the working pressure of 80 bar is, therefore, only about 35 bar. If the theoretical degree of retention is 99% then the permeate will still contain 0.6 g/l salt [33] and a second, uneconomical reverse osmosis step would be required.

Ultrafiltration allows the treatment of more concentrated waters (e.g. 50%) at lower working pressures, as long as blockage can be avoided and the necessary flow velocities can be maintained [33]. Adsorption, extraction, stripping, precipitation or flocculation or biological treatment are held to be more economical for more dilute solutions. The use of ultrafiltration is justified economically if the permeate or the concentrate can be reutilized or if emulsions have to be broken.

As mentioned above the universal application of membrane separation is limited by temperature sensitivity. Thus, cellulose acetate has no long-term resistance to temperatures above 30 °C; other polymers are better in this respect. The hydrolysis sensitivity of cellulose acetate at high and low pH expresses itself by an increase in permeability accompanied by a loss of retentivity. The rate of hydrolysis of cellulose acetate is least at pH 4.5 [33] when its working life is longest. Membrane clogging is a further problem, which can prevent or significantly reduce transfer. These phenomena are caused by the deposition of hydroxides for example, particularly in colloidal form. Prefiltration is, therefore, essential. Cellulose acetate membranes are also sensitive to microbial attack; they must be protected from drying out. Concentration polarization also limits transfer. A layer of higher concentration than that of the bulk is created at the interface between the membrane and the untreated solution, which lowers the theoretical flow rate. This phenomenon can be limited by turbulent flow.

The widespread use of membrane filtration for industrial waste water purification is not feasible. The process is, however, of importance as a stage in the solutions-specific problems and pollutant concentrations. Its economics must be considered for each individual case. Some examples of the application of membrane techniques to industrial waste waters are listed in Table 3.3.

Table 3.3. Examples of the application of membrane filtration processes [35]

Branch of industry	Type of effluent	Recoverable constituents
Chemical industry	process water	catalysts, noble metals
Pharmaceutical industry	process water	proteins, antibiotics, hormones
Galvanizing plants	electrolysis baths, rinsing baths	noble metals, nonferrous metals
Textile industry		
– Viscose manufacture	spinning acids	copper
– Wool scouring	scouring waters	lanolin
– Dye works	dye baths	dyes, pigments
– Sizing	wash water	starch, proteins
Photographic industry	fixing baths	silver

Up to now ultrafiltration has played by far the most important role in waste water purification, where it is included with utility as a stage in certain purification processes and is directed at material recovery. Here are some examples: Waste waters from electro-dip finishing are always ultrafiltered to produce a paint-free wash liquid from the dipping bath that is used for rinsing the completed workpiece. The used rinse water loaded with excess paint is returned to the dipping bath in an almost ideal closed circuit [36]. Ultrafiltration has become competitive with respect to other processes in the last few years for the splitting of oil emulsions, particularly boring emulsions in the metal manufacturing industry [37, 38]. After prefiltration the emulsions containing ca. 1 % oil are concentrated in a circulatory system to about 50 %, the calorific values of 3000 to 7000 kcal/kg (12 560–29 308 kJ/kg) are achieved for the concentrate [33]. The emulsion concentrates produced are usually unstable and form two phases on standing. The aqueous phase can be returned to the circuit, while the remaining phase only has a water content of 20 %–30 %. This emulsion-breaking process has the particular advantage that no chemical additives are required to achieve the splitting, that high temperatures are not required and that a high degree of purification of the filtrate is achieved at relatively low total cost.

The whey produced in cheese making can be concentrated to a 20 %–30 % protein solution by ultrafiltration. The lactose contained in the permeate can be concentrated tp 20% lactose solution by a following reverse osmosis plant. How these concentrates can be employed has not yet been satisfactorily explained.

The so-called diafiltration of polymeric complex formers constitutes a special case of ultrafiltration and is particularly applied for the separation of reutilizable or toxic heavy metals [34] from electroplating waste waters, waste waters from chlorine-alkali electrolysis, the photographic industry and other areas. For this process highly polymerized, water-soluble complex formers, which are selective for the metal to be recovered, are added to the waste water and react in homogeneous solution rapidly and quantitatively with the metal to form large aggregates (see Fig. 3.5), that can no longer pass through the membrane. The permeate can be led into the drain, the partially loaded, complex-forming mixture is led back into the ultrafilter and the volume of the permeate is made up with raw water. When the polymer is completely loaded it is concentrated in the ultrafiltration plant and can be further processed. A whole range of metals: Hg, Cd, Cr, Pd, Pt, Au, Ag, Cu, Zn etc. can be enriched from very high dilution by this method. The complex former need only be highly selective if a particular metal is to be recovered from a mixture of metals.

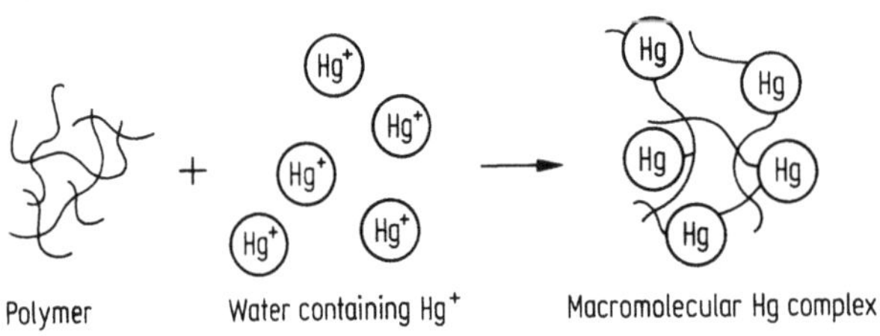

Fig. 3.5. The formation of larger aggregates in diafiltration

As previously mentioned reverse osmosis has not yet found wide application in waste water treatment, even though many investigations are under way, e.g. for mercury-containing waste waters, electroplating effluents, photographic effluents etc. [28, 39]. The pilot plant experiment for the recovery of glycerol from petrochemical waste waters containing other organic and inorganic substances can be mentioned as another example [41]. In metal recovery the separation efficiency for Cu^{2+} amounts to 96%–99% (with a maximally achievable concentration of 8%–10%); for Ni^{2+} 97%–99% (10%–12%); for Cd^{2+} 95%–98% (8%–10%); CrO_4^{2-} 90%–98% (8%–12%) and for cyanide 90%–95% (4%–12%, pH-dependent) [227].

Mercury salts are well retained by membranes, but it is not possible to reduce the residual concentrations below that required for waste waters so that further treatment would be necessary.

The concentration of inorganic salts in electroplating effluents is limited by the osmotic pressure. For solutions containing mixtures of inorganic and organic substances, account must be taken of the fact that low molecular weight organic substances can readily penetrate the membrane, particularly as the concentration increases. Certain organic substances are also soluble in cellulose acetate [40].

3.1.5 Electrodialysis

Electrodialysis is employed on a large scale chiefly for the desalination of brackish waters and sea-water; thus, for example, 13% of Japan's salt requirement is met by sea-water desalination [42]. The process is also finding increasing interest for the processing of sulphite lyes, for NaOH recovery from hemicellulose-containing waste lyes in the viscose-manufacturing industry and for electroplating and radioactive waste waters [42]; the limiting factors are concentration polarization and membrane fouling. Pilot plants are in operation for the recovery of carboxylic acids (salts), chromic acid solutions, steel pickling solutions and for the recovery of toxic metals.

An electrodialysis plant consists of alternating anion and cation-permeable membranes between Pt electrodes [43]. Under the influence of the electric field the anions migrate to the nearest cation membrane and are retained there and the reverse process happens to the cations. Thus, compartments are either enriched or depleted of salts. As the spaces are depleted of ions the electrical resistance finally increases and becomes so high that further electrolysis would be uneconomical. The back diffusion as a result of the drop in concentration is largely avoided by the choice of specially designed membranes [44]. The economic desalination of aqueous and other solutions is only possible at concentrations of from about 5–10 g/l [45]. The principle of the method is illustrated in Fig. 3.6.

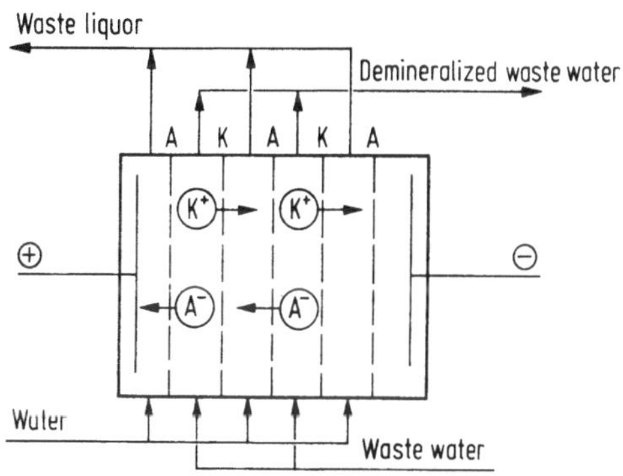

Fig. 3.6. The principle of electrodialysis

3.1.6 Electrophoresis

Colloidal pollutants can, as an alternative to flocculation, be removed electrophoretically. In an electrical field the charged colloidal particles migrate to the electrode bearing the opposite charge, where charge neutralization occurs. This causes the formation of larger aggregates, which are subject to "flotation" by the gases produced at the electrodes (cathodically H_2, anodically O_2 or if chloride is present some Cl_2). Because of their powerful oxidizing properties nascent oxygen and chlorine have a deodorizing and detoxifying action. Electrophoresis is rarely employed in waste water purification except for the treatment of oil-water emulsions and for the removal of proteins from solution.

3.1.7 Adsorptive Processes

Specific adsorption waste water purification of *mixed* organic waste waters is not widely practised at present [46], however, adsorptive elimination processes play a part in many mechanical, chemical and biological detoxification methods. Adsorption becomes economical and effective for some waste water flows with relatively uniform impurity content at not too high a concentration, while very heterogeneous waters can scarcely be detoxified satisfactorily because of the differences in the adsorbability of their components [47, 48, 49].

Examples of application to waste water purification are the removal of dyestuffs from dyeing liquors and the removal of odorous materials (phenols, sulphur compounds, mineral oils) from potable waters, when some compounds (e.g. H_2S, Cl_2) can undergo chemical reaction. Active carbon adsorption can be usefully employed after a biological purification plant. Active carbon adsorption is particularly recommended for the following waste waters [56]:

- The removal of dyestuffs from textile and dyeing industry waste waters.
- The removal of pesticide residues from the waste waters from pesticide manufacture.
- Improvement of the purification of oil refinery fluid effluents.
- Phenol and polyolefin removal in heavy organic manufacture.
- Purification of communal and paper-manufacturing effluents.

The removal is based on the fact that the pollutant in aqueous solutions is concentrated on the surface of the adsorbent and can be removed from the system with the adsorbent (i.e. as adsorbate) by filtration for example. The loaded adsorbent must either be regenerated and the adsorbate destroyed, e.g. by catalytic or thermal oxidation, or regeneration is not attempted and the adsorbent plus adsorbate is incinerated. Recovery of adsorbed material is occasionally possible. The important adsorbents are active carbon (granulated or powdered) for hydrophobic organic solutes in water and to a lesser extent aluminium oxide for hydrophilic organic solutes (e.g. carboxylic acids, alcohols, amines) and synthetic polymers, which, depending on their composition, have adsorptive properties for one or other of these categories. The adsorptive properties of flocculating agents such as iron salts, aluminium salts, sodium silicate and organic polyelectrolytes is discussed on p. 196. Trials are also under way employing superfluous sludges from biological sewage treatment for the adsorption of poisons such as heavy metals ("biosorption") [51].

If the hydrophilic or hydrophobic nature and the electrical charge of the pollutant are taken into account it is also possible to find suitable adsorbents not only for pollutants in true solution but also for dispersed and emulsified materials. Active carbon, aluminium oxide and synthetic polymers are almost exclusively employed for adsorption from true solution; a sufficient removal of solids and suspended material, which would reduce the adsorptive capacity by occupying the adsorbent surface, is achieved by prefiltration.

Here we wish to limit ourselves exclusively to the discussion of adsorption of active carbon. A good active carbon should be distinguished by the following properties:

- High adsorption capacity for long operating life or low circulating mass in entrained bed adsorbers.

— Rapid adsorption kinetics because of large pore size adsorbents, suitable particle size and chemical surface properties.
— High mechanical stability for transport to and from regeneration and for the regeneration itself.
— Unreactivity or slow reaction with the regenerating agent, e.g. steam.

At the same molecular weight aromatics are more powerfully adsorbed than aliphatics. Hydrophilic substitutents reduce adsorption on active carbon; this is illustrated by the virtual impossibility of adsorbing lower aliphatic alcohols, carboxylic acids and sulphonated aromatics in contrast to aromatics, phenols, nitroaromatics or aromatic amines. Other important pollutants such as amino acids or carbohydrates show only a slight tendency to adsorption [52].

Just as for adsorption from gases for pollutants in true solution the loading capacity at first increases with chain length within a homologous series until it reaches a maximum caused by the increasing steric hindrance to diffusion to the adsorption sites and begins to decline.

One of the factors affecting the economics of active carbon adsorption is the concentration of the pollutant in the water. Advantages over biological treatment include the small amount of space required, the costs, the recovery of water and valuable solutes, the degree of purification and the freedom from odour of the treated water.

The adsorption of substances in true solution can usually be described in terms of the Freundlich and Langmuir adsorption isotherms. However, the theoretical treatment of the total adsorption process is complicated by the fact that it is composed of a number of parallel and linked sub-processes such as diffusion of the particle to be adsorbed to the site of adsorption, actual adsorption, capillary condensation and the phenomena of chemisorption. The Langmuir adsorption isotherm yields the amount of substance adsorbed per unit surface area as a funcrion of the concentration. As the concentration is increased the amount adsorbed increases at first until finally a saturation value is arrived at with the formation of a monomolecular layer, i.e. at low concentrations the amount adsorbed is proportional to the concentration while at higher concentrations it is constant.

The unit processes utilized in waste water purification are the addition technique and the percolation technique. In the addition technique active carbon in powder form is stirred into the waste water, when a residual concentration in comparison to the loading of the adsorbent is set up according to the Freundlich isotherm [46]. If a great deal of adsorbent is added, which is very uneconomical, the residual concentration can be reduced at the expense of a low degree of loading. The percolation technique is more economical; here the waste water is passed through a column filled with granulated active carbon. When the flow is upwards a moving bed is spoken of — with downward flow a solid bed. In moving bed columns the active carbon adsorbent moves slowly down the column countercurrent to the waste water, so that heavily loaded particles are continuously removed for reneration and fresh or regenerated carbon is added at the top. Solid bed adsorbers can either be connected in series, when regeneration is performed countercurrent to the direction of waste water flow, or in parallel, when columns are regenerated as they become fully loaded. Solid bed adsorbers should only be employed for wastes containing low concentrations of organic contaminants. The maximum loading of the adsorbent depends on the concentration; with mixtures account must be taken of the fact that less well adsorbed substances are displaced by more strongly adsorbed pollutants (chromatographic effect). The dependence of the adsorption on the pH of the waste water must also be borne in mind. For instance, active carbon adsorbs phenol better from acidic waters and aniline from basic ones [47, 48].

Biological side reactions can be expected to occur in the percolation method, such as are intentionally exploited in the biological processes for producing potable water but which, on the other hand, can cause clogging of the active carbon. In such cases the columns should be operated with an upward flow, in order to wash out the bacterial flora.

When the loading capacity of the carbon is reached, it must be regenerated either thermally or by solvent elution followed by purification of the exhaust gases or the eluate. Experience has shown that wet chemical (eluational) regeneration leads to complete exhaustion of the carbon after only a few cycles, while thermal regeneration at elevated temperature in the presence of an oxidative regenerating agent always regenerates the original adsorption activity. The process is best carried out in a fluidized bed at about 800 °C.

A few examples of active carbon adsorption from industrial practive will now be described. Research and prototype plants have been described in the literature for the adsorptive purification of highly polluted coke plant waste waters using the entrained bed process, with continuous regeneration in a fluidized bed furnace at 750°–900 °C [53]. It was demonstrated in a continuous trial of 7 months' duration that the purified effluent had been reduced in total oxidizable carbon content (TOC) by 90%–95% (The residue was mainly carbon-containing salts, which were also included in the determination.) The $KMnO_4$ consumption was reduced by 88%–100%. The benzene-caustic process, in contrast, only reduces the total oxidizable carbon by 65% at the most. The phenol removal is reported to be at least 99% for the adsorptive process, that for the benzene-caustic process 92–94% [53]. The purified coking plant waste water is suitable for instance for use in coke quenching. It is possible to reduce the phenol content of coking plant effluent waters to 1 mg/l with granulated active carbon, which is a better result than can be obtained by biological purification [54]. One advantage of the process is its insensitivity to other toxic constituents and to variations in pH, temperature and pollutant concentration. That the uptake capacity is rapidly diminished by the concentration of high molecular weight materials is a disadvantage. (It has also been reported that multicyclone dust produced by Winkler generators is a good phenol adsorbent, water which has been pretreated to reduce its phenol content to 2 g/l is mixed with 6% multicyclone dust and run into settling ranks. The disadvantage is that the phenol cannot be recovered. Power station fly ash can be employed in a similar manner [54].)

Another example is the removal of traces of metals by the BSM process which operates using active carbon coated on the surface with sulphur compounds [55]. Pretreatment to bring the Hg content down to 0.1 mg/l is necessary. In some cases processes are employed involving chemisorption where actual chemical bonds are formed. The use of chemically modified cellulose containing free NH and SH groups is an example; by their use more than 98% of Hg^{2+} can be removed from industrial waste waters (starting concentration 1000 mg Hg^{2+}/l) [56].

Porous organic polymers are employed for various purposes in waste water treatment. Thus, porous hydrophobic resins of crosslinked polystyrene are very effective in removing chlorinated pesticides such as endrin, DDT, 2,4-D, toxaphen and polychlorinated biphenyls. The removal of phenol, chlorophenols, p-nitrophenol and aromatic hydrocarbons has also been described [57]. The compounds recovered can be reutilized.

Combined processes are of some importance. An example is the treatment of waste water from the electrolytic production of chlorine and caustic by the amalgam process having a Hg content of 2–20 mg Hg/l (as Hg, HgO, Hg_2Cl_2, $HgCl_2$) [58]. The first step is chlorination of the water with the aim of converting all the Hg components into $HgCl_2$. In the next step the water is dechlorinated and the major portion of the $HgCl_2$ is removed by adsorption on active carbon. The resulting water is then "polished" by treatment with the ion exchanger wofatit EA-60, when the residual Hg concentration is reduced to <0.1 mg/l. The elution of both the carbon and the ion exchange resin is performed with conc. hydrochloric acid. The $HgCl_2$ loaded acid is added to the dilute brine before the dechlorination stage; if the Hg content of the brine rises above 5 mg/l electrolytic reduction to mercury metal occurs.

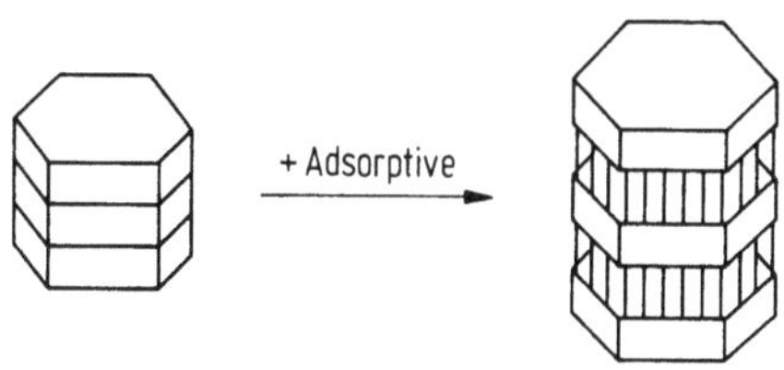

Fig. 3.7. Sorption by tetracalcium aluminate hydrate; the diameters of the hexagonal crystals are about 5×10^{-5} m [59]; the adsorbate particles can form various structures (represented by the lines)

The expandable stratified crystals will now be described more fully. They are particularly important substances for the binding and detoxification of highly poisonous compounds (e.g. chemical warfare agents), which lie between catalysts and ion exchangers. Certain types of natural and synthetic layer compounds are able to take up ions or neutral molecules from the gas or the liquid phase and to store them between their crystal layers, when the individual crystal layers are forced apart by the molecules that have been taken up (unidimensional swelling) (see Fig. 3.7) [59]. Since the adsorbed molecules are packed in dense layers the uptake capacity is very large. The unidimensional expansion is crystallographically very strictly defined. The thickness of the elementary layer of the non-expanded and the loaded crystals can be determined by X-ray crystallography. In contrast to the zeolites, the unidimensional expansion of the layered crystals offers a flexible storage procedure independent of the size of the molecules of the adsorbed substance which exploits the volume actually created during the adsorption process. Permanent gases cannot be captured between the layers. Water can penetrate and, in some cases, crystallographically well defined hydrates can be produced, however, the binding is very loose compared to that of acidic or basic adsorbents, so that the water can easily be displaced by organic molecules or it only occupies unutilized space. In contrast to other adsorbents mentioned the adsorption process with layered crystals is mainly chemisorption (hydrogen bridges, ionic or covalent bonds). A distinction must be made between two types of these compounds:

— Swellable natural layered crystals (e.g. the clay mineral montmorillonite, which is the major component of bentonites).
— Tetracalcium aluminate hydrate (TCAH).

Montmorillonites of different compositions and TCAH have structures consisting of incompletely neutralized macro-ion sheets, which are electrically neutralized by counter-ions in between the sheets, in the case of the clay minerals they are counter-cations such as Mg^{2+} or K^+, in the case of TCAH OH^- acts as counter-anion; these are able to participate in ion exchange processes.

Expandable silicate layers preferentially take up basic cationic compounds (particularly onium complex-forming substances such as mustard gas and the nitrogen mustards, Tammelin's ester and some psycho-active substances (LSD, phenylethylamine derivatives, indolylethylamine derivatives)). The synthetic TCAH

$$(4\,CaO \cdot Al_2O_3 \cdot x\,H_2O \text{ or } /Ca_2Al(OH)_6/OH \cdot n\,H_2O/)$$

preferentially takes up acid compounds by a rather different mechanism. It also takes up neutral compounds rather more readily than the natural layered crystals. The OH^- ions can be stoichiometrically replaced by virtually any anions and n can have the values 0, 2, 2.5, 3 or 6, which in the oxide formula corresponds to a water content of x = 7, 11, 12, 13 or 19. TCAH can be produced cheaply and with constant properties (important for process control); it can, however, be transformed into inactive carbonate hydrate by atmospheric carbon dioxide. This means that when it is employed in protective masks a soda lime CO_2 prefilter must be employed. Temperature increase also leads to the formation of $3\,CaO \cdot Al_2O_3 \cdot 6\,H_2O$ with loss of activity. Dosch [59] has studied, amongst other things, the removal of fluoroacetic acid derivatives and highly toxic phosphate esters from solutions.

For example ethyl fluoroacetate is sorbed with hydrolysis to the acid without the addition of a suspending agent. The alcohol formed is also complexed.

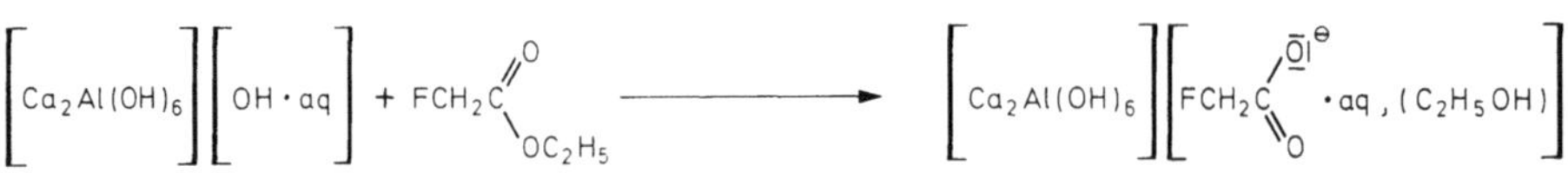

Phosphate esters are rapidly saponified by the action of the OH^- ion within the inorganic lattice and again chemisorbed in the form of the cleavage products. TCAH, thus, acts as a model esterase. The detoxification of organophosphorus esters cannot, therefore, be regarded as a simple adsorption process. Dorsch regards the first step of the sorption as a TCAH-catalysed ester cleavage, whose mechanism has not yet been completely clarified. Thiolphosphate esters are more slowly attacked and thionothiolphosphates even more slowly. Organophosphorus esters of the Sarin and Tabun type are rapidly and effectively detoxified by TCAH (see also Fig. 3.8). Other military substances such as mustard gas, nitrogen mustards, VX agents are not adequately attacked. The rate of detoxification of the more easily hydrolysed phosphate esters increases with decreasing TCAH crystal size (optimum ca. 10 µm). Increase in temperature and the presence of certain solvents or suspending agents (such as n-hexane or water) accelerate the ester hydrolysis, while acetone or methanol do exactly the opposite (incrustation; epitaxial growth of cleavage products on the TCAH surface; adsorption and reaction with solvent molecules [60]).

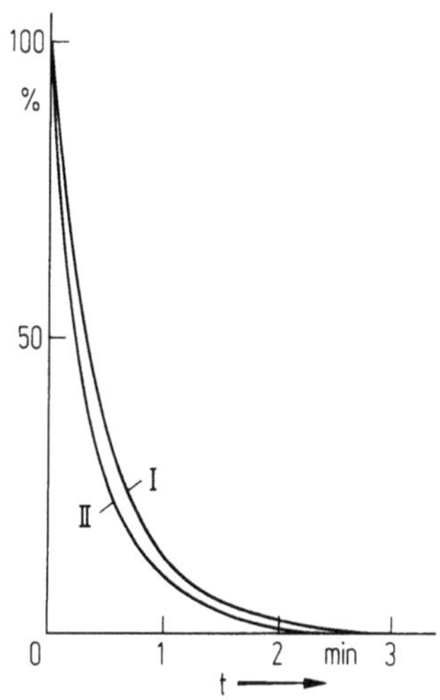

Fig. 3.8. The time course of the detoxification of Tabun (I) and Soman (II) by tetracalcium aluminate hydrate [60]

The following results were obtained for organohalogen compounds: p-halophenols and thiophenols are adsorbed without change, while for p-halobenzoic acids and α,ω-halo-carboxylic acids the course of sorption is dependent on the agent in which they are suspended and on its participation. Thus in the presence of water the halogen is more or less completely removed with the formation of two complex types $[Ca_2Al(OH)_6]$ [halogen aq.] and $[Ca_2Al(OH)_6]$ $[R{-}CO_2^- \cdot aq.]$. In anhydrous suspending media the hydrolysis is repressed, however, acid chlorides such as thionyl chloride, acetyl chloride and ethyl chloroformate are hydrolysed no matter what the suspending agent. Here too halogen-containing complexes are formed, however, the organic residue is not generally sorbed.

It is more difficult to employ TCAH for adsorptions from the gas phase, since the necessary energy for enlarging the layers (in solution: energy of solvation + energy of adsorption) is not available. In particular, this makes the adsorption of larger molecules more difficult. Nevertheless, descriptions have been published [59] of attempts to utilize sub-stoichiometric organic complexes of TCAH in which the layers had already been separated by several Å, so that gas molecules were able to penetrate the remaining volume.

Dosch [59] has published a review — with no pretention to comprehensiveness — of the sorbability of various classes of organic compounds. The following are classified as easily sorbable:

Aliphatic monohydric alcohols, phenyl alcohols, diols, sugar alcohols, sugars and sugar-like substances, glycosides, mercaptans and dithiols, aromatic and aliphatic aldehydes, aldoximes and ketoximes, carboxylic acids and dicarboxylic acids, phenyl carboxylic acids and cyclic polycarboxylic acids, sulphonic, arsonic and phosphoric acids, fatty alcohol sulphates and sulphonates, oxy acids, amino acids, phenols, aliphatic amines and diamines, phenylamines and substituted anilines, heterocyclics such as pyridine, all mono and polybasic inorganic anions.

While the following are only sorbable to a limited extent:

Ketones, acid amides, thioethers and thioureas.

Compounds that are not immediately sorbable include:

Hydrocarbons, halogenated hydrocarbons, nitriles, nitrocompounds, onium compounds.

The following classes of compound are decomposed by TCAH:

Acid halides and other halogen-donating compounds, esters (saponified), dialkyl sulphates (form alkoxylated TCAH). The substances taken up by TCAH to a limited extent or not immediately sorbable, are taken up by TCAH when they are in the presence of readily sorbable substances.

The stability of the adsorptive combination depends principally on the acidity of the sorbed molecule. Binding stability is reported to increase with decreasing pK_S values, e.g. in the following order [59]:

Methylamine — water — methanol — ethylene glycol — formaldehyde — thiophenol — acetic acid — benzoic acid — malonic acid — hydrochloric acid.

Now to consider the naturally occurring clay minerals. They are capable — particularly after modification — of taking up molecules that can form onium complexes. Ion exchange allows the replacement of the counter-cations by protons that yield the so-called H clays, which are particularly suitable for the chemisorption of the onium-forming mustard gas, nitrogen mustard, Tammelin esters and VX substances. Nitrogen mustards and VX compounds may be taken as examples [60]:

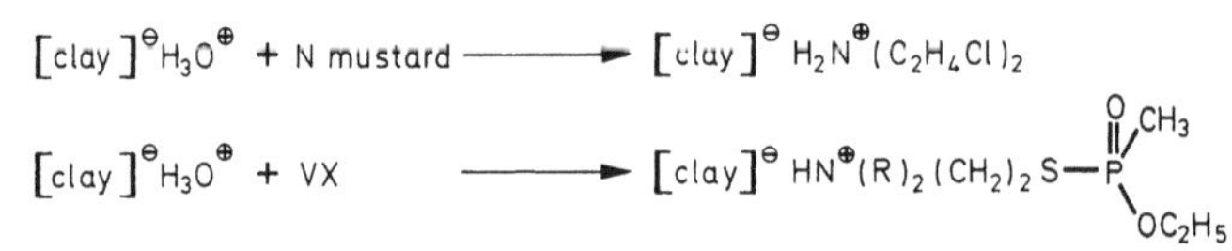

The toxins are incorporated in the lattice as cationic onium complexes. Another possibility for the modification or activation of clays for particular toxic substances is to exchange the surface counter-ions for heavy metal ions such as Ag^+, when very active sorbents are obtained, for instance for the removal of the pesticides Metasystox or DDVP. The rate of elimination of the toxic substance is dependent on the particular heavy metal ion used (see Fig. 3.9). Silvercontaining montmorillonite

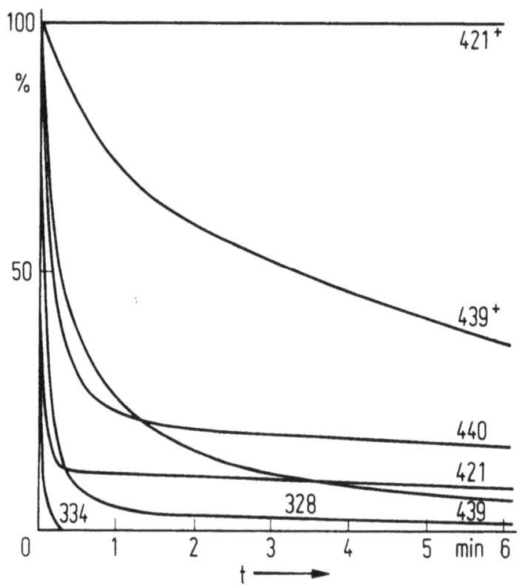

Fig. 3.9. The adsorption curves of various heavy metal-treated montmorillonites for Metasystox (Thiolform +) and DDVP. Metasystox: 421: Zn, Cd, Sn^{2+}, Co; 439: Cu^{2+}, Hg^{2+}, Fe^{3+}, Ni; 334: Ag. DDVP: 440: Zn; 328 Ag [60]

has produced the best results for the removal of all the organophosphorus insecticides and also of mustard gas [60]. The rapid mustard gas adsorption results in the formation of a physiologically inactive addition complex that cannot be extracted with nonpolar solvents even on heating [60]. Of the polar solvents only pyridine is capable of extracting it and that only within 10 minutes of uptake. After that the sorbed mustard gas is increasingly decomposed by the water contained in the montmorillonite, probably with the formation of 2-hydroxy-2′-chlorodiethyl sulphide (GC) [60]. Organophosphorus esters sorbed onto silver-impregnated montmorillonite are not eluted by nonpolar solvents, but by polar solvents (as are thiol esters). Thiono and thionothiol esters are not extracted by polar solvents either. The investigations were carried out at a toxin: detoxifying agent ratio of 1:20. Animal experiments have revealed a very good detoxifying activity for Ag-montmorillonite powders against mustard gas, so that other possibilities for personnel decontamination (e.g. sprays etc.) are opened up. The alternate use of TCAH and H or heavy metal-montmorillonites enables the detoxification of elimination of the whole spectrum of classical war gases and a whole range of other toxic substances (e.g. organophosphorus pesticides). Expandable layered crystals have only found limited application to date in waste water purification.

3.1.8 Flocculation Processes

Flocculation processes are becoming more important as treatment steps in the armoury of waste water purification technology. Particles in waste water with colloidal oı coarse disperse dimensions remain so if they possess an electrical charge. Usually the adsorption of anions (or other processes) endows the particles with a negative charge. It is assumed that an electrical layer is formed around the particles composed of these anions and cations as counter-ions, a double layer that is sometimes closely adherent to the particle surface and sometimes lies diffusely in the surrounding liquid, whereby a potential difference, the zeta potential, is set up, which determines the likelihood of the particles adhering to each other. The stability of a colloidal system is determined by the sum of the van der Waals attraction forces between the particles and the electrical repulsion from the like charges resulting from the electrical double layer. As a consequence of the particles being surrounded by counter-ions in the double layer this repulsion cannot be calculated simply in terms of Coloumb's law. The larger the repulsion forces compared to the Van der Waals forces the greater the stability of the colloidal system.

The destabilization of a colloidal system and hence its flocculation or coagulation can be achieved by suppression of the negative charge, i.e. the zeta potential. This is achieved in practice by the addition of various types of flocculating agents [61]. Sometimes these are electrolyte such as iron or aluminium salts, which act as carriers of positive charge to compress the diffuse double layer and restrict the range of the electrostatic repulsive forces. The reduction of the surface potential can also be achieved by chemisorption of the counter-ions. It should be noted that an overdose can reverse the carge of the colloid, i.e. restabilize it. Flocculation for waste liquid effluent purification can often be achieved very economically by utilizing industrial waste iron salts (e.g. ferrous sulphate in the form of green vitriol, ferric chloride sulphate, red mud), but aluminium salts such as $Al_2(SO_4)_3 \cdot 18\ H_2O$ are also employed. There is an interesting Soviet electroflocculating process where the effluent is placed in a direct current flowing between two electrodes, the ferrous metal anode is slowly dissolved thereby and the iron oxide floc formed acts to flocculate the impurities [62].

The formation of the cationic charge carrier from the added iron or aluminium salts will be considered now. If ferrous salts, green vitriol for instance, are employed then they are first atmospherically oxidized then precipitated in a neutral or alkaline environment with the formation of aquo and hydroxoaquocomplexes depending on the pH, e.g.

$$[Fe(H_2O)_6]^{3+},\ [Fe(H_2O)_5(OH)]^{2+},\ [Fe(H_2O)_4(OH)_2]^{+},$$
$$[Fe_2(H_2O)_8(OH)_2]^{hx}$$

Complex mixtures are produced at higher pH by further reaction and condensation. The final product in this condensation series is $Fe_2O_3 \cdot x\ H_2O$. A similar process occurs with aluminium salts, which also form an increasing number of hydroxoaquocomplexes as the pH increases. At high pH this can lead to the production of negative soluble aluminate ions, which are no longer capable of the coagulation of colloidal particles. From this it can be seen that the pH is of decisive importance for the flocculation process [63]. The six or eight nuclear complexes produced between pH 4–5 are of particular importance, since these are in equilibrium with neutral $Al(H_2O)_3(OH)_3$ and can be incorporated into its lattice, endowing it with a positive charge which enables it to neutralize the negative charge of the colloidal particles [64] (cf. Fig. 3.10).

A further flocculation method, which is particularly employed in waste water treatment, is inclusion flocculation. The added flocculating agent (Fe or Al salt) is flocculated out by hydrolysis and during this process, leading from true solute via colloid to solid, components of the waste water are also carried down with it. The process does not involve a change in the forces of repulsion or attraction. In practice the two types of flocculation described above can scarcely be distinguished and take place more or less simultaneously. Magnesium chloride is also available in some countries for the precipitation of dissolved or colloidal impurities in the waste water from sulphate pulp manufacture, because it is produced in large quantities domestically by the potash industry as a waste product. The magnesium hydroxide floccules that are produced in alkaline medium take up lignin aglomerates and thus sediment them. The sedimentation process can be significantly improved by the addition of slaked lime or suspensions of lime sludge [65].

The flocculation of colloidal particles by the addition of organic polymers is another method of destabilization [66, 67]; they do not operate via the Van der Waals forces but by the formation of macromolecular bridges. The bridges are formed by reactive groups of ionogenic or non-ionogenic character. Examples of non-ionogenic flocculating agents include polysaccharides, polacrylamides and polyethylene oxides. While examples of ionogenic polymers include anionic polyacrylic acids and cationic polyvinylbenzeneammonium compounds. It is important that the organic flocculating agents are not employed in excess, thus ensuring that they are completely precipitated during flocculation, so as not to cause secondary pollution of the water.

Flocculation adsorption is a special case of adsorption. Here the necessary adsorb-

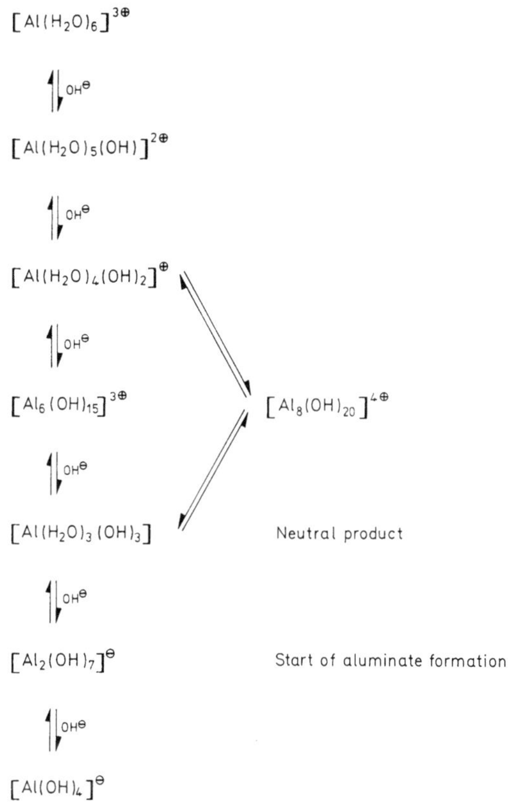

Fig. 3.10. The formation of hydroxoaquo complexes of aluminium salts in flocculation processes

ing surface is created by the hydrolysis of the flocculating agent (Fe and Al salts, silicates). When choosing the flocculating agent care should be taken that the surface of the adsorbent to be created has the opposite charge to the colloidal particles to be removed from solution. Some true solutes can also be co-flocculated by chemisorption in appropriate pH ranges, this, however, is virtually a transition to precipitation; an example is the removal of dyestuffs containing HSO_3 groups by sorption to ferric hydroxide in the alkaline range. The precipitates have to be sedimented and dewatered; following this they are either dumped or precipitated [68]. One example of application is the breaking of emulsions. In particular used oil emulsions are acidified by the addition of acidic ferrous chloride and broken by heating. The oil released is removed by centrifugation. Then neutralization causes metal(III) hydroxide to precipitate, which binds the residual traces of oil adsorptively. The oil-containing hydroxide sludge can be re-added to the used oil emulsion after it has been reacidified.

Highly disperse hydrophobic silicic acids are also employed as components of some breaking agents [69], where they act as adsorbents for the oily components of the waste water. After the emulsion has been broken by a water-soluble separating agent the oil is bound by the silicic acid to form an easily filterable, crumbly oil cake. The amount of salt added in the separating agent is relatively small; normally separating agent amounting to 40% by weight of the oil content is added, thus 1 m^3 of 2% oil emulsion would require 8 kg separating agent. The reaction time is between 5 and 10 minutes and the residual oil concentration below 10 mg/l. A series of specific process combinations is under trial in model investigations, at the pilot plant level and on the industrial scale. It is reported that lipophilic organic substances, such as oil products, halogenated hydrocarbons, chlorinated insecticides, halogenated aliphatic ethers, polycyclic aromatics etc. can be economically removed by the addition of water-moistenable polyethylene powder, followed by flocculation [70]. The polyethylene chains contain hydrophilising carboxylic acid amides in a ratio of 300:1.

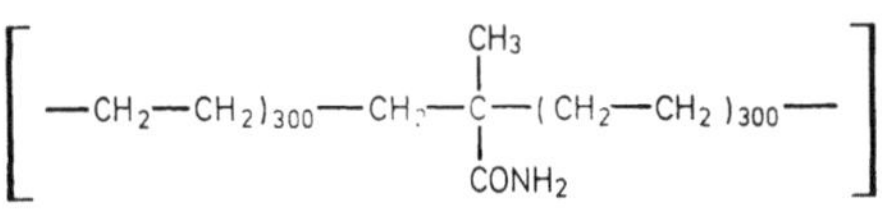

Molecular weight 15000–30000

For example the efficiency of separation of decane is maximally 98%–99% on conventional flocculation. This means that the decane residues are too high for potable water. Conventional flocculation, preceded by the addition of 10 mg/l polyethylene, reaches efficiencies of 99.85–99.95%. The removal of polycyclic aromatics also plays an increasing role in the production of potable water. Conventional flocculation does not meet with success. Chlorination can slowly yield substances such as 5-chloro-3,4-benzpyrene and 3,4-benzpyrene-5,8-quinone from benzpyrene. Chlorine dioxide reacts even more rapidly to produce seven identified products. Ozone only has an effect in the absence of other organic traces, but the formation of carcinogenic epoxides cannot be excluded. Model experiments with water containing 250 ng/l polycyclic demonstrated that the addition of 50 mg/l polyethylene followed by flocculation was able to reduce the polycyclics to less than 1 ng/l [70]. The investigation was made using TLC. Since polar substances are scarcely sorbed by polyethylene a combined agent has been developed in which active carbon and polyethylene are epitactically bound to each other [70], so that dissolved hydrophilic components can be simultaneously removed by adsorption.

3.2 Biological Waste Water Treatment

There is only enough space here to sketch out the basic principles [3]. Biological sewage treatment processes are widely used for the treatment of communal sewage and are being increasingly employed for industrial problems in combination with other processes. In biological or biochemical purification processes the unwanted materials present in the waste water are "offered" as nutrients to chosen bacterial cultures or mixed populations for inclusion in their metabolism. To put it in a simplified manner two natural self-purification processes then run side by side, degradation (dissimilation) processes and synthesis (assimilation) processes; the first as energy metabolism, the second as anabolism. This means that, on the one hand, degradation occurs to yield the simplest least toxic end products (CO_2 and water in the case of anaerobic conditions: mineralization) or the pollutant is incorporated into the cell substance (in the form of amino acids and proteins, of sugars and polysaccharides, of nucleic acids etc.), i.e. the micro-organisms grow and multiply so as to form a separable solid phase (biomass). The dissolved substances are first adsorbed by the micro-organisms and then taken up into the cells (resorbed), there they are either stored or immediately incorporated into the metabolic process. Suspended substances are processed analogously under anaerobic conditions; first they are adsorbed onto the cell surface and then enzymically processed by exoenzymes to yield resorbable molecules. Finally, the degradation of water pollutants by micro-organisms occurs in many small stages and an energy barrier has to be overcome at each of them. In order for this to occur the substance must be transformed into a condition in which it can react; a process which is assisted by substrate-specific biocatalysts (enzymes) produced by the organisms. Whether the dissimilation or the assimilation process predominates depends on the residence time of the waste water in the biological purification process and, in the case of aerobic processes, on the intensity of aeration. As has already been mentioned several times biological degradation can occur anaerobically as well as aerobically. Anaerobic degradation is only of utility for suspended substances (mainly sludges) [4], whereby it is of advantage to increase the temperature. The carbon compounds are then mainly transformed to methane and CO_2 (along with H_2S and NH_3) in the dissimilation process.

Aerobic degradation with the addition of air or oxygen is of much greater importance in waste treatment; here the pollutants, insofar as they are not incorporated into the cell substance, are completely decomposed to CO_2 and water. The "consumer" of the oxygen is the energy metabolism, while catabolism determines the formation of excess sludge. Ammonia is produced from the nitrogen-containing portion of the water's content; this is further oxidized to nitrite and nitrate by nitrifying bacteria and the sulphur content is oxidized to sulphate. The energy which is produced in individual steps is stored in the form of energy-rich, very reactive phosphorus compounds. The living cell is unable to utilize energy produced by the direct oxidation of carbon. The CO_2 produced in energy metabolism is not a product of C oxidation. In strictly anaerobic bacteria sulphate, carbonic acid or reducible organic compounds, for example, serve as acceptors, in facultative anaerobics it is often nitrate or nitrite. Such denitrifying bacteria, which are always in those zones of a biological purification plant which are almost free from oxygen, utilize nitrate or nitrite instead of oxygen as the final hydrogen acceptor in their respiratory chain. The major product is nitrogen. The waters either contain nitrate initially or it is produced microbially from NH_4^+. These phenomena are also employed for the deliberate removal of NH_4^+ and of NO_3^-. The enzyme-catalysed, intracellular, biochemical processes occurring during biological degradation are largely unknown.

The conditions for the application of aerobic purification are:

- An adequate supply of dissolved oxygen for the micro-organisms. The O_2 demand is the higher the greater the content of microbially degradable organic substances; the BOD is a measure of this requirement.
- The maintenance of the pH in the 5–9 range and the avoidance of sudden pH variations.
- Sufficient nutrients, i.e. alongside the C source in particular N and P plus certain trace elements must also be available in the right proportions. The nitrogen is required by the micro-organism for the production of cellular protein, the

phosphorus is of primary importance for the dissimulation process in the energy metabolism.

- A final condition is that no poisons (bactericidal or bacteriostatic) or inhibitors are present in the waters. The adaptation to be observed in the micro-organisms also favours degradation, that is when new chemical substances are introduced enzymes are produced to facilitate their degradation (enzymic adaptation) and those species, that can process the new substances more efficiently, select themselves by multiplying more rapidly (ecological adaptation); this is the basis for the degradation of industrial water pollutants.
- The rate of degradation can be increased by raising the temperature, but this lowers the rate of oxygen supply. It has been established that the rate of degradation in well managed treatment plants is approximately the same both in summer and in winter.
- The rate of degradation is governed by Michaelis and Menten type kinetics (see an appropriate textbook).

Organic water pollutants are biodegradable to different degrees [5]. Mangold has presented some results of the investigation of the connection between biodegradability and chemical constitution [6].

- The introduction of functional groups into the carbon skeleton increases biodegradability, when the molecular weight then ceases to play an important role.
- In homologous series (e.g. alcohols) the biodegradability decreases as the carbon chain increases.
- Chain branching in aliphatic compounds usually only reduces the degradation rate to a limited extent.
- The introduction of double bonds has no effect (insofar as the relationship between saturated and unsaturated bonds is large).
- Introduction of substituents into a benzene ring improves the biodegradability, depending on the type of the substituents. For example, the degradation rate increases in the order $-CH_3$, $-OOC \cdot CH_3$, $-CHO$, $-CH_2OH$, $-CN$, $-NH_2$, $-COOH$, $-OH$, $-SO_3H$.

The various technologies employed in biological waste water treatment cannot be dealt with here in detail [7–10]. All the processes produce large quantities of activated sludge (biomass), which has to be dewatered and processed further (detoxified if necessary) and either tipped or incinerated. Thus the processes that occur in biological water treatment are, in principle, the same as those that occur in the biological purification of water; simplifying the process it can be said that organic pollutants are transformed into inorganic products (CO_2, H_2O and mineral salts) and insoluble bacterial mass (new cells). The bacterial mass and other insoluble materials form agglomerates, that either collect at the surface of the water or sediment out, which guarantees their separation from the purified water. At the same time a range of purely physico-chemical processes occur simultaneously in a biological water treatment plant (e.g. coagulation, adsorption, flocculation, precipitation), which also contribute to the reduction in the concentration of the pollutants.

The important biological processes include, on the one hand, the large-scale natural processes and, on the other, the artificial processes. The natural processes, whose space requirement is generally ca. 100 times greater, include land treatment, waste water sprinkling and pond processes. Unpleasant smells are unavoidable when they are employed. In the artificial processes the natural purification mechanisms are exploited in an optimally controlled manner in specially designed plant. The most important of these are the contact beds and the activated sludge processes. Contact beds contain packing to which the bacteria can adhere, in the activated sludge process the bacteria are present in suspended floccules [3]. The natural process is accelerated mainly by increasing the concentration of aerobic bacteria under optimal metabolic conditions with recycling of the biomass. The necessary air is supplied with aerating systems of many different types. The degree of purification achieved by a mechanical-biological plant can be up to 95%, but the water is by no means free from germs and pollutants, since not all pollutants are sufficiently biodegradable. Biological purification processes are impossible

— for most inorganic materials,
— for materials that are very biologically stable, such as halogenated hydrocarbons,
— with highly polluted waters.

It is also impossible to avert the danger of phosphate eutrophication by biological purification. Only physico-chemical or chemical processes can achieve this.

Nevertheless, for many (pure) chemical compounds it is possible to find a suitable bacterial strain, which utilizes the compound in its metabolism under suitable conditions. Even low concentrations of phenol and formaldehyde can be bacterially degraded under suitable conditions [13]. The degradation of phenol probably proceeds as follows:

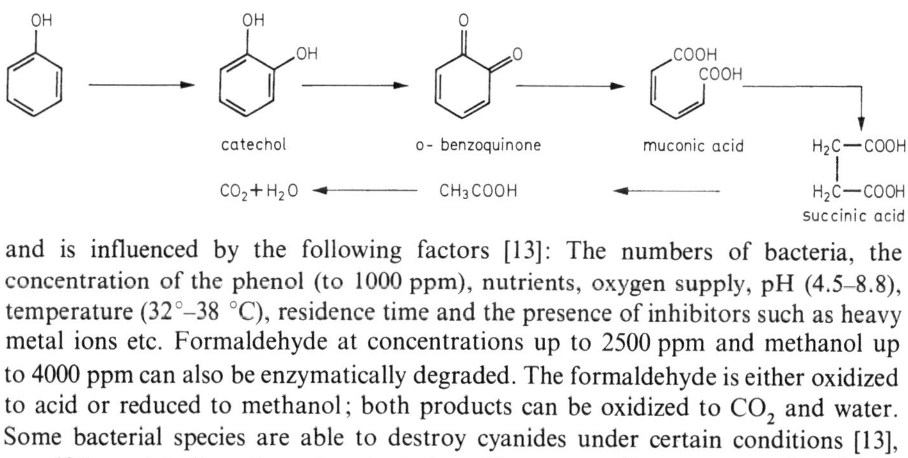

and is influenced by the following factors [13]: The numbers of bacteria, the concentration of the phenol (to 1000 ppm), nutrients, oxygen supply, pH (4.5–8.8), temperature (32°–38 °C), residence time and the presence of inhibitors such as heavy metal ions etc. Formaldehyde at concentrations up to 2500 ppm and methanol up to 4000 ppm can also be enzymatically degraded. The formaldehyde is either oxidized to acid or reduced to methanol; both products can be oxidized to CO_2 and water. Some bacterial species are able to destroy cyanides under certain conditions [13], even if the metabolic pathway is not yet clear. The same applies to a range of pesticides and herbicides including butylphenols, chlorophenols, nitrophenols, thiolcarbamates, chlorophenylacetic acid, benzoates etc. However, partial biodegradation can lead to the formation of new persistent substances; for instance, the nonionic detergent polyethylene oxide with a 4-nonylphenylene end group is microbially degraded to the persistent 2-(4-nonylphenoxy)ethanol [232].

Now a few words concerning the handling of sludge, which is troublesome because of the 95% water content. One possibility is anaerobic sludge digestion. First of all acid bacteria reduce the pH with the production of CO_2 and evil-smelling products such as lower fatty acids, H_2S etc. A second bacterial stage results in the formation of mainly CO_2 and CH_4 as odorless gases (ca. 70% methane, 30% CO_2, calorific value 6000 kcal/m^3 (25 121 kJ/m^3)) [11, 12]. Offensive odours are avoided by performing both steps under controlled conditions.

The methane bacteria are only active under very specific conditions however, such as in the complete absence of oxygen, at a pH between 6.9 and 8.0, at constant temperature (optimally 30°–40 °C), with the availability of specific nutritional conditions, in the absence of poisons and inhibitors (e.g. heavy metal ions; hence unsuitable for most industrial treatment plants). This process is carried out nowadays in closed spaces (vessels) and the residence times of 15 to 30 days are normal.

Aerobic digestion is also practised. The digested sludge can be used for fertilization purposes in the liquid state or, after dewatering, it can be tipped or incinerated.

The range of detoxification processes will be enriched in the future by the addition of targeted biological degradation. Even today bacterial cultures are available for various industrial biotechnical purposes. An example is the extraction of waste rock containing small quantities of copper, if the copper is present in the form of the sulphide [213]. This takes place under the influence of Thiobacillus bacteria — a sequence of interrelated chemical and biochemical oxidations results in the formation of dissolved copper sulphate. The copper is then most easily precipitated by sheet iron. The term biotechnology expresses the fact that biological processes are utilized on a technological scale. This principle is being exploited more and more in effluent and waste treatment. Even today there are processes that are employed to produce basic chemicals from waste, for example, the microbial production of methane from waste in digestion towers.

The main object of biotechnology is to harmonize the microbial reactions with the conditions in the reactors. Many classical processes, in food manufacture for example, (e.g. cheese, alcoholic drinks, silage) are based on the action of bacteria, fungi or yeasts. One development that has occurred in the last few decades is the direct production of lipid or protein (biomass) from organic carbon sources (e.g. methanol, petroleum oil, cellulose) by micro-organisms with the employment of suitable nutrient media [214]. Enzymatic methods, which occur without increase in the temperature or the pressure will certainly play their role in detoxification reactions as well in the future. Alongside the great activity enzymes are also distinguished by their high specificity. This can be used as a basis for dividing them into the following groups [215]:

- oxoreductases catalysing oxidations and reductions,
- transferases catalysing the transfer of groups,
- hydrolases catalysing hydrolyses,
- lyases cleaving groups with the formation of double bonds or adding groups to double bonds,
- isomerases catalysing isomerization,
- ligases joining groups with the elimination of a pyrophosphate bond.

Temperature and pH are of enormous importance for enzymic reactions. Instant deactivation and denaturation has to be expected outside the normal range for each particular enzyme. Enzymes can be extracted from plant or animal tissues industrially. Isolated enzymes are employed in various large-scale industrial processes, where they have the following advantages over classical chemical and fermentation techniques

— mild reaction conditions
— few side reactions
— the use of water as a solvent.

Good results have already been obtained in the elimination of phenols and aromatic amines from industrial effluents using vegetable peroxidases and hydrogen peroxide (up to 99.9%) [230]. It is particularly important for practical application that the enzyme can be reused several times by exploiting specific immobilization techniques. This is achieved in practice [216] by:

— adsorption on surfaces,
— incorporation in crosslinked gels,
— by micro-encapsulation,
— by covalent coupling to polymeric carriers,
— by enclosure within semipermeable membranes.

Some of the other advantages of immobilized enzymes are the possibility of realizing continuous processes, the ease of following the reaction process and the absence of enzyme from the product.

An example of enzyme application that will certainly be of importance in the future is the degradation of cellulose wastes to D-glucose. Thus 1 t of waste paper can be used to produce ca. 1/2 t of glucose, which on fermentation yields ca. 365 l ethanol [215]. A calculation made in America estimates that alcohol produced in this manner will cost about one third as much as that produced from ethylene. The glucose can also be converted to fructose or employed directly as valuable tertiary livestock nutrient [215]. Ground newspaper containing 70% cellulose is completely hydrolysed within 48 h. The residue consisting mainly of lignin can be incinerated or utilized as the starting material for other chemicals.

Investigations have also been reported of the use of immobilized enzymes to detoxify pesticides [217]. A micro-organism (Pseudomonas fluorescens) immobilized in soft PVC granules has also been employed for the elimination of heavy metals from waste waters [218]. Biological processes have recently been employed for the reutilization of polystyrene wastes. For this purpose the polystyrene (dissolved in ethyl acetate and mixed with water) is employed as a nutrient for reeds (Schoenoplectus lacustris L.) and sedge (Phragmites communis Trin.). An anaerobic process for the production of edible protein from these plants is reported [219].

3.3 Chemical Methods of Treating Waste Waters

3.3.1 Waste Water Neutralization

Neutralization processes play an important part, on the one hand, in the treatment of acidic or alkaline effluents and, on the other, in the disposal of used acid and alkali solutions and residues [71]. The aim is to adjust the pH to one that is "safe" or to the legally regulated limit for aqueous effluents and waste chemicals led into them. With concrete drainage systems damage must be expected at below pH 5 [72]. Concrete is stable at elevated pH, but deposits can occur in the drainage system.

The use of a basic neutralizing agent often not only neutralizes the acidity but brings about precipitation of undesired cations as hydroxides or basic salts. This, however, is not invariably the case, since the pH range for hydroxide precipitation varies greatly from metal to metal ion. Furthermore, it is often a great distance from the neutral point. For some metals the optimum precipitation range lies above the maximum pH laid down, while other metals redissolve in this range as soluble hydroxy complexes.

The neutralization can be performed in stationary or flow-through plant; with stationary plants it is recommended that two basins be employed, one with alkaline and one with acid neutralizing agents. If the pH of the input varies sharply then the first basin can be used for preliminary neutralization and the second for fine adjustment.

Firstly we will concentrate primarily on acid neutralization; alkaline neutralization and precipitating agents will be dealt with in the next section; The neutralizing agents employed in practice for the neutralization of acidic effluents are [6]:

- slaked lime } good solubility
- caustic soda, potash and their solutions } good solubility
- soda ash Na_2CO_3
- limestone, $CaCO_3$ and magnesite, $MgCO_3$
- calcined magnesite, MgO
- dolomite, mainly in the form of partially calcined decarbolite, $CaCO_3 \cdot MgCO_3$ or $CaCO_3 \cdot MgO$.

The carbonates have their effect because of the partial hydrolysis occurring in aqueous medium, e.g.

$$Na_2CO_3 + 2\,H_2O \rightarrow 2\,NaOH + H_2CO_3$$

Similarly the oxides react in aqueous medium to yield hydroxides e.g.

$$MgO + H_2O \rightarrow Mg(OH)_2$$

Smaller quantities of acid neutralizing agent are required for waste water treatment than basic ones, since only a few effluents are strongly basic and waters with pH between 5.5 and 9 are held to be relatively safe for discharge into the drainage systems or into rivers and lakes [73]. Thus, only strongly alkaline wastes have to be at least partially neutralized.

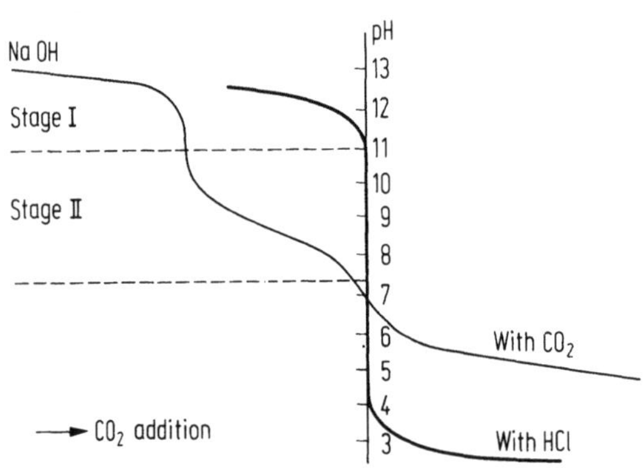

Fig. 3.11. Titration curves for the neutralization of caustic soda by carbonic acid and by HCl

Sulphuric and hydrochloric acids are economical and widely employed. However, there are limitations to the employment of sulphuric acid, since the sulphates produced on neutralization have to be limited in concentration in the neutralized water to ≤ 300 mg/l because of their corrosive effect on the drainage and other concrete systems. Sulphates form very hygroscopic compounds with components of the concrete (ettringite), which reduces its strength so that concrete structures crumble. This problem does not occur with hydrochloric acid, but the extra chloride load in the already highly polluted water is a significant problem, so that chloride concentrations should not be allowed to rise above 500 mg/l.

Carbonic acid i.e. carbon dioxide is the third economically employable acid; in water it dissociates as follows

$$CO_2 + H_2O \rightleftharpoons H_2CO_3 \overset{KD_1}{\rightleftharpoons} H^+ + HCO_3^- \overset{KD_2}{\rightleftharpoons} 2\,H^+ + CO_3^{2-}$$

and yields carbonates and bicarbonates that are relatively harmless to the environment. The solubility of carbon dioxide can be increased almost according to Henry's law by increasing its partial pressure up to 5 bar. Carbonic acid is an acid of medium strength, the pH of its 0.1 N solution is 3.73 at 18 °C. The neutralization with carbon dioxide takes place in two stages and over-acidification is virtually excluded (maximal variations of 0.1–0.2 pH units from the desired value), which is not always possible in the case of stronger acids such as hydrochloric acid and sulphuric acid. In the pH range 3–10 small excesses of these bring about large alterations to the pH (see Fig. 3.11) which makes it necessary to employ an expensive control system.

The first stage in the neutralization of, for example, NaOH is the formation of sodium carbonate down to a pH of about 11, which is then converted to sodium bicarbonate in the next stage extending down to pH 7.5 (see Fig. 3.11). The addition of further CO_2 causes dissolution of CO_2 and the pH drops slowly into the acid range [74].

$$2\,NaOH + CO_2 \rightarrow Na_2CO_3 + H_2O \text{ (pH down to ca. 11)}$$

$$Na_2CO_3 + CO_2 + H_2O \rightarrow 2\,NaHCO_3 \quad \text{(pH down to 7.5–8)}$$

Good mixing must be ensured between the CO_2 and the product to be neutralized. This and hence the creation of the necessary absorption area is effected in industrial practice by recirculation of the liquid by pumping. Return to the reactor or release into the drain is controlled by an automatic pH sensor. Other processes employ jets, with whose aid alkaline waters are sprayed with CO_2 so that a large absorption surface is created with a low expenditure of energy and the pH is adequately reduced. The chemical reaction occurs in a reactor (see Fig. 3.12). The CO_2 addition can be made with relatively simple equipment.

Industrial neutralization with carbon dioxide has the particular advantage that environmentally safe salts are formed with a relatively high buffering capacity, so that only 70%–80% of the stoichiometric quantity of CO_2 need be employed (e.g. for the neutralization of 10 m^3 waste water containing 1% NaOH to pH 8.5 requires in practice only 88 kg CO_2 instead of the calculated 110 kg. The temperature increase on neutralization stays within limits. For 1% NaOH it amounts to about 4 °C.

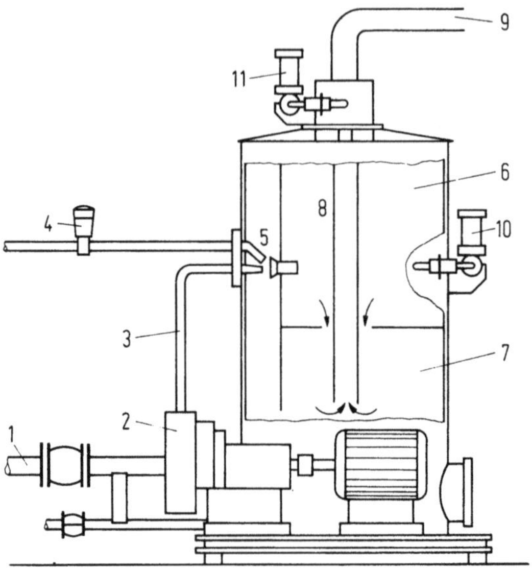

Fig. 3.12. CO_2 neutralization reactor [75]. 1. feed-pipe, 2. reactor pump and motor, 3. pressure line, 4. CO_2 line with control valve, 5. jet with diffuser, 6. reaction chamber, 7. postreaction chamber, 8. riser, 9. outlet, 10. pH monitor (outlet), 11. pH monitor (inlet)

Flue gases containing CO_2 which are produced during many processes can also be very profitably employed as neutralizing agent [76]. As well as being neutralized by the contained CO_2 certain effluents (e.g. from textile processing) are additionally purified to a large degree by the froth fotation of colloidal impurities [77].

3.3.2 Precipitation

Precipitation reactions have the effect of immobilizing dissolved waste water contaminants, i.e. they are transformed into less reactive, less dangerous substances by the addition of suitable oppositely charged ions, when — as with flocculation — intermediate stages such as colloid formation, destabilization, micro and macro-flocculation occur [78]. Precipitation can also be effected in certain cases by other means, e.g. by altering the oxidation state as by the reduction of cupric ions to cuprous ions or metallic copper. Silver ions can be precipitated as metallic silver (slightly contaminated by silver oxide) by the addition of H_2O_2. This present section is not, however, concerned with such processes.

Precipitation is often designated as a method for the "further purification" of water, i.e. a process additional to the normal mechanical-biological treatment; it can be performed either before, during or after the other purification processes depending on when the precipitant is added [79]. The removal of unwanted or toxic metal ions from effluents is one particular application of precipitation, the removal of unwanted anions (such as phosphate by calcium salts in neutral or alkaline medium) is less frequent.

Like all other purification processes precipitation should be formed as close as possible to the site of production of the effluent and before it is diluted by admixture with other waste waters, since additional contaminants (e.g. foreign salts, complexing agents, protective organic colloids) can increase the residual solubility.

The precipitation reactions discussed in this section are characterized by the formation of an ionic crystal lattice. The solubility product of the salt or hydroxide formed can be employed for the quantitative description of the process. From this it can be seen that there is, in principle, an equilibrium which can be set up between the solute and precipitate so that if one of the ions is in large excess the one of the opposite charge is completely driven from the solution into the precipitate. The residual metal concentration of the treated solution can be calculated at constant temperature and in dilute solution from, to take the example of neutralization precipitation, the solubility of the precipitated hydroxide. In this case it falls with increasing pH [78] (see Table 3.4). Prolonged contact of the sediment with the mother liquor can cause changes to occur in the residual metal concentration. Complexes occur frequently in waste waters, which can have the effect of bringing heavy metals into solution or keeping them there [80] and thus preventing precipitation. If the substances only form weak complexes, then the metals can usually be precipitated at higher pH i.e. in excess alkali, e.g. copper ions from solutions of pyrophosphoric acid by means of an excess of milk of lime. Interfering ammonium ions can be bound with some success by the addition of formaldehyde [1] thus allowing the precipitation of the metal.

Oxidative destruction of some special complexing agents is possible with an excess of chlorine or sodium hypochlorite with long reaction times (several hours or even days). There is no satisfactory method of destroying complexing agents such as ETDA, ethylene diamine, triethanolamine, citric acid or tartaric acid etc. The problem of treating metal complex-containing effluents has been reviewed [81].

3.3.2.1 Neutralization Precipitation

Neutralization precipitation with NaOH, lime or soda ash is the most commonly employed precipitation process in waste water treatment. Its advantages are the ease

Table 3.4. Hydroxide precipitation at various pH values [78].

Compound	Solubility product (room temperature) (mol/l)	pH of precipitation	Theoretical solubility (mg/l metal)		
			pH = 7	pH = 8	pH = 9
$Fe(OH)_3$	$8{,}7 \cdot 10^{-38}$	2,8–4,0	$4{,}9 \cdot 10^{-12}$	$4{,}9 \cdot 10^{-15}$	$4{,}9 \cdot 10^{-18}$
$Al(OH)_3$	$2{,}0 \cdot 10^{-32}$	4,3–5,0	$5{,}4 \cdot 10^{-7}$	$5{,}4 \cdot 10^{-10}$	$5{,}4 \cdot 10^{-14}$
$Cr(OH)_3$	$6{,}0 \cdot 10^{-29}$	5,5–6,5	$3{,}0 \cdot 10^{-6}$	$3{,}0 \cdot 10^{-9}$	$3{,}0 \cdot 10^{-12}$
$Cu(OH)_2$	$2{,}0 \cdot 10^{-19}$	5,8–8,0	1,3	$1{,}3 \cdot 10^{-2}$	$1{,}3 \cdot 10^{-4}$
$Pb(OH)_2$	10^{-17}	7,0–9,5	$2{,}1 \cdot 10^{2}$	2,1	$2{,}1 \cdot 10^{-2}$
$Zn(OH)_2$	$4{,}0 \cdot 10^{-17}$	7,6–8,3	$2{,}6 \cdot 10^{2}$	2,6	$2{,}6 \cdot 10^{-2}$
$Co(OH)_2$	$2{,}0 \cdot 10^{-16}$	7,0–9,0	$1{,}2 \cdot 10^{3}$	12	0,12
$Ni(OH)_2$	$5{,}8 \cdot 10^{-15}$	7,8–9,3	$3{,}4 \cdot 10^{4}$	$3{,}4 \cdot 10^{3}$	3,4
$Cd(OH)_2$	$1{,}3 \cdot 10^{-14}$	9,1–10,0	$1{,}5 \cdot 10^{5}$	$1{,}5 \cdot 10^{3}$	15
$Mn(OH)_2$	$4{,}0 \cdot 10^{-14}$	7,0–9,0	$2{,}2 \cdot 10^{5}$	$2{,}2 \cdot 10^{3}$	22
$Fe(OH)_2$	10^{-13}	6,0–10,0	$5{,}6 \cdot 10^{5}$	$5{,}6 \cdot 10^{3}$	56

[1] Range from commencement of precipitation to quantitative precipitation

of handling and dosing the precipitating agent and the relatively easy removal of the precipitates. The use of solutions of caustic soda as the precipitating and neutralizing agent can pose certain storage and dosage problems because of crystallization. A 18%–22% solution has the best storage properties, since this solidifies below −20 °C [1]; 40% caustic soda solidifies at about +17 °C, a 46% solution, on the other hand, only at ca. +5 °C. Impurities in technical grade caustic solutions can lower these temperatures a great deal. Care is necessary in dissolving or diluting large quantities of sodium hydroxide because of the high heat of solution.

Table 3.5. The price of important precipitants in DM/t (1976) [78]

NaOH (50%)	ca. 300
Quick lime	ca. 100
Sodium carbonate (soda ash)	ca. 300
Thus the price of chemicals is about the same for precipitation with caustic soda and with soda ash, while it is only a third as high when lime is employed.	
Aluminium sulphate ($17\%–18\%$ Al_2O_3)	ca. 300
Mixed aluminium and ferrous sulphate ($18\%–20\%$ Al_2O_3 and Fe_2O_3)	ca. 200
Ferrous chloride solution (ca. 40%)	ca. 200–240
Ferrous sulphate (wet)	ca. 3
The low price of ferrous sulphate can make it attractive for treating large quantities of waste water, even though the necessary oxidation is expensive.	

Milk of lime is more economical (see Table 3.5), but it is not so easy to store (danger of caking) or to prepare the 6%–12% suspensions [1]. Because of the natural carbonate hardness of the water and particularly because of the effect of atmospheric CO_2, there is a tendency for calcium carbonate to be precipitated from milk of lime. Neutralization precipitation is valuable for acidic waste waters because precipitation and neutralization are accomplished simultaneously, with the formation of a voluminous — slow settling — precipitate.

$$Me^{3+} + 3\,NaOH \rightarrow Me(OH)_3 + 3\,Na^+$$

$$Me^{2+} + Ca(OH)_2 \rightarrow Me(OH)_2 + Ca^{2+}$$

Quantitative theoretical predictions of the actual quantities of metals remaining after precipitation are scarcely possible, since, on the one hand, there are differences in the solubility products quoted in the literature and then there are many other factors that can influence the result. The precipitation of Cr(III), Fe(III) and Al(III) ions is reported to be quantitative in the acid region. The residual solubility increases in the following order amongst the bivalent metal hydroxides [78]; increasingly higher precipitation pHs are needed here:

$$Cu < Pb < Zn < Co < Ni < Cd < Mn < Fe$$

The amphoteric nature of some metals is the cause of some important reactions that occur during precipitation. As the pH is increased the solubility of most metal hydroxides goes through a minimum (=precipitation) then as a consequence of the formation of hydroxo salts it climbs again. The precipitation process must be precisely monitored particularly for chromium, zinc and aluminium hydroxides. Soluble complexes can be formed if excess alkali is employed, e.g.

$$Zn(OH)_2 + 2\ NaOH \rightarrow Na_2[Zn(OH)_4]$$

In the case of chromium and zinc this danger can be avoided if milk of lime is employed for the precipitation rather than caustic soda, since the complex calcium salts then formed are also very insoluble.

$$Zn(OH)_2 + 2\ Ca(OH)_2 \rightarrow Ca[Zn(OH)_4]$$

Since the solutions treated contain not just metallic cations but also their anions, the precipitating agent can also cause the formation of basic salts, mainly of bivalent metals and at low pH. In the case of the sulphates and the chlorides, however, their behaviour is like that of the pure hydroxides.

Concentration precipitation of $CaSO_4 \cdot 2\ H_2O$ (solubility 2 g/l at 20 °C) can occur in the treatment of waste waters containing large quantities of sulphates with milk of lime [78].

The course of neutralization precipitation can be followed potentiometrically. On addition of a basic neutralizing agent any mineral acid present is first neutralized (see Fig. 3.13), only then does the actual precipitation take place. The precipitation range for the strongly basic cadmium with caustic soda or milk of lime is very high at pH 9–9.9. The use of soda ash is to be recommended in this case, when the

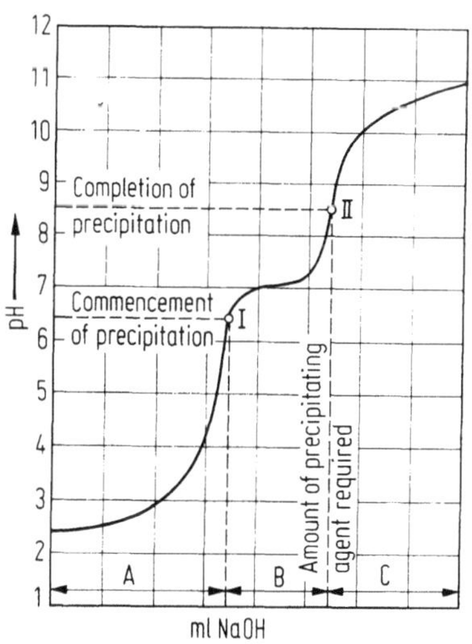

Fig. 3.13. The titration curve of an acidic solution containing heavy metals (neutralization precipitation) [82]

precipitation is already complete by about pH 7.5 and the cadmium comes out of solution as the pure carbonate. A similar, but much smaller, lowering of the precipitation is only known in the case of zinc. The general advantage of carbonate precipitation is the crystallinity of the precipitate that is formed, which guarantees its easy filtration and that the resulting filter cake has a lower residual water content. The theoretical residual concentrations for bivalent metallic ions are virtually all lower than is the case for hydroxide precipitation. Hydroxides rather than carbonates are precipitated in the case of trivalent metal ions. The formation of double salts such as $2\,ZnCO_3 \cdot 3\,Zn(OH)_2$ also occurs when precipitating with soda ash. For mixtures of metal salts the use of a combination of caustic soda and soda ash can lead to similar removal rates at low pH. The use of soda ash is unsuitable for precipitations at high pH since large excesses are required. Table 3.6 summarizes the solubility products of the most important carbonates. Soda ash is unsuitable for the precipitation of some metals such as chromium because of the formation of easily soluble carbonate complexes. Precipitation with NaOH can be employed successfully here in spite of possible complex formation, since the colloidal hydroxide ages relatively quickly.

Table 3.6. Carbonate precipitation [78]

Compound	Solubility product (room temperature) (mol/l)	Theoretical solubility (mg/l)	
		pH = 7	pH = 8
$CoCO_3$	$1{,}6 \cdot 10^{-13}$	$1{,}1 \cdot 10^{-4}$	$1{,}1 \cdot 10^{-6}$
$PbCO_3$	$8{,}0 \cdot 10^{-14}$	$1{,}8 \cdot 10^{-4}$	$1{,}8 \cdot 10^{-6}$
$CdCO_3$	$5{,}0 \cdot 10^{-12}$	$6{,}3 \cdot 10^{-3}$	$6{,}3 \cdot 10^{-5}$
$ZnCO_3$	$1{,}6 \cdot 10^{-11}$	$1{,}2 \cdot 10^{-2}$	$1{,}2 \cdot 10^{-4}$
$FeCO_3$	$2{,}5 \cdot 10^{-11}$	$1{,}6 \cdot 10^{-2}$	$1{,}6 \cdot 10^{-4}$
$MnCO_3$	$6{,}3 \cdot 10^{-11}$	$3{,}8 \cdot 10^{-2}$	$3{,}8 \cdot 10^{-4}$
$CuCO_3$	$2{,}5 \cdot 10^{-10}$	0,18	$1{,}8 \cdot 10^{-3}$
$NiCO_3$	$1{,}3 \cdot 10^{-7}$	85	0,85

The best precipitation conditions will now be described for a range of heavy metal ions [82]. Copper ions are equally well precipitated by both caustic soda and milk of lime. At a minimum pH of 7.5 the solubility of the hydroxide is well below 1 mg/l. Precipitation with soda ash is not recommended since the basic copper carbonate formed [$2\,CuCO_3 \cdot Cu(OH)_2$] possesses a higher solubility. The solubility of copper hydroxide is not increased by complex formation.

Nickel ions can be precipitated at pH 9.5 with about the same efficiency by all three precipitating agents.

For the precipitation of zinc hydroxide with caustic soda or milk of lime the pH should be raised above 8.5. With caustic soda at above pH 11 the soluble zincate is formed. With lime the insoluble calcium zinate is formed, with sodium carbonate the product is the insoluble basic zinc carbonate when the pH is lowered to 8.

In order to precipitate cadmium ions with caustic soda or lime pH values higher than 9.5 are required, whilst as mentioned previously the precipitation begins at pH 7.5 with soda ash, this is because of the particular insolubility of cadmium carbonate which is precipitated in pure form here.

Trivalent metal ions are less basic in character than their bivalent counterparts, in general this is reflected in a lower precipitation pH. Thus chromium(III) ions can be very completely precipitated at between a pH of 6.5 and 9, although the solubility increases at higher pH because of the formation of chromite anions. This is avoidable by precipitating with lime, since then insoluble calcium chromite is formed at higher pH. Sodium carbonate cannot be employed because of the formation of easily soluble carbonates.

Ferrous iron is precipitated above pH 3.5 by all three precipitating agents. If the iron is present in the bivalent state it must first be oxidized by atmospheric oxygen or some other oxidizing agent. In practice the precipitation should be performed at about pH 7 because the precipitated ferric hydroxide sediments with difficulty in the acid region.

Trivalent aluminium ions are well precipitated by all three agents in the 5–8.5 pH range. Above pH 8.5 the aluminium hydroxide begins to go into solution because of the formation of soluble aluminates even when lime is employed. The above exposition applies to the precipitation of the hydroxides from solutions of the pure metal salts. The relationships are scarcely comprehensible in complicated waste water mixtures, particularly since the presence of complexing agents (e.g. ammonium ions, cyanides, organic complexing agents) has such a great effect that hydroxide precipitation does not occur. Nevertheless the presence of complexing agents can also be of advantage for the separation wished for. It has been found in practice that mixtures of various metals ions can often be precipitated better to leave lower residual concentrations or at lower pH values; this is a result of mutual adsorption and mixed crystal formation (e.g. between bivalent ions of very similar radius) and the formation of mixed salts and complexes. Exactly reproducible results cannot be obtained because of the many factors affecting the precipitation process. Hence for waste waters of complicated composition the optimal pH cannot be chosen for each metal ion. Practical investigations must be made to find the best "compromise" pH for a particular effluent; account must also be taken of the local regulations concerning what may be led into the drains, because further expense for additional neutralization would be uneconomical. A whole range of other factors must be taken into account. For instance a subsequent drop in the pH is to be expected when precipitating trivalent metal ions. Large quantities of iron or aluminium salts immediately increase the efficiency of precipitation of bivalent metals etc.

The precipitation of metals as the pure hydroxides with caustic soda is to be recommended if the hydroxides are then going to be further utilized. If they are to be deposited as wastes, then the use of milk of lime is recommended since the pH range for precipitation can be widened and the precipitates that are formed are usually very easily filtered.

The precipitation ranges are summarized again below [83].

Al^{3+}/NaOH: 4.9– 8.5
Cd^{2+}/NaOH: 9.5–13.0 } milk of lime: 6.4–13.0
Cr^{3+}/NaOH: 6.4– 9.0

Cu^{2+}/NaOH: 7.2–13.0
Fe^{3+}/NaOH: 3.3–13.0
Ni^{2+}/NaOH: 9.5–13.0 } milk of lime: 8.3–13.0
Pb^{2+}/NaOH: 7.0– 8.5
Zn^{2+}/NaOH: 8.3–11.0

The sedimentation process, another aspect of hydroxide precipitation, will only be dealth with briefly here. Often the hydroxides do not settle quantitatively but remain in part suspended as tiny floccules. In such cases the addition of organic flocculating agents is to be recommended. The very different effects of neutral salt (NaCl), calcium and polyelectrolyte concentrations and combinations of these factors on the sedimentation properties of various precipitated metal hydroxides has been described by Jola [84]. Under optimal pH conditions the filterability drops in the order Cd > Al > Fe > Cu > Zn > Ni > Cr [85].

3.3.2.2 Sulphide Precipitation

Although sulphide precipitation demands a relatively high capital investment and a complicated process technology (particularly when H_2S is used), the relatively low cost of the chemicals consumed makes it of interest for large-scale operation and laboratory use in some specific cases. Acid precipitation with H_2S either from pressure cylinders or generated on the site from iron sulphide and HCl has the advantage of not introducing additional cations into the water. Alkali metal sulphides or ammonium sulphide are employed for precipitation in alkaline solution. Hydrogen sulphide and sulphide ions produce insoluble precipitates (atomic numbers: 23, 25–34, 42–52, 74–85, 92) with numerous metal ions. Depending on the precipitating agent and the ions to be precipitated a wide pH range can be employed. When precipitating with H_2S in acid solution the active species is the HS^- ion, since the sulphide ion concentration is tiny because of the extremely small dissociation of HS^-. Because of dissociation hydrogen sulphide and hydrosulphide anions are also present when aqueous sodium sulphide is employed as the precipitating agent.

$$S^{2-} + H_2O \rightleftarrows HS^- + OH^-$$

$$HS^- + H_2O \rightleftarrows H_2S + OH^-$$

Often the sulphide precipitation does not take the form of a simple rapid process. Delays occur in precipitation, the formation of addition compounds, coprecipitation of readily soluble sulphides etc. Complete precipitation cannot be guaranteed without further ado. In the precipitation of Ni, Co, Zn and Mn sulphides it must be taken into account that each of them can occur in two modifications. Apart from the case of manganese sulphide the solubility products of the sulphides are so low that the foreign salts do not have the effect that they have in hydroxide precipitation (see Table 3.7). Thus the precipitation of sulphides is usually possible in the presence of complexing agents and good separations can be obtained even in the acid region. Ammonial baths containing cupritetramine complexes are employed in the production of printed circuit boards. The only method of removing the copper in this case is by precipitation as the sulphide.

Mercury, copper and lead sulphides are only soluble in oxidizing acids, so that

Table 3.7. Sulphide precipitation [78]

Compound	Solubility product (room temperature) (mol/l)	Theoretical solubility (mg/l)	
		pH = 5	pH = 7
HgS	$4 \cdot 10^{-53}$	$1 \cdot 10^{-35}$	$1 \cdot 10^{-39}$
CuS	$8 \cdot 10^{-37}$	$6{,}5 \cdot 10^{-20}$	$6{,}5 \cdot 10^{-24}$
PbS	$3{,}2 \cdot 10^{-28}$	$6{,}6 \cdot 10^{-11}$	$6{,}6 \cdot 10^{-15}$
CdS	$1{,}6 \cdot 10^{-28}$	$2{,}4 \cdot 10^{-11}$	$2{,}4 \cdot 10^{-15}$
γ-NiS	$2 \cdot 10^{-24}$	$1{,}5 \cdot 10^{-9}$	$1{,}5 \cdot 10^{-13}$
α-NiS	$3{,}2 \cdot 10^{-19}$	$2{,}4 \cdot 10^{-2}$	$2{,}4 \cdot 10^{-6}$
β-CoS	$2 \cdot 10^{-25}$	$1{,}5 \cdot 10^{-8}$	$1{,}5 \cdot 10^{-12}$
α-CoS	$4 \cdot 10^{-21}$	$3 \cdot 10^{-4}$	$3 \cdot 10^{-8}$
ZnS (Sphalenit)	$1{,}6 \cdot 10^{-24}$	$1{,}4 \cdot 10^{-7}$	$1{,}4 \cdot 10^{-11}$
ZnS (Wurtzit)	$2{,}5 \cdot 10^{-25}$	$2{,}1 \cdot 10^{-5}$	$2{,}1 \cdot 10^{-9}$
MnS green	$2{,}5 \cdot 10^{-13}$	$1{,}8 \cdot 10^{4}$	1,8
MnS pink	$2{,}5 \cdot 10^{-10}$	$1{,}8 \cdot 10^{7}$	$1{,}8 \cdot 10^{3}$

(Sphalinite) (Wurtzite) green pink

under certain conditions they are suitable for disposal by dumping. The solubility of cadmium sulphide is also so low compared with that of the hydroxide and the carbonate, that precipitation as the sulphide has been recommended for the removal of cadmium [78]. However, long delays are observed in cadmium precipitation. The solubility of the sulphides is significantly temperature-dependent. Some examples of sulphide precipitation will now be described: mercurous salts in hydrochloric acid yield a black precipitate of mercuric sulphide and mercury. Mercuric salts also precipitate quickly and quantitatively to the sulphide in hydrochloric acid (up to 5.2 N). At higher concentrations of HCl the yellowish mercuric sulphide chloride $Hg_3S_2Cl_2$ is precipitated. The lower limit for the sulphide precipitation is reported as being 5 mg Hg/l. The black sulphide modification is unstable and converts spontaneously to the red on standing. It should be remembered during precipitation that mercuric sulphide forms a complex anion with excess sulphide ions, which forms soluble salts with alkali metal and ammonium salts.

The precipitation of arsenic and antimony as the sulphides is also a suitable method of removing them. Arsenic(III) ions can be precipitated quantitatively as yellow As_2S_3 in the presence of any concentration of HCl, however delays occur in the precipitation of five-valent arsenic. With antimony care should be taken not to work at too low a pH; the sulphide precipitates well in HCl concentrations up to 4 N.

Thus sulphide precipitation can be successfully employed for the removal of several metal ions, but other methods for rendering the amounts remaining in solution harmless should be employed in combination.

Sulphide precipitation has not been widely employed in practice because the high toxicity of H_2S and its low odour threshold render expensive safety and control systems necessary. All fumes must be exhausted and rendered innocuous by combustion or in some other way. The filtrate must also be treated to destroy residual H_2S, e.g.

oxidatively with H_2O_2. Problems can occur if the sulphides precipitate in colloidal form.

A more recent process for the precipitation of heavy metals employs sulphur-lime solution, which consists principally of calcium polysulphide (CaS_4), which, because of its high alkalinity, can be employed for the treatment of very acid effluents without the necessity for further neutralization [86, 87]. The residual heavy metal concentrations after treatment are reported to be in the ppb range.

The solution of CaS_4 decomposes in the presence of air or CO_2 as follows:

$$CaS_4 + 3/2\,O_2 \rightarrow CaS_2O_3 + 2\,S$$

$$CaS_4 + CO_2 + H_2O \rightarrow CaCO_3 + H_2O + 3\,S$$

Heavy metal ions are precipitated as their sulphides by calcium thiosulphate as well as by H_2S.

$$Me^{2+} + CaS_2O_3 + H_2O \rightarrow MeS + CaSO_4 + 2\,H^+$$

$$Me^{2+} + H_2S \rightarrow MeS + 2\,H^+$$

Since the sulphide is unstable chromium(III) ions can be precipitated as $Cr(OH)_3$ because of the high alkalinity of the solution, while chromium^{VI} is first reduced to Cr^{3+} by H_2S or CaS_2O_3.

$$Cr^{VI} + 3/2\,S^{2-}(H_2S) \rightarrow Cr^{3+} + 3/2\,S$$

$$Cr^{VI} + 3/2\,CaS_2O_3 + 3/2\,H_2O \rightarrow Cr^{3+} + 3/2\,CaSO_4 + 3/2\,S + 3\,H^+$$

The solid sulphur that is precipitated and the calcium salts from the decomposition of the CaS_2O_3 have a positive effect on the conditions for the flocculation of the metal sulphides and hydroxides.

As well as the cations described, precipitation can also be used for the removal of certain anions during water purification. The precipitations of phosphate, cyanide, fluoride and numerous organic acids are of importance, e.g. as sparingly soluble calcium salts. Hence precipitation as a method of immobilization is not limited to heavy metals. For example, hydrofluoric acid and soluble fluorides can most economically be rendered harmless by treatment with milk of lime. The fluoride concentration is thereby reduced to about 15 mg F^-/l (the theorectical solubility of CaF_2 is 8 mg F^-/l) [1, 88]. A pH of 8.5 is necessary in order to avoid the formation of a soluble complex.

In the laboratory polyethylene bottles should be employed as reaction vessels, since glass apparatus is attacked immediately. In idustrial applications appropriate acid-resistant construction materials are necessary. Dissolved H_2S can be removed from effluents by the addition of ferrous sulphate as FeS (pH 7–9) and conversely. The FeS precipitate redissolves below pH 4.5. Since ferrous ions also react with carbonate, phosphate, oxygen and organic complexes, more than the stoichiometric quantity is usually required. However, experimental investigations have demonstrated that the formation of ferrous sulphide is the preferred reaction [89].

3.3.2.3 Phosphate Precipitation

As well as their employment as flocculating agents as described earlier iron and aluminium salts can also be utilized for precipitation reactions e.g. for the removal of dissolved phosphates [90–95].

The following processes occur simultaneously during the elimination of phosphates by the addition of ferric salts [96]:

- the precipitation of phosphate as $[Fe(PO_4)_x(OH)_{3-x}]$
- the adsorption of phosphate on the surface of precipitated in part colloidal ferric oxides or oxide hydroxides
- the coagulation of suspended particles by polynuclear Fe(III) hydrolysis products (kinetic intermediates in the transition from metal ions to insoluble hydroxo-oxo metal polymers).

Investigations and plants have been described for the removal of dissolved phosphates from mechanically biologically treated waters by the addition of $FeCl_3 \cdot 6\,H_2O$, $FeCl_3 \cdot 6\,H_2O + Ca(OH)_2$, $AlCl_3 \cdot 6\,H_2O$ and $Al_2(SO_4)_3 \cdot 18\,H_2O$ [97]. The amount of precipitating agent added is adjusted depending on the total phosphorus content, taking the stoichiometric proportion of phophorus to iron ($FePO_4$) or to aluminium ($AlPO_4$) as a basis. When 150% of the stoichiometric quantity of iron or aluminium was employed in the form of either $FeCl_3$ or $AlCl_3$ hexahydrate, then more than 90% of the dissolved phosphate was eliminated. The same result was also obtained by the employment of twice the stoichiometric amount of $Al_2(SO_4)_3 \cdot 18\,H_2O$. The employment of a mixed precipitating reagent consisting of ferric chloride hexahydrate and calcium hydroxide, with 80% of the stoichiometric amount of iron with the addition of 100 mg $Ca(OH)_2$/l of water treated, led to a greater than 90% elimination of phosphate. Such precipitations are of particular interst because of the possible employment of the precipitation products as fertilizer.

Metal salts also have the advantage when employed as phosphate precipitating agents that the precipitates conglomerate to large sized particles. The chemistry of the precipitation reaction has been described in more detail [98].

The effect of the precipitating agent depends on the pH. Thus when precipitating orthophosphate with iron salts the best pH for the reaction is 5, with aluminium salts it is pH 6 and for lime above pH 10. These values also correspond to the minimum solubilities of the pure phosphates. However, satisfactory precipitation is also possible at other pH values [98].

Technical scale plants have been described which achieve residual phosphate concentrations of 4 µg P/l in continuous operation [96].

3.3.2.4 Complex Formation and Precipitation of Cyanides

The precipitation of cyanides by iron salts is one of the oldest methods of detoxifying cyanide-containing waste waters. Even today when performed as a stationary process it remains of importance for the detoxification of concentrates. Ferrous salts and cyanide ions react in alkaline medium at ca. pH 8.5 to form the nontoxic ferrocyanide ion; the simple ferrous cyanide — $Fe(CN_2)$ — does not exist.

$$6\,CN^- + Fe^{2+} \rightarrow [Fe(CN)_6]^{4-}$$

The formation of this complex anion (as the sodium or potassium salt) is not sufficient of itself to ensure proper detoxification, since photolytic decomposition easily occurs.

$$[Fe(CN)_6]^{4-} + H_2O \xrightarrow{\text{light}} [Fe(CN)_5(H_2O)]^{3-} + CN^-$$

The corresponding ferricyanide produced on reaction with ferric ion is too toxic for employment in detoxification. The solution to this problem is achieved by precipitating the ferrocyanide anions as the colourless ferrous ferrocyanide [99].

$$[Fe(CN)_6]^{4-} + 2\,Fe^{2+} \rightarrow Fe_2[Fe(CN)_6]$$

A small amount of Prussian blue ($Fe_4[Fe(CN)_6]_3$) is always produced too. The precipitation of ferrous ferrocyanide is to be preferred to the precipitation of Prussian blue (ferric ferrocyanide), since the ferrous salt can be regarded as being stable within a pH range of 3–9 while Prussian blue is completely hydrolysed above pH 8, that is ferrocyanide ion is present which could under certain conditions be photochemically cleaved to cyanide. This situation could occur, for example, in contact with metal hydroxide sludges, which are virtually always present at chemical waste disposal sites, and result in the release of free cyanide. The same applies to the waste water, since this is usually adjusted to be slightly alkaline before it is released into the drains. The substance must, therefore, be regarded as an unstable end product rather than as an adequate means of detoxification [99].

The step involving the formation of complex ferrocyanide ions (reduction in cyanide concentration) can be followed photometrically [100]. The increase in potential occurring (ca. 700 mV) can be employed to control the automatic addition of ferrous salt. Since the precipitation process is principally employed for concentrated cyanide solutions anodically polarized iron electrodes should be employed rather than ones constructed from gold which would gradually disolve [100]. The second stage of the process, the precipitation of ferrous ferrocyanide, cannot be rigorously differentiated from the first step, since some of the ferrocyanide ion formed immediately reacts with the added $FeSO_4$. The second stage is electrode-inactive. Almost quantitative precipitation of ferrous ferrocyanide should be achieved, if the pH is adjusted to 3.5 after the addition of the correct quantity of ferrous salt. The precipitant must be added in solution and not as the solid. Nevertheless there remains a danger that unreacted cyanide will be occluded in the precipitate and could perhaps be eluted at the disposal site. The process is employable for cyanide concentrates (1–100 g CN^-/l), hardening salt wastes and the more unstable zinc and cadmium complexes. In the latter cases the metal hydroxides are also precipitated. The reactions with the more stable copper and nickel complexes are more complicated. Here ferrous ferrocyanide is not produced quantitatively. The ferrous salts of complex cyanides of copper and nickel and cuprous cyanide are formed as minor products so that the precipitate sludge produces problems of its own. The small amounts of residual cyanide present after precipitation (1–2 mg/l) can be quantitatively destroyed by oxidation with hydrogen peroxide. However, higher residual concentrations (up to 20 mg/l) have been described in the literature. The precipitation of ferrous ferrocyanide is a purely stationary process.

The handling of sludges, in particular those obtained in flocculation and precipitation, is a constituent part of waste water treatment. The methods used depend on the identities and concentrations of the metals present. Sedimentation produces sludges containing a maximum of 3% solids — dilute sludges — which must be dewatered, in filter presses for example. Sludges containing several percent of one or more metals should be recovered. The sludges from communal sewage treatment plants can be anaerobically fermented and employed as compost if they do not contain too high a concentration of heavy metals. Disposal by dumping should be considered for slidges that cannot be leached; those that cannot be dumped must be incinerated in specially equipped plants, taking into account the problems posed by their metal content. Metal-containing sludges always belong on a chemical waste disposal site.

3.3.3 Ion Exchange

Alongside the already described physicochemical methods of reducing unwanted salt loads in waste waters and removing toxic constituents, ion exchange is a suitable chemical method for low concentrations, cf. [101–104] for example. If the salt concentration is greater than 10 meg./l then the economics of the method have to be justified in the particular case; but the method only ceases to be economical in principle at concentrations above 100 meg./l. The position of the chemical equilibrium can also be displaced in the desorption direction (the counterion effect) [105]. As well as finding their most important areas of application in the production of very pure water and in the recovery of raw materials, ion exchange processes can also be employed for the detoxification of dilute ionic solutions; detoxification and recovery are usually coupled [106].

In practice ion exchange units are employed which range in size from those of the smallest dimensions (ca. 60 cm^3) to plants containing 90 m^3 of resin [107]. Ion exchange is employed in conjunction with other methods for the purification of waste waters. The synthetic ion exchange resins employed consist of a stable matrix onto which acidically or basically reacting residues are attached, which are capable of exchanging cations or anions respectively (hence cation and anion exchangers) (cf. Table 3.8). The most usual starting materials for the manufacture of ion exchange resins are styrene and acrylic acid derivatives, which are polymerized after the addition of crosslinking agents such as divinylbenzene. Polycondensation resins are also employed.

Table 3.8. Important ion exchange active groups of the most useful ion exchanger resins

Cation exchangers		
Strong acid $-SO_3^-$		weak acid $-COOH$
Anion exchangers		
Very strongly basic	strongly basic	weakly basic
$-CH_2-\overset{+}{N}(CH_3)_3$	$-CH_2-\overset{+}{N}(CH_3)_2(CH_2CH_2OH)$	$-CH_2-\overset{+}{N}(CH_3)_2H$

The functional groups active in exchange are mainly introduced into the matrix after polymerization by means of various chemical reactions, e.g. by sulphonation for the manufacture of acidic cation exchangers and by chloromethylation followed by amination for the manufacture of basic anion exchangers. A wide range of specialized resins also exists which are designed for the solution of specific problems (see Table 3.9).

The choice of the particular ion exchange resin depends on the pH of the solution to be handled. Anion exchangers can take up ions in the strongly acidic to the weakly basic range (strongly basic up to about pH 10, weakly basic to about pH 8). Strong acid exchangers operate within the pH range 1–13, weakly acid ones in the range 4–9 and chelating resins operate within the pH range 1.5–10. The latter in particular have a powerful pH-dependent selectivity for heavy metal ions, (the pH ranges mentioned are not identical with the stability ranges of the resins, which should be checked in specific cases [101]).

The capacity of the exchange resins is also limited in the case of rinsing waters, so that — like other adsorbents — they have to be regenerated after a certain operation time. Thus, in general, two exchanger units are required to guarantee continuous operation. Recent continuous processes exploit the fall in loading between the inlet and the outlet. For example, if an exchanger column is washed through from bottom

Table 3.9. Specialized ion exchangers [107]

Specific ion exchangers
— Binding of silicic acid by two ortho phenolic hydroxyl groups
— Binding of boric acid by N-methylglucosamine groups
— Binding of nickel by 1,2-dioxime groups
— Binding of potassium by resins containing dipicrylamino groups
Selective ion exchangers
— Separation of heavy metals by aminoacetic acid groups
— Binding of copper to amino groups of anion exchangers
— Separation of heavy metals from salt solutions by carboxyl resins
— Binding of chlorocomplexes to strongly basic anion exchangers
Amphoteric ion exchangers
— Separation of dissolved materials by ion exchange resins containing both cation and anion active groups
— Decolorization of solutions
Redox exchangers
— Oxidation and reduction reactions at groups derived from quinones, malachite green or thionine dyestuffs
Optically active exchangers
— The separation of isomeric compounds at optically active groups
Adsorbent resins
— Macroporous synthetic resins containing small numbers of ion exchange groups or non-specific groups, whose purpose is merely to confer some hydrophilic character to the resin
— Binding of cationic, anionic or nonionic organic substances such as dyes, humic acids and detergents

to top then after a short time the most heavily loaded resin at the bottom of the column can be removed and replaced at the top by regenerated material. The eluates produced on regenerating the resins have to be rendered innocuous by another method. Electrolytic processes are being increasingly employed for the recovery of metals. Further the temperature sensitivity of exchange resins must be allowed for in practice. Acid exchangers are stable up to about 120 °C, weakly basic ones up to ca. 100 °C and strongly basic ones only up to about 70 °C [105].

The reaction velocity and reaction equilibrium are decisive in ion exchange — as in all chemical reactions. The reaction velocities are determined by the process of diffusion of the dissolved ion to the exchanging groups of the gel. The exchange then proceeds according to the law of mass action until an equilibrium is set up, in practice a whole range of thermodynamic parameters such as affinity (e.g. number of positive or negative charges and the basicity or acidity of the ions), contact time, temperature etc. ensure that the system is a long way from equilibrium [108]. The parameters of the process, i.e. the loadability, are determined by the concentration of immobilized ions, that is the number of groups participating in exchange per unit volume. The correct ion exchanger with the most appropriate properties must be chosen for each particular problem; for example the valency and the binding types of the ions allow the construction of affinity series which simplify the choice of the resin (see Table 3.10). These have been reviewed [101].

Table 3.10. Affinity series of some ions [108]

Exchanger type	Affinity series
Cation exchanger:	
strongly acidic	$Me^{4+} > Me^{3+} > Me^{2+} > Me^{+} > H^{+}$
weakly acidic (carboxylic acid type)	$H^{+} \gg Me^{2+} > Me^{+}$,
Anion exchangers:	
weakly basic	$OH^{-} \gg$ anion active metallic complexes CrO_4^{2-} SO_4^{2-} Cl^{-}
strongly basic	anion active metallic complexes $\gg CrO_4^{2-} > Cl^{-} > CN^{-} > OH^{-}$

Strongly acidic cation exchangers (sulphonic acid groups) take up heavy metals without any particular selectivity in the acidic and neutral ranges. It is a disadvantage here that the calcium which occurs in many waste waters is also bound just as well so that the retention of heavy metals cannot be expected after the breakthrough of Ca^{2+}. The order of affinity on normal polystyrene sulphonic acid resins is as follows [107]:

$$Li^{+} > Na^{+} > NH_4^{+} > K^{+} > Cs^{+} > Mg^{2+} > Ca^{2+} > Ba^{2+}$$

Weakly acidic exchangers (carboxylic acid resins) based on acrylates have a higher capacity, but are only effective in the weakly acid to neutral range and Ca^{2+} interferes. The chelating resins (e.g. the iminodiacetic acid types) are particularly effective for the absorption of heavy metals, since they possess a graduated affinity, which can be exploited by the choice of the optimal pH. At equal concentrations the following order of selectivity obtains [228]:

$$Cu^{2+} > UO_2^{2+} > Pb^2 > Ni^{2+} > Zn^{2+} > Cd^{2+} > Co^{2+} > Fe^{2+} > Mn^{2+} > Ca^{2+} > Mg^{2+}$$

At pH values below 4 the Ca^{2+} binding is almost completely suppressed, so that these types of ion exchanger are suitable for the selective removal of heavy metal ions from waste waters. In general a preloading with Na^+ amounting to half the total exchange capacity is necessary (Cu^{2+} and UO_2^{2+}, vanadium and lead can be absorbed by the iminodiacetic acid type in the hydrogen form).

Negative ions (such as the chlorocomplexes of metals in hydrochloric acid solution) are bound without change by anion exchangers. Since the chloride complexes have differing stabilities, they can be eluted to some extent separately with stepwise varying concentrations of HCl. The chlorocomplexes of zinc and ferric iron can be eluted with water. Only nickel is not taken up by anion exchangers because it does not form a negatively charged complex of sufficient stability with chloride. Noble metals too can be very firmly bound by strongly basic anion exchange resins. The following selectivity obtains for chlorocomplexes at equal concentration [228]:

$$Co^{2+} > Hg^{2+} > Fe^{2+} > Cu^{2+} > Ni^{2+} > Ag^{+} > Au^{+} > Zn^{2+} > Cd^{2+}$$

Elution is performed with 4% caustic soda. Complex cyanides are bound very tightly by anion exchangers and cannot be eluted with NaOH. Weakly basic anion exchangers with a macroporous resin structure are recommended for removal from neutral to alkaline solution (max. pH 8). The chloride form is employed which takes up the complex ions more or less in the order of their stability [228].

Anion exchangers exhibit in general the following range of affinities [107]:

$$\text{Acetate} < F^- < Cl^- < SO_4^{2-} < NO_3^- < CrO_4^{2-}$$

with hydroxide taking up a position on the far left for strongly basic ion exchangers and far to the right for weakly basic ones [107]. The selectivity for H^+ and OH^- is determined by the strength of the acids or bases formed by reaction of the immobilized ions in the exchangers with H^+ or with OH^- ions respectively. Weakly acidic and weakly basic exchangers are selective for H^+ and OH^-, since the acids and bases formed are only slightly dissociated and swell only slightly. The theory has been well reviewed [107]. The basic methodological possibilities for the employment of ion exchangers are shown in Fig. 3.14.

Because of their microporous structure or their chemical constitution ion exchangers are capable, to a certain limited extent, of taking up nonionogenic organic compounds by adsorption or by hydrogen bonding [108].

As has been mentioned their most significant application in waste water purification is decontamination combined with recovery of raw materials, whether it be recovery of rinsing water for recirculation, the recovery of metals from rinsing waters (e.g. gold, silver, coinage metals, heavy metals, chromic acid) or the removal of organic bases (e.g. quinine). The advantages that ion exchange resins offer for such applications are listed in Table 3.11.

The recovery of chromic acid from rinsing water will be described as an example of practical application [108]. When the amounts of rinsing water are small

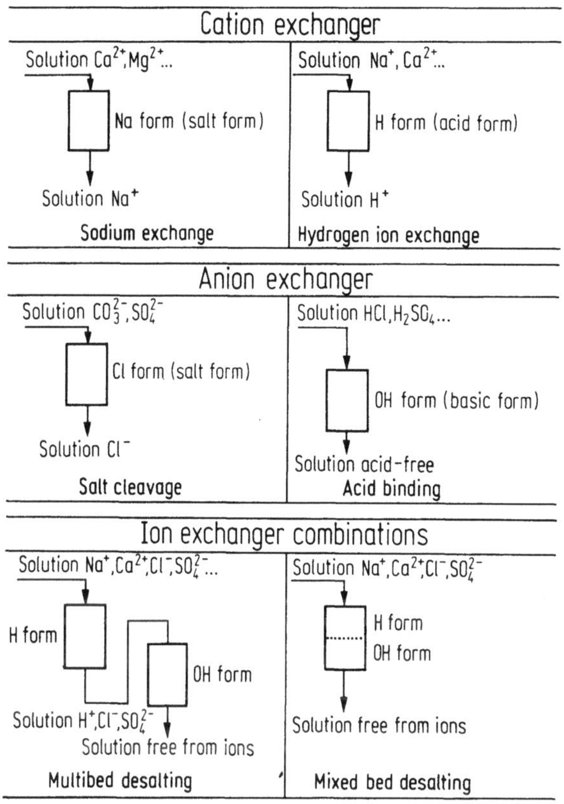

Fig. 3.14. Basic ion exchange processes [105]

Table 3.11. The advantages of ion exchange processes for material recovery [108]

High throughput possibilities
Spherical, highly stable particles and macroporosity allow throughputs of dilute solutions of 20–40 m^3/m^3 resin/hour; specialized resins allow throughputs of up to 100 m^3/m^3 resin/hour. Even for concentrated solutions the throughput can be as high as 5–10 m^3/m^3 resin/hour
High usable capacity
On average 1.0 eq./l for strongly acid cation exchangers 1.5–2.0 eq./l for weakly acid cation exchangers 0.5–0.6 eq./l for strongly basic anion exchangers 1.1–1.2 eq./l for weakly basic anion exchangers
Purity of the eluate

the process is relatively simple. The rinsing water is evaporated. The condensate is re-employed for rinsing, the concentrate added to the chromium electrolyte. In this form, however, the process has the decisive disadvantage if continually enriching the impurities (e.g. Fe^{3+}, Cr^{3+}, calcium compounds etc.) by recycling all the rinsing residues quantitatively. This can be avoided by the pre-employment of a strongly acid ion exchanger (H form,), which binds all the interfering cations. The process of recovery of chromic acid from larger quantities of rinsing water is more complicated. Here the individual stages are first a strongly acid cation exchanger (H form) for the removal of metallic impurities, then a weakly basic anion exchanger (OH form), that binds the chromic acid as chromate. The water is returned to the rinsing circuit. The anion exchanger is regenerated with caustic soda. The initial red-yellow portion of the eluate has a pH of about 7 and contains 50–60 g/l chromate (calculated as CrO_3) [108]. This is led into a further strongly acid cation exchanger (H form), when chromic acid is formed which has an average concentration of 50 g/l in the eluate [108]. The next strongly alkaline yellow fraction contains ever larger concentrations of chromate. It is employed for the pre-regeneration of the next batch. The final wash water, which only contains small amounts of chromate, is fed into the raw water holding vessel [108].

Ionic mercury can be removed from water by the following processes [110]: cementation (i.e. as the metal [74]), sulphide precipitation, extraction with complexing agents and ion exchange. Of these processes ion exchange is gaining more and more attention. The commercially available acid ion exchange resins have the disadvantage of only being able to treat mercury-containing solutions in the almost complete absence of chlorides, since in the presence of chloride the slightly dissociated $HgCl_2$ and in the presence of excess chloride the stable anion complex $[HgCl_4]^{2-}$ are formed which do not interact with the exchanging groups [110]. It has been reported by Oehme [111] that the tetrachloromercuriate complex is sorbed by anion exchangers (styrene/divinylbenzene matrix, quaternary ammonium or tertiary amine groups) in the chloride form, with a saturation concentration of 100–200 g Hg/l resin. The disadvantage of strongly basic ion exchangers is that only up to 33% and often nothing at all can be eluted from them.

The Dutch Akzo Zout Chemical Company has developed a mercury-selective polymeric resin containing free sulfhydryl residues (Imac TMR) [112, 113], for the removal of mercury from the effluents of the Solvay process for the electrolysis of brine and, in principle, for the removal of mercury from other waste waters. The resin binds mercury ions (Hg^{2+} or $[HgCl]^+$) even in saturated solution [112]. In doing this it displaces the equilibrium for the formation of $[HgCl_4]^{2-}$, which is normally completely on the right-hand side, completely over to the left.

$$2\,RSH + Hg^{2+} \rightarrow R{-}SHgS{-}R + 2\,H^+$$

$$RSH + [HgCl]^+ \rightarrow R{-}SHgCl + H^+$$

$$Hg^{2+} + 4\,Cl^- \rightleftarrows [HgCl]^+ + 3\,Cl \rightleftarrows HgCl_2 + 2\,Cl^-$$

$$[HgCl_3]^- + Cl^- \rightleftarrows [HgCl_4]^{2-}$$

The capacity of the resin is 240 g Hg/l resin [110]. It can be regenerated using concentrated hydrochloric acid.

One specific application area of strongly acid and strongly basic ion exchangers ought to be mentioned briefly; their employment as catalysts, which can be used to prevent the formation of unwanted products and hence contribute right from the start to effluent purity. The same reactions can be catalysed as are catalysed by H^+ and OH^- ions in homogeneous phase, e.g. hydrolysis and esterification, the elimination of water and hydration, the aldol reaction, condensation, oligomerization and polymerization, rearrangements, alkylations etc. [107]; there are, however, some advantages over homogeneous catalysis [107].

The development of exchange resins is progressing towards the development of more selective exchangers. Chelating resins (cation exchangers incorporating complex-forming groups such as iminodiacetic acid) are being employed more and more in specific applications.

Finally, there are the liquid ion exchangers (e.g. amines, alkylphosphoric acids), one of whose applications is the regeneration of HCl-containing pickling acids (the removal of enriched iron salts). The process is an extractive one, which removes the material to be extracted by reaction with the extracting phase. Certain secondary amines, e.g. N-dodecenyl-N-trialkylmethylamine [114] can be employed to remove ferric ions. The "carrier" is not a polymer in this case but the diluent (xylene is favoured). The ion exchange takes place very rapidly at the interface between the two liquid phases (in a water-free droplet). The selectivity is determined by the nature of the functional and substituting groups and of the diluent. The removal of iron salts from spent HCl pickling acids [114] first necessitates the oxidation of the Fe^{2+} with chlorine or some other agent, when the following reactions occur:

$$FeCl_2 + 1/2\,Cl_2 \rightarrow FeCl_3$$
$$FeCl_3 + HCl \rightarrow HFeCl_4$$
$$HFeCl_4 \rightarrow H^+ + [FeCl_4]^-$$
$$R_2NH + HCl \rightarrow R_2NH \cdot HCl$$

The next step is the extraction of the Fe^{3+} by the secondary amine. The amine is finally regenerated by treatment with water.

$$R_2NH \cdot HCl + FeCl_3 \rightarrow R_2NH_2FeCl_4$$
(in concentrated HCl)
$$R_2NH_2FeCl_4 \xrightarrow{(H_2O)} R_2NH \cdot HCl + FeCl_3$$

3.3.4 Reduction Processes

Reductive processes are employed to solve some specific waste water purification and detoxification problems. Base metals and inorganic or organic reducing agents, such as SO_2, sodium sulphite and hydrogen sulphite, sodium thiosulphate, sodium sulphide, ferrous sulphate, $NaBH_4$ or formaldehyde, are employed. Electrolytic reduction is possible (e.g. the cathodic reduction of chromate to Cr^{III}, but it is very cost-intensive. For economic reasons the base metal usually employed is iron, in order to avoid significant additional pollution of the waste water. The unused dissolved Fe^{II} can be atmospherically oxidized afterwards to ferric iron which is precipitated [115]. When other reducing agents are employed provision must be made for their removal. Organic reductants, such as formaldehyde or formic acid, can be employed to advantage if a biological purification step is to follow. Reductive processes are employed to a limited extent for the removal of metallic ions, mainly chromate, and further for specific purposes such as the reduction of free chlorine to chloride by thiosulphate, the reduction of nitro compounds with nascent hydrogen or by iron filings. The laboratory and industrial destruction of nitrite by reductive reaction with sulfamic acid is also known; the nitrite nitrogen is transformed into elementary nitrogen [116]:

$$NO_2^- + NH_2SO_3H \rightarrow HSO_4^- + N_2 + H_2O$$

Because of the heat of reaction concentrates have to be diluted 1:1. The necessary acidification is normally achieved by the reagent itself, exceptionally waste sulphuric acid can be added to reduce the pH to 3.8 [116]: When this pH has been achieved the destruction of nitrite is complete without the intervention of any side reactions to speak of, so that neutralization can begin.

Peroxides and peroxy acids can also be easily destroyed by the employment of reducing agents.

$$ROOH + NaHSO_3 \rightarrow ROH + NaHSO_4$$
$$ROOH + Fe^{2+} \rightarrow ROH + Fe^{3+}$$
$$H_2O_2 + H_2SO_3 \rightarrow H_2O + H_2SO_4$$

The most important reduction process that is employed in the laboratory and in industry alike is the transformation of hexavalent chromium to the trivalent form, this will be described in a little more detail. In order to decontaminate waters from dissolved chromates, they must first be transformed to the trivalent state. Hydrogen sulphite solution is usually employed below pH 2.5 [1]:

$$4\,CrO_3 + 6\,NaHSO_3 + 3\,H_2SO_4 \rightarrow 2\,Cr_2(SO_4)_3 + 6\,H_2O + 3\,Na_2SO_4$$

As can be seen from Fig. 3.15, the reduction is dependent on pH and time. Gaseous SO_2 can be formed as a by-product, so that the reaction must be performed in an efficient fume cupboard or the chromate reduction bath must be equipped with exhaust facilities. If large quantities of chromate are to be destroyed then it becomes economical to employ SO_2 itself as the reagent. Ferrous sulphate has also been sucessfully employed in the stationary detoxification process:

$$2\,CrO_3 + 6\,FeSO_4 + 6\,H_2SO_4 \rightarrow Cr_2(SO_4)_3 + 3\,Fe_2(SO_4)_3 + 6\,H_2O$$

The chromium^{III} produced remains in solution and is precipitated by the addition of alkali after the reduction. The sulphur compounds most usually employed for the detoxification of chromic acid and chromosulphuric acid (sulphite, SO_2, dithionite) all have the disadvantage that malodorous lower sulphur compounds are formed in the reaction. More recent processes for the reduction of chromic and chromosulphuric acids exploit the reaction with hydrogen peroxide [117]. As is well known hydrogen peroxide can exhibit both oxidative and reductive properties; if a system with

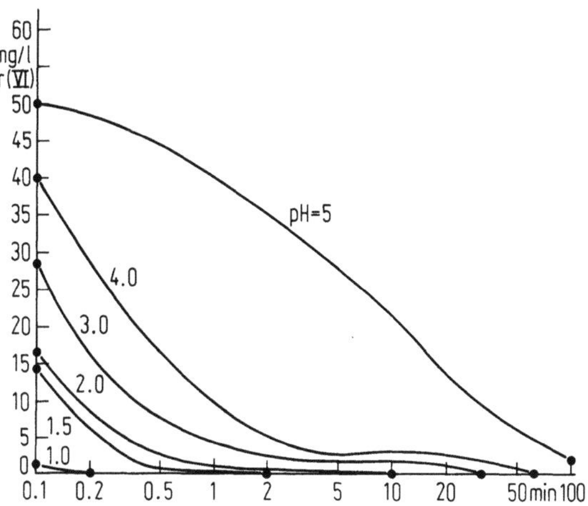

Fig. 3.15. Rate of reduction of chromium (III) compounds as a function of pH [1]

a higher redox potential is present it can be employed for the reductive destruction of oxidizing agents, when it is itself oxidized to water and oxygen.

As an oxidizing agent H_2O_2 has a redox potential of +1.77 volt:

$$H_2O_2 + 2\,H^+ + 2\,e^- \rightarrow 2\,H_2O$$

As a reducing agent H_2O_2 has a redox potential of +0.68 volt:

$$H_2O_2 \rightarrow O_2 + 2\,H^+ + 2\,e^-$$

If 2 mol H_2O_2 in sulphuric acid (2 mol H_2SO_4 not necessary for chromosulphuric acid) are reacted with 1 mol chromic acid, then firstly an intermediate deep blue coloured chromium pentoxide is formed that rapidly loses oxygen to yield the green to violet (hydration isomerism!) chromium (III) salt:

$$Cr_2O_7^{2-} + 4\,H_2O_2 + 2\,H^+ \rightarrow 2\,CrO_5 + 5\,H_2O$$

Overall reaction

$$Cr_2O_7^{2-} + 3\,H_2O_2 + 8\,H^+ \rightarrow 2\,Cr^{3+} + 3\,O_2 + 7\,H_2O$$

Since an alkaline precipitation has to follow the reaction, the quantitative removal of the excess H_2O_2 is absolutely essential in order to obviate the re-oxidation of the Cr^{III}. The hydrogen peroxide can be destroyed catalytically with active carbon, suitable metal or by reaction with manganese dioxide with the formation of water and oxygen and a bivalent derivative. The detoxification of chromate waste with hydrogen peroxide is employed industrially both for the stationary detoxification of concentrates and for the decontamination of apparatus and work pieces in the metal processing industry by immersion. A decontaminating dip of the following composition has proved valuable: water 900 cm^3, 50 cm^3 35 % w.w. H_2O_2 and 50 cm^3 concentrated sulphuric acid (for chromosulphuric acid no additional acid need be employed!). This solution is relatively stable even when it has been used. When reducing large amount of concentrates the process is only economic below pH 1. The necessary amounts of hydrogen peroxide (1.7 cm^3 35 % w.w. per gram CrO_3) may only be added in small aliquots because the reaction is exothermic. If the solutions to be destroyed contain more than 30 g CrO_3/l then the reaction should be carried out in several batches or in parallel. It should be noted that excess H_2O_2 can effect the re-oxidation of the chromium^{III} to chromate after the neutralization step; this does not usually occur because the reaction takes several hours.

In some cases electrolytic methods can be employed for the reductive treatment or removal of harmful and toxic substances from waste waters, if the salt concentration is high enough (1–5 g/l) to ensure sufficient conductivity [43]. The topic has been thoroughly reviewed with a large number of examples [118].

Alongside anodic oxidation of unwanted substances, which will be dealt with later, the second basic electrochemical method is the cathodic reduction or removal of toxic heavy metals, particularly Hg, Cd, Pb, Cu, Zn, Cr etc. It is a method of recovering valuable metals. Exhausted photographic laboratory fixing baths contain 3–6 g/l silver for example, that can be recovered chemically, or more elegantly, electrochemically. The chemical recovery process is very complicated, labour-intensive and malodorous (H_2S!), so that electolytic methods for the removal of the silver are superior [119]. The necessary currents and electrolysis times can be calculated from the silver content of the starting solution. Electrolytic processes for the removal of metals are interesting for the future because lower residual metal concentrations are realizable than is the case for hydroxide precipitation. The levels required for leading the water into the drains cannot be achieved by hydroxide precipitation. However, the elctrolytic removal of metals is only economic if high volume-time yields can be achieved. This is made possible by the employment of large specific electrode surfaces, i.e. with solid bed or fluidized bed electrodes [120]. Pilot scale investigations are under way.

3.3.5 Oxidation Processes

Oxidation processes are important wet chemical detoxification processes for many classes of compound in aqueous solution and also for chemical wastes in nonaqueous solution, whereby a whole range of oxidizing agents finds application. The spectrum ranges from the frequently observed nonspecific decomposition to the directed oxidation of a particular portion of a molecule or the formation of precisely defined products. However, chemical oxidation is costly because of the large amounts of relatively expensive reagents that are required. The aim is either the complete mineralization or, if that is not possible, the conversion into partial oxidation products which are nontoxic or can be more easily eliminated from the waste waters in further (biological for instance) processes. Complete oxidation, which has the advantage that no by-products are produced that require further treatment with the exception of a solution of mineral salts which may in some circumstances be reusable, can only rarely be performed purely chemically, because either the necessary reaction times are too long and/or the oxidizing agents are too expensive or materials are present that are very resistant to oxidation. Oxidizing agents such as potassium permanganate or chromate, which form polluting products, should be avoided on the industrial scale, this, however, does not necessarily apply to some laboratory scale oxidations. The application of chemical oxidation processes has therefore concentrated on the following reagents: oxygen (particularly at high temperatures or in combination with γ-radiation (see p. 254), ozone, hydrogen peroxide, chlorine or hypochlorite solution (possibly in combination with UV irradiation (see p. 236, 245, 249). Other oxidizing agents have, in comparison, a very limited importance. The course of many of these reactions is extremely complicated and their mechanisms have not been satisfactorily elucidated. For instance, concentrated nitric acid, chromic acid and chromosulphuric acid do not generally produce defined reaction products but uncharacterized mixtures. The nitrating properties of nitric acid should be remembered, particularly for aromatic compounds. Chromosulphuric acid contains the carcinogenic potassium dichromate and should therefore be avoided as far as possible.

3.3.5.1 Waste Water Incineration

The incineration of waste water [121] achieves the complete oxidation of the waste water components to relatively innocuous products. The main criterion for deciding whether incineration is possible is a solids content of $\geqq 10\%$ by weight. Just as with liquid chemical wastes waters with a calorific value of 2000–3000 kcal/kg (8374 to 12560 kJ/kg) can be directly combusted with a burner, while waste waters of lower calorific value are either injected into the oil or solvent-fired combustion chamber or burned as a mixture with waste solvent at 800°–1200 °C with a multipurpose burner.

3.3.5.2 Wet Oxidation

By wet oxidation is meant the decomposition of organic compounds with atmospheric oxygen or pure oxygen at 150°–370 °C and 10–220 bar. Extensive decomposition occurs yielding CO_2 and water as the final products. With problematical materials flue gas scrubbing is necessary. Depending on their chemical structures organic waste water components are decomposable to differing degrees at the temperatures employed of between 100° and 370 °C. Extending the reaction time by 1/2 hour has scarcely any effect on the result [122–125].

Local authority sewage sludge may be taken as an example; the following rates of decomposition are obtained:

175 °C	200 °C	225 °C	250 °C	300 °C
10–15%	35–45%	65–70%	75–80%	90–95%

The basic methodological possibilities are the low pressure process (below 250 °C; partial decomposition) and the high pressure process (over 250 °C and higher pressure; almost complete decomposition), so that the process can be performed using a wide range of temperatures and pressures [126], e.g.:

- as high pressure oxidation with air to complete oxidation of all organic components or the partial destruction of the organic components,

— as medium pressure oxidation with oxygen leading to complete destruction of all organic material or its partial destruction,
— as medium pressure oxidation with air in the presence of catalysts leading to partial oxidation of the organic content.

Regarding the mechanism, it is reported that the organic substance is first carbonized [127, 128]. The dissolved oxygen reacts catalytically on the surface of the carbon so as to yield hydrogen peroxide which then decomposes to form oxygen and hydroxyl radicals. These radicals then react with the carbon to yield carbon dioxide (catalytic autoxidation). Accordingly substances that do not carbonize below 300 °C (such as acetic acid) are not degraded by wet oxidation, in contrast to substances that carbonize readily (such as sugars) which are very completely degraded. Another investigation indicates that acetic acid is oxidized very slowly above 300 °C [129].

Examples of wet oxidation in industrial practice are the treatment of sulphite liquors and local authority sewage sludge. The organic materials (e.g. proteins, lipids, carbohydrates, starches, cellulose) are degraded, high molecular weight substances are first cleaved to lower molecular weight substances, then progressively to carbon dioxide and water. At high temperatures only very stable hydrolysis and oxidation products remain (e.g. acetic acid, simple amino acids). Apart from the treatment of waste liquors from the paper and pulp industries experiments have been made with cyanide-containing wastes, with the purification of refinery rinsing waters and waste disposal with simultaneous chromium recovery [122]. At the present time wet oxidation is also being tested for the purification of effluents containing large amounts of organic pollution, such as those produced for example in the manufacture of acrylonitrile [122]. Some of the principal materials and effluents that can be treated by wet oxidation are listed in Table 3.12. No cases are known as yet of the commissioning of large scale wet oxidation plants for industrial waste water purification.

Table 3.12. Constituents of aqueous industrial effluents degradable by wet oxidation (selected examples) [122]

Type of material	Pressure (bar)	Temperature (°C)	Conversion rate (%)
Phenol	40	200	95
2,4-Dimethylphenol	210	275	99
2-Chlorophenol	210	275	99
4-Nitrophenol	210	275	99
Methanol	40	200	77
Formaldehyde	40	200	93
Refinery effluents			
Cyanides and nitriles	10	180	99
			99.7
Dyestuffs			
Nitrocellulose and other explosive materials			
Effluents from paraffin oxidation			

In general wet oxidation of effluents can be recommended where their combustion would be uneconomical, but when they are, on the other hand, too polluted to be treated in a biological treatment plant. Amongst the factors determining the economics of the process are whether or not reutilization of inorganic constituents is possible and whether the energy necessary for the process can be derived from the water purification process [131]. For example, a plant for the wet oxidation of organically polluted black soda liquor from the production of wood pulp, that was started up in Tasmania in 1966, has been the subject of a report [130]. Oxidation at 300 °C and 200 bar yields an almost organic-free sodium carbonate solution, which can be recycled as soda liquor after caustification. The Zimpro process [132] is recommended for the degradation of DDT and of alkyline cyanide from electroplating effluents.

Further specific applications are described in the patent literature. Thus the presence of adsorbents (preferably active carbon) is reported to lower the reaction temperature. The regeneration of loaded active carbons from waste water purification under conditions of mild wet oxidation is an example of the exploitation of this phenomenon [122].

Reports have also been made of investigations into the catalytic effect of added metal ions on the reaction; copper ions, in particular, have a significant catalytic effect. Thus the copper and ammonium ion-catalysed wet oxidation of cyanide-containing waste waters has been described, which would, however, be uneconomic in practice because of the expensive or not yet realizable removal of the copper ions afterwards. Mn-V catalysts and activated chromium compounds have also been investigated [126]. The lower limit for the calorific value of the effluent to be treated is at least 100 kcal/kg (419 kJ/kg). The possibility of catalyst poisoning causes problems, as does the inadequate oxidation of suspended particles leading to "catalyst clogging" and the very corrosive effects particularly of salt-containing effluents. When mixed effluents are treated it must be expected that not all the organic components will be oxidized and that organic residues will remain, not all of which will be biodegradable.

A specific catalyst system based on acidic solutions of bromide, nitrate and manganese ions, having a broad spectrum of effectivity, has been developed in the USA. For example, Atrazine, butyl phthalate, chloroaniline, diphenyl hydrazine, ethylene dibromide, Malathion and pentachlorophenol are degraded to the extent of more than 60% at temperatures of 165°–200 °C with reaction times of less than one hour. Other compounds (e.g. acetonitrile, chloroanthracene, DDT, hexachlorobutadiene, nitrobenzene, trichloropropane, TCDD) require higher temperatures and longer reaction times for less degradation [229].

In contrast to the wet oxidation described so far the Katox process [27] consists of catalytically induced oxidation employing particular contact media whose optimal effectivity is at pH 7. The rate of reaction is directly dependent on the concentration of dissolved oxygene, so that the salt concentration of the solution treated should not be too high since this lowers the solubility of the oxygen. Oxygen-enriched air or ozonized oxygen can also be employed as oxidizing agents instead of atmospheric oxygen. The process is independent of temperature, fluctuations in the amount of pollution and intermittent operation and is reported to remove cyanide, sulphide, nitrite and other poisonous contaminants to the extent of 90% [27, 133]. The

Katox precipitation process is a further development of this method [134], which follows the catalytic oxidation by a directed chemical precipitation. The catalytic oxidation takes place in three tanks laid out in series with a supply of air and thorough mixing of added active carbon, impregnated active carbon or of peat coke. The residence time is normally 4 hours. The oxidizable constituents of the water are bathed in oxygen-containing water and oxidized at the surface (in particular at the internal surface) of the catalyst; the products obtained are then desorbed (see also Fig. 3.16). The reaction method has not yet been elucidated and biological processes can also occur. The catalyst can be de-activated by Van der Waals adsorption of colloids or disperse water constituents, by chemisorption and by coating with inactive materials (see Fig. 3.16). However, it is not necessary to interrupt the process, since the constant turbulence caused by the introduction of the air continually regenerates the catalyst particles by abrasion and breakdown. In the second stage the unoxidized dissolved and colloidal materials are removed by neutralization or by the addition of precipitating or flocculating agents. The typical areas of application are highly organically polluted waters not suitable for biological purification. The total costs lie, depending on the composition of the water and the amount of purification desired, between 0.70 and 1.00 DM per m^3 effluent [134]. The flotation of active carbon mixtures is also employed for catalysing the atmospheric oxidation of cyanides, nitrites and hydrazines [135]. More than 80% of these substances can be removed from the waste water in this manner. The remainder is then detoxified by the use of chlorine. For example, nitrite (up to 125 mg NO_2/l) can be reduced to a concentration of less than 2 mg/l between pH 6 and 7 by the action of atmospheric oxygen followed by chlorine.

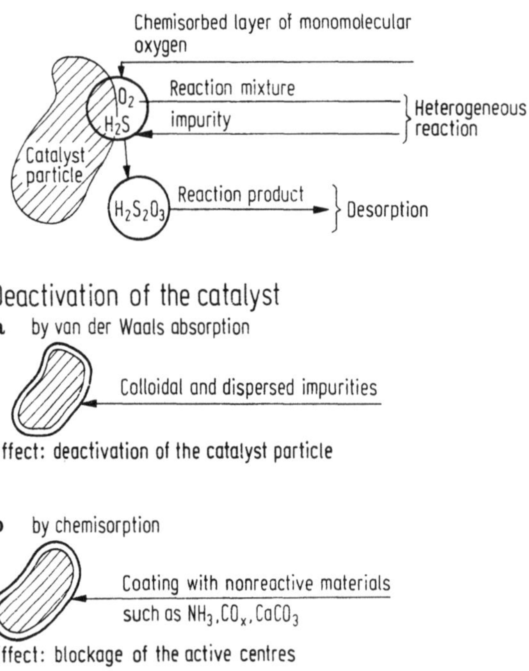

Fig. 3.16. Model for the catalytic oxidation of waste water pollutants [134]

3.3.5.3 Chlorinating Oxidizing Processes

In principle there is no difference whether a chlorinating oxidizing water treatment is performed with active chlorine in the form of bleach (hypochlorite) or of chlorine gas (from gas cylinders or electrolytically generated), since hypochlorite is formed initially in both cases.

$$Cl_2 + 2\,OH^- \rightleftharpoons Cl^- + OCl^- + H_2O$$

The basis of every chemical chlorination of substances in water is a series of chemical equilibria, which depend on the pH of the solution and its temperature [136]:

$$Cl_2 + H_2O \rightarrow HClO + H^+ + Cl^-$$

$$HClO \rightarrow H^+ + ClO^-$$

$$ClO^- + H_2O \rightarrow HClO + OH^-$$

$$Cl_2 + 2\,OH^- \rightarrow ClO^- + Cl^- + H_2O$$

It can be seen from Fig. 3.17 that at pH 8 and 20 °C 33 % of the chlorine compounds are present in the form of a highly reactive hypochloric acid, on reducing the pH to 7 at the same temperature then 80 % is present in this form [136]. That is, the higher the pH the more the equilibrium is displaced in the direction of the less active hypochlorite anion. The hypochloric acid present decomposes slowly with the formation of hydrochloric acid and nascent oxygen,

$$HClO \rightarrow HCl + O$$

which is the basis of the oxidizing action of aqueous chlorine and hypochlorite. Alongside this some chloric acid is also formed depending on the pH.

$$3\,HClO \rightarrow HClO_3 + 2\,HCl$$

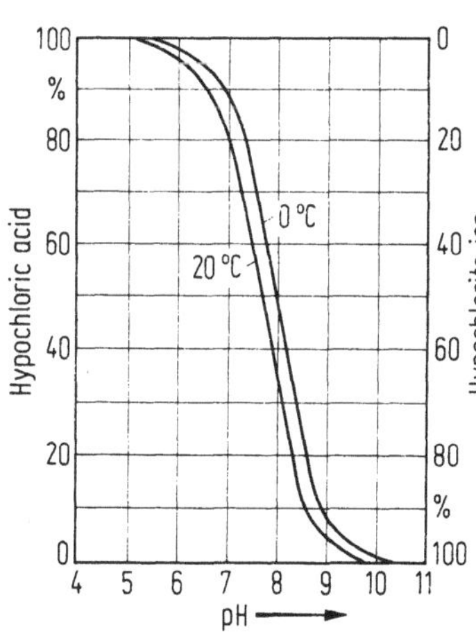

Fig. 3.17. Proportions of hypochlorite anion and hypochloric acid in water at 0 °C and 20 °C at various pH values [134]

When hypochlorite is employed in the form of the calcium salt dichlorine monoxide can also be formed.

$$Ca(ClO)_2 + H_2O \rightarrow Ca(OH)_2 + Cl_2O$$

$$Cl_2O + H_2O \rightarrow 2\ HClO$$

Alongside its oxidizing effect on chemical compounds chlorine is also an excellent disinfecting agent; this effect is also a result of the oxidizing action of hypochloric acid. This acts either by chlorination of organic compounds or by inhibiting enzymes by the addition of chlorine to double bonds [136].

When treating water and waste water with chlorine it should be borne in mind that chlorine impedes the access of oxygen, that corrosive HCl is produced and that any biological degradation is impossible at less than pH 5.5. Furthermore the decision to employ chlorine for the treatment of aqueous wastes and waste waters also depends on the type of compounds present, since chlorination can occur as an unwanted side reaction. Thus, the chlorination of phenol yields toxic chlorophenols which smell and taste up to a 1000 times more powerfully [136]. Carcinogenic waste water constituents such as polycyclic aromatics are also scarcely likely to be rendered harmless by chlorination. Several authors have established that the chlorination of 3,4-benzpyrene yields products, some of which at least are carcinogenic. 5-Chloro-3,4-benzpyrene and 3,4-benzpyrene-5,8-quinone were identified. Other authors, however, believe that 3,4-benzpyrene can be degraded to the extent of 70%–90% with the formation of innocuous decomposition products by reaction with either chlorine gas or ClO_2 [136].

The chlorination of waters containing ammonia leads to the formation of chloramines, which then participate in the reaction scheme. Ammonium compounds which can form the highly explosive nitrogen chloride on reaction with excess chlorine are examples of incompatible substances.

$$3\ Cl_2 + NH_4Cl \rightarrow NCl_3 + 4\ HCl$$

Excess detoxifying agents must themselves be detoxified after the reaction is complete, e.g. with sodium thiosulphate or more elegantly with H_2O_2. Active chlorine is employed industrially for cyanoxidation and for the treatment of sulphur compounds (mercaptans to disulphides, sulphides to sulphones, disulphides to sulphonic acids or sulphonyl chlorides). For example dissolved H_2S in waste waters can be oxidized to sulphur or sulphuric acid depending on the dosage of chlorine or hypochlorite.

Alongside alkaline solutions of chlorine, various bleaching solutions are employed (Javelle water, potassium hypochlorite solution and bleach, sodium hypochlorite solution) with active chlorine contents of 12%–15%, as well as calcium hypochlorite and chloride of lime. Chloride of lime has the additional advantage when employed for detoxification that is releases chlorine in acid solutions. Particular care is necessary with application of dry chloride of lime, which bursts into flames on reaction with mustard gas and countless other substances. The use of a high percentage of chloride of lime results in the formation of a whole range of toxic substances from mustard gas (from chlorinated sulphides and sulphoxides to chloral, chloroform and their hydrolysis products), however they do not possess the character of weapon gases. Nitrogen mustards only react with hypochlorites to form the highly toxic N-chloramines, so that this cannot be considered as a detoxification reaction.

Because of its reactivity, when chlorine dioxide, ClO_2, is employed it is prepared on site; there are two possible routes, both starting from 24% sodium hypochlorite solution. This is reacted either with hydrochloric acid or with chlorine gas.

$$5\,NaClO_2 + 4\,HCl \rightarrow 4\,ClO_2 + 5\,NaCl + 2\,H_2O$$

$$2\,NaClO_2 + Cl_2 \rightarrow 2\,NaCl + 2\,ClO_2$$

Electrolytic generation is also possible, as is production by the reduction of chlorates. Chlorine dioxide is employed in some instances for oxidative detoxification. Thus, in the pH range 5–9 sulphides can be oxidized to sulphates; careful control of the reaction conditions avoids the formation of colloidal sulphur. Simple cyanides are oxidized to cyanates in the neutral to weakly alkaline range and above pH 10 to CO_2 and N_2. Between pH 5 and 9 mercaptans react to form sulphonic acids or sulphonates and malodorous amines are deodorized within the same pH range. In the neutral to weakly alkaline range phenol in oxidized to benzoquinone and above pH 10 to mixtures of lower nonaromatic carboxylic acids [137]. A recent Degussa process employs chlorine dioxide synthesized from sodium chlorite and formaldehyde for the oxidation of effluents containing 100 to 1000 ppm phenol, neither phenol nor formaldehyde are detectable after the reaction has been performed. This process ought to be particularly valuable for the synthetic resin industry, since the effluents already contain formaldehyde [138]. Its main area of application to date is in bleaching, in the disinfection and removal of unpleasant flavours from potable waters. In general it is employed in excess [137].

One of the advantages of the employment of chlorine dioxide, alongside its high oxidizing activity, is in some cases the fact that some organic compounds, that are chlorinated by chlorine to yield potentially carcinogenic substances, are stable towards ClO_2 [220, 221]. This also reduces the consumption of oxidizing agent as compared to chlorine. There are, however, possible risks when ClO_2 comes into contact with other substances, for instance ClO_2 reacts explosively with elementary sulphur. Sodium thiosulphate (4.6 kg for 1 kg 100% chlorine dioxide) can be used for the detoxification of ClO_2.

Alkali and alkaline earth chlorites (particularly sodium chlorite) have also found employment recently in acid medium for the oxidation of mono and polyphenols and hydrocarbons, which are rapidly and quantitatively destroyed [139].

The oxidation of cyanide with active chlorine ranks amongst the classical detoxification processes. In general it is sufficient to oxidize the cyanide ions to cyanate and then to release the water into the treatment plant for further processing. If the effluent is to be passed directly into the drains further reaction to CO_2 and N_2 is necessary, since the cyanate is known to be toxic to aqueous life (fish food) [140]. Oxidation with chlorine or hypochlorites is of great technical importance for both variants. At pH $\geqq$ 10 the first stage involves reaction to form toxic cyanogen chloride and some cyanate directly [141]:

$$CN^- + OCl^- + H_2O \rightarrow CNCl + 2\,OH^-$$

This reaction takes place rapidly and almost independently of pH and can be followed with suitable electrodes (e.g. gold versus calomel) by means of the decrease in cyanide and finally the first excess of active chlorine and can be automated with a control system [100]. In acid solution HCN and cyanogen chloride would be evolved, while immediate hydrolysis to cyanate is guaranteed above pH 12 (see Fig. 3.18) [142]. This reaction step is strongly pH-dependent; small amounte of hypochlorite effect a catalytic acceleration.

$$CNCl + 2\,OH^- \rightarrow CNO^- + Cl^- + H_2O$$

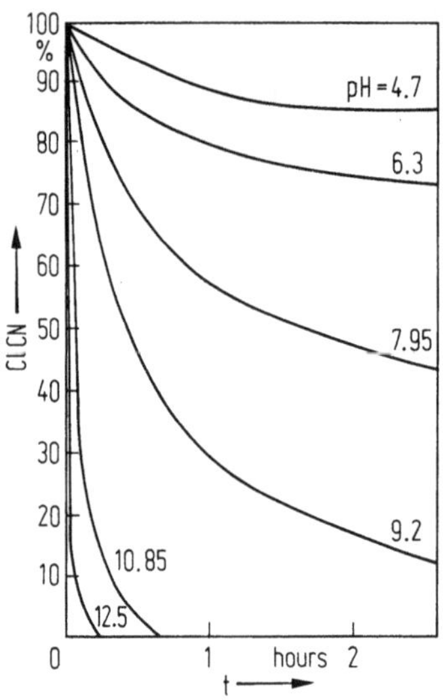

Fig. 3.18. Dependence on pH of the hydrolysis rate of cyanogen chloride [142]

The necessary hydrolysis time for the quantitative conversion of the cyanogen chloride to cyanate after the addition of hypochlorite has been reported by Stumm et al. for various pH values [142]: pH 8 = 20 h, pH 9 = 12 h, pH 10 = 4 h, pH 11 = 1 h, pH 12 = 1/4 h. The effect of hypochlorite is only an oxidation in the formal sense since the cyanate is actually produced by the hydrolysis of the intermediate cyanogen chloride. This hydrolysis of the cyanogen chloride is electrode-inactive. The limiting cyanide concentrations can be readily controlled by circulation through a conductivity measuring device [100]. Since the cyanogen chloride formation proceeds exothermically, it is necessary to limit active chlorine oxidation to solutions containing a maximum of 1000 mg/l CN^- for continuous technical scale processes [100]. At higher concentrations temperatures are achieved which expel the intermediate cyanogen chloride from the solution. The problem can be avoided by continuous addition of water. Some authors [141] are of the opinion that other detoxification methods should be chosen for higher concentrations, since the addition of water increases the costs significantly and requires considerably larger reaction plant. On the laboratory and technical scale the above concentration limit can be exceeded 10-fold [100], if the detoxifying agent is added suitably slowly. The use of efficient fume removal apparatus is essential. The cyanate solutions obtained must be adjusted to a pH of below 9 before being washed down the drain [142], the acid used should not be too concentrated and should be added slowly with efficient mixing to avoid the release of ammonia from the cyanate. Cyanate hydrolyses under neutral to acid conditions to ammonium hydrogen carbonate, which finally decomposes to carbon dioxide and an ammonium salt.

$$CNO^- + H_2O \rightarrow (NH_4)HCO_3$$

If total oxidation is necessary, the cyanate solution obtained is acidified and further oxidized at pH 4–6 [100] by the addition of hypochlorite to yield N_2 and CO_2, which, of course, are the preferred products from the environmental point of view.

$$2\,OCN^- + 3\,ClO^- + 2\,H^+ \rightarrow N_2 + 2\,CO_2 + 3\,Cl^- + H_2O$$

Thus in acid medium further hypochloric acid is formed, which releases nascent oxygen as the oxidizing agent. The reaction occurs very slowly and only goes to completion in the presence of an excess of hypochlorite [142]. The simplified equations for the total reaction emphasize that 2.5 times the stoichiometric amount of oxidizing agent is required for the total oxidation as for the partial oxidation.

$$1\text{ mol } CN^- + 1\text{ mol } OCl^- \rightarrow 1\text{ mol } CNO^- + 1\text{ mol } Cl^-$$

$$2\text{ mol } CN^- + 5\text{ mol } OCl^- \rightarrow 1\text{ mol } N_2 + 2\text{ mol } CO_2$$

Von Beckerath et al. point out that, in the acid region, the estimatable free chlorine in bleach solution varies between 12% and 14% and is all effective, which cannot be automatically assumed for alkaline medium [143]. Investigations at pH 12 indicated amongst other things that the quantitative oxidation of 2.55 g/l CN^- requires stoichiometrically 3.4765 g chlorine which would amount to 25.37 cm^3 of the 13.7% bleach solution employed; in practice, however, 97 cm^3 of this bleach solution were required. This demonstrates that, in practice, a 13.7% bleach solution only liberates 3.62% free chlorine for reaction at pH 12 [143]. Therefore, the amount of hypochlorite required for a particular detoxification reaction should previously be determined experimentally.

Stumm et al. have published detoxification guidelines for cyanide concentrations up to 1000 mg/l which are valid for laboratory and technical scale detoxification [142].

When waste water contains heavy metal ions — as in electroplating effluents — then complex cyanide ions of varying stability are formed. Such complex cyanides are not detected potentiometrically, so that the aparent cyanide concentration sinks. Although the reaction velocity for free cyanide is independent of pH value or chlorine or hypochlorite excess, a dependence has been established for complex cyanides of heavy metals [141]. In general the opinion is held that active chlorine only attacks free cyanide, so that this must be continually produced from complexes by complex dissociation [100]. This means that the reaction times are too long to be of practical value in the case of the more stable complexes; on the other hand, the reaction times of the less stable complexes approach those of the free cyanides.

Thus, Zn and Cd complexes are oxidized within a few minutes with a slight excess of active chlorine. For copper cyanide a reaction time of 20–30 minutes is necessary with a chlorine excess. Silver, gold and nickel cyanides are more difficultly oxidizable. In the literature it is reported that the oxidation of weaker complexes is sometimes slower than that of stronger complexes [144]. Thus, according to their stability constants, copper is the more stable and nickel the more labile, although the practical reaction times give the opposite result. This emphasizes that a distinction must be made between a stationary (tabulated) stability constant and a dynamic stability constant (effective in the reaction). This question has not been completely clarified. If ferrocyanates are treated with hypochlorite it should be remembered that the oxidation leads primarily to the stable ferricyanide complex. When carrying out the detoxification of cyanides with active chlorine it should be remembered that the effluents concerned often contain other pollutants that hinder the detoxification or even render it impossible. Such substances include, for instance, aldehyde-bisulphite adducts or gelatine [145]. In such cases the alteration of the reaction conditions does not lead to the expected success. The detoxification of cyanides may be used as an example for the citation of the disadvantages of oxidation with active chlorine. Alongside the formation of the toxic cyanogen chloride that has already been discussed they are:

- The formation of an equivalent amount of chloride ion causes an additional salinization of the waste water (e.g. just the oxidation of 1 kg CN^- to cyanate produces 5 kg of NaCl from the bleaching solution).
- The excess hypochlorite must be destroyed after the oxidation is complete.
- The hypochlorite solution used cannot be stored.

Chlorine acts as a significantly more powerful oxidizing agent when chlorine and short wavelength UV radiation are employed simultaneously [146] (see p. 227).

Various chloramines are employed for specific detoxification processes, particularly for 2,2′-dichloroethyl sulphide. The divalent organic sulphide is thereby converted to the tetravalent state, with the formation of sulphimine compounds. The cost of chloramines is too high for their widespread practical application, so that only small quantities can be detoxified and their application can only be considered in a catastrophe or militarily. Here chloramines are mainly employed for decontaminating the skin. The presence of incompatible substances should also be excluded prior to the employment of chloramines. They include NH_3, NH_4^+, urea and similar NH compounds.

3.3.5.4 Oxidation with Potassium Permanganate

Potassium permanganate has been known for a very long time as an oxidizing agent possessing certain advantages (easy to handle technically and in dosing, odourless and tasteless). The manganese dioxide formed as a result of the oxidation can also be effective in promoting flocculation. The high fish toxicity (maximum concentration ca. 5 mg/l) is a significant disadvantage of permanganate. The maximum concentration for manganese dioxide is reported as 1.3 g/l. The cost of potassium permanganate often inhibits its employment industrially; on the other hand, it remains perfectly practical for many laboratory detoxifications (e.g. cyanide, low molecular weight sulphur compounds etc.). The reactions are strongly dependent on pH. In principle, permanganate can undergo two redox reactions, with the acceptance of 5 electrons below pH 3 to form the Mn(II) ion and between pH 3 and 11.5 with the acceptance of 3 electrons and the formation of insoluble manganese dioxide.

The oxidation of cyanide will again be taken as an example [147]. There is no uniform mechanism for this, since, depending on the pH, differing reactions yielding various products take place. Since hydrocyanic acid is a weak acid below pH 9 the whole of the cyanide is present as HCN. Below pH 5 there is practically no reaction between cyanide and permanganate which leads to the conclusion that hydrocyanic acid does not react with permanganate. As the OH^- concentration increases various reactions begin to occur. From pH 6 to 9 cyanogen is formed quantitatively, while between pH 9 and 12 several reactions occur simultaneously leading to cyanogen, cyanate and carbon dioxide. Above pH 12 cyanate is produced quantitatively.

$$2\,MnO_4^- + CN^- + 2\,OH^- \xrightarrow{pH>12} 2\,MnO_4^{2-} + CNO^- + H_2O$$

The oxidation of cyanides by permangantes has to be accelerated by the addition of catalytic quantities of copper ions. The addition of 3 mg Cu^{2+}/l brings about the oxidative reduction of 20–50 mg CN^-/l to 0.1 mg/l in about 10 minutes under laboratory conditions, and in the same time to 1.4–0.6 mg/l in industrial waste water containing a great deal of iron (complex iron cyanides are not attacked by potassium permanganate) [147]. The reaction of permanganate with cyanides and, in particular, its behaviour towards complex cyanides is not fully understood.

3.3.5.5 Oxidation with Hydrogen Peroxide and with per Compounds

Hydrogen peroxide has a special status amongst oxidative detoxifying agents, since it also possesses reducing properties alongside its powerful oxidizing effect and further the perhyrdoxyl ions so formed in aqueous solution can lead to the rapid hydrolysis of some pollutants. Its application, particularly for the oxidative destruction of toxic waste chemicals, to the detoxification of aqueous effluents, to disinfection and deodorization can lead to environmentally satisfactory solutions to a multitude of problems [148, 149]. It is an advantage of H_2O_2 that it exhibits high reactivity over a wide pH range, that it does not produce toxic reaction products, increase the salt load of the water or affect its pH. In addition it is significantly more stable than other oxidizing agents. The reaction can be satisfactorily controlled and regulated by means of redox electrodes, pH electrodes and suitable controllers. The potential "corrosive effect" of unused or excess H_2O_2 is only a short-term danger, since water impurities rapidly decompose the compound. Its relatively high price is a limiting factor for its large-scale industrial application. H_2O_2 is employed on the laboratory, technical and industrial scale especially for the effective detoxification and deodorization of a wide range of concentrated solutions such as fixing, developing and clearing baths, cyanide wastes, hardening salt wastes (cyanide and nitrite), formaldehyde-containing water and other substance. It is employed industrially for the treatment of aqueous effluents containing cyanide and large quantities of other dissolved organic substances, which are simultaneously oxidized [150–152]. A wide range of inorganic and organic sulphur compounds, which are distinguished by their noxious odours, can be elegantly immobilized by treatment with H_2O_2. Even sulphur dioxide is directly and quantitatively oxidized to sulphuric acid [150]. Certain problems concerning recycling can also be solved by the application of H_2O_2. Thus silver dissolved as its thiosulphate complex is produced in photographic laboratories and can be precipitated with H_2O_2 as an insoluble mixture of silver oxide, bromide and sulphide [153, 154]. In the disposal of industrial wastes H_2O_2 is employed for the intensive supply of oxygen to activated sludge plants and for the treatment of foamy sludges. Investigations are also being made into the application of Fenton's reagent, consisting of a mixture of H_2O_2 and ferrous salt, that energetically oxidizes many organic substances as a consequence of the production of hydroxyl radicals.

$$Fe^{2+} + H_2O_2 \rightarrow Fe^{3+} + OH^- + {}^{\cdot}OH$$

The ferric ions can either react slowly with the formation of hydroperoxide radicals

$$Fe^{3+} + H_2O_2 \rightarrow Fe^{2+} + H^+ + \dot{O}OH$$

or they can be reduced again by organic intermediates. At the moment this method is restricted in its application to specific cases of detoxification [155]. There are good reviews of its application to oxidative waste water treatment [156, 231].

Since hydrogen peroxide reacts rapidly and sometimes with vigorous decomposition with various impurities, particularly with metals and their salts, with reducing agents and easily oxidable organic compounds, used and contaminated residues must on no account be returned to the stock of pure compound. H_2O_2 affects the skin and mucous membranes (nose, throat and eyes) "corrosively". Hence protection such as goggles and rubber gloves should be worn when handling it. Spilled H_2O_2 and residues should be washed down the drain immediately with a great deal of water. It should be noted that H_2O_2 can react with various substances (e.g. aliphatic ketones) to produce explosive peroxides as by-products.

Organic peroxides are almost never employed for detoxification purposes [157]. Inorganic peroxides such as the alkali metal and alkaline earth peroxides, urea peroxide and peroxyphosphoric and sulphuric acids, on the other hand, find technical application. However, only permonosulphuric acid (Caro's acid) and persulphuric acid and their salts have a certain industrial importance. In the future it is to be expected that urea peroxide will gain in importance for disinfection as will peracetic application. However, only permonosulphuric acid (Caro's acid) and persulphuric acid and their salts have a certain industrial importance. In the future it is to be expected that urea peroxide will gain in importance for disinfection as well peracetic acid [153], which exhibits a broad antimicrobial spectrum. Ion exchangers can also be disinfected (water bacteria, metabolic and degradation products) economically by means of highly dilute peracetic acid solutions.

Cyanide Oxidation

In contrast to active carbon hydrogen peroxide oxidizes cyanide ion directly to cyanate in an exothermic reaction, so that there are no dangers from toxic intermediates; the cyanate can be hydrolysed by raising the temperature:

$$CN^- + H_2O_2 \rightarrow CNO^- + H_2O$$

However, the rate of cyanate production is too slow for practical application but can be significantly increased by the addition of heavy metal catalysts, e.g. 200 mg $CuSO_4$/l solution [100]. (In contrast the presence of copper during the hypochlorite oxidation of cyanide carries with it the danger that insoluble copper cyanide will be precipitated before the last free cyanide is reacted and thus escape adequate detoxification [158].) The reaction takes place, in principle, over a wide pH range from 3–12, with an optimal rate between pH 4 and 5. The reason that, in practice, a pH of 10–11 [158] is employed is because the 50% equilibrium for the reaction

$$KCN + H_2O \leftrightharpoons KOH + HCN$$

is at pH 9.1; this means that 50% of the cyanide is present as free hydrocyanic acid at this pH. The formation of ammonia as a by-product must be expected in the alkaline range:

$$CNO^- + 2\,H_2O_2 \rightarrow NH_3 + CO_2 + OH^-$$

H_2O_2 is generally employed in excess in a stationary process, when the excess reagent is catalytically decomposed at the surface of precipitated heavy metal hydroxides. The oxygen bubbles formed in this process aid the expulsion of the ammonia formed in the basic range.

Cyanide concentrations of between 500 and 2500 mg/l are reported to be detoxifiable. Concentrations of less than 500 mg/l require reaction times that are too protracted. The method is recommended for the detoxification of small volumes of cyanide concentrates, which allow dilution to the recommended concentration range. At a cyanide concentration of 2000 mg/l and at 25 °C a reaction time of 1 hour has been recommended. At higher concentrations and at pH values below 10 considerable quantities of HCN are released even at 25 °C [100]. For the decomposition of higher concentrations of cyanide the addition of 10–20 cm^3 waterglass per litre solutions is recommended to protect the H_2O_2 from unwanted partial decomposition. According to the literature heavy metal complexes are also attacked (to varying extents however) [100, 158]. The oxidation of cyanide by hydrogen peroxide is significantly temperature-dependent (see Fig. 3.19); even at 10 °C reaction times occur which prohibit its employment [100]. In practice hydrogen peroxide is added until cyanide can no longer be detected. The amount required is 2 mol 100% H_2O_2 per mol CN^- and at low cyanide concentration (< 1 g/l) 3 mol. However the measurement method (e.g. silver against calomel electrodes) does not — as is the case with hypochlorite oxidation — indicate the presence of excess oxidizing agent, which is why stationary treatment is recommended with excess peroxide. Hydrogen peroxide also interferes with the usual photometric methods for the determination of cyanide. Oehme et al. discovered that the end point of the reaction could be determined readily by the temperature increase (thermodiode) caused by the catalytic destruction of the excess H_2O_2 (see Fig. 3.20) [100]. The ammonia released with the exhaust gases should be neutralized in a gas wash bottle or, on the technical scale, led through an absorber containing sulphuric acid; the excess sulphuric acid can then be employed for the neutralization of the alkaline reaction solution after the oxidation is complete.

The disadvantage of the H_2O_2 process for its widescale industrial application is that it only proceeds at a satisfactory rate in the presence of cupric ions and that 25%–45% of the cyanide is transformed directly into ammonia, which is just as toxic and also forms stable amine complexes with the cupric ion. Copper salts also have the disadvantage that they catalyse the destruction of H_2O_2, so that the consumption of the oxidizing agent can be considerably increased. Recently, however, a new — but still secret — catalyst indicator system has been developed that is claimed to ease the technical application of H_2O_2 detoxification, particularly for hardening salts [159, 160].

Hardening salts contain nitrites and barium salts as well as cyanides. Hypochlorite solutions are only suitable for the detoxification of highly dilute solutions, since an acid medium is necessary for a satisfactory reaction time, so that not just the release of HCN from the cyanide portion but also the formation of toxic oxides of

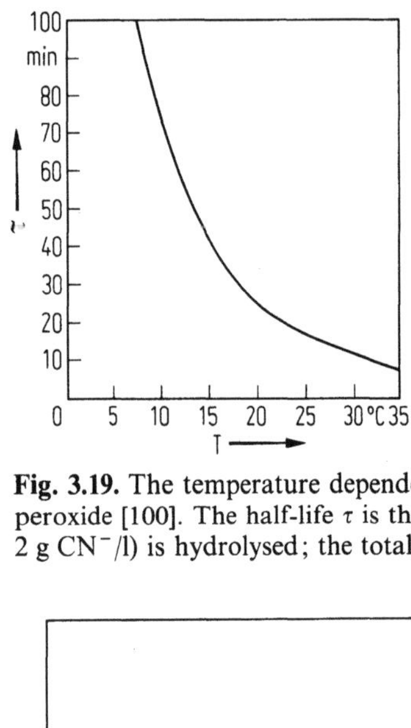

Fig. 3.19. The temperature dependence of the reaction rate of cyanide oxidation by hydrogen peroxide [100]. The half-life τ is the time in which half of the cyanide originally present (here 2 g CN^-/l) is hydrolysed; the total time of reaction is approximately 3τ)

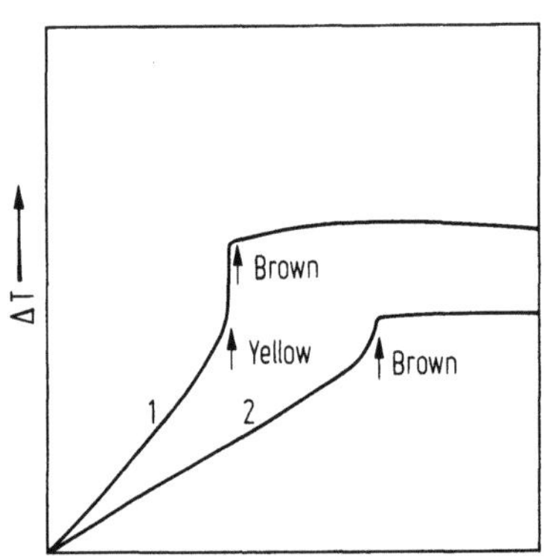

Fig. 3.20. Variation in temperature during the oxidation of cyanide by hydrogen peroxide. The total increase in temperature during the reaction amounts to ca. 80 °C. At the end of the reaction the commencement of the decomposition of excess H_2O_2 is signalled by an obvious temperature increase. Conditions: Cyanide concentration 2 g CN^-/l; 1. Addition of 3.5 cm^3 30% peroxide and 50 mg Cu^{2+}/l as catalyst 2. Addition of 2 cm^3 and 5 mg. The temperatures were measured by means of a thermodiode

nitrogen has to be expected. The employment of hydrogen peroxide produces more favourable results [161]. Not only is the cyanide oxidized, but the nitrite is quantitatively converted to nitrate, along with this, reaction of nitrite with cyanide leads to the formation of some nitrogen. It is possible to work at a uniform pH, even though the optimum pH for the oxidation of nitrite lies between pH 4 and 5. The hardening salts are first dissolved to a concentration of 50 g/l (calculated as CN^-) in water and H_2O_2 is added gradually at pH 11.5–12, so that the temperature does not rise above 75 °C. As usual the course of the reaction is monitored with pH and redox electrodes. After all the oxidizing agent has been added the solution is allowed to stand for 1/2 hour. The total reaction time amounts to 3 hours. For each 1 kg cyanide 2.5 kg of 35% by weight H_2O_2 are consumed, while each 1 kg nitrite (calculated as NO_2) consumes a further 2 kg H_2O_2. After the reaction is completed the alkaline solution must be neutralized, at best with sulphuric acid, since then insoluble barium sulphate is formed. The sludge must be disposed of at a sterile site since micro-organisms posses the ability to transform $BaSO_4$ into soluble species.

The detoxification of cyanides with Caro's acid (and persulphuric acid) also leads to the formation of cyanate:

$$CN^- + SO_5^{2-} \rightarrow OCN^- + SO_4^{2-}$$

$$CN^- + S_2O_8^{2-} + 2\,OH^- \rightarrow OCN^- + 2\,SO_4^{2-} + H_2O$$

The reaction occurs sufficiently rapidly at pH 9–10 at room temperature and all the cyanides and complexes attacked by chlorine are also oxidized by Caro's acid [158]. Catalytic quantities of copper ions once again have a pronounced acceleration effect. Dosage and reaction can be monitored by redox measurements. However there are also disadvatages in comparison with hydrogen peroxide; apart from the high price there is also the increase in the salt load of the water, this time with sulphate. The amount employed is also substantially larger than is the case for the hydrogen peroxide detoxification (each kg CN^- requires 5.85 kg $KHSO_5$ or 9.16 kg $Na_2S_2O_8$ in comparison to 1.3 kg H_2O_2). Additionally as an acidic oxidizing agent Caro's acid greatly reduces the quality of the effluent water, which can even lead to displacement into the acid range with highly concentrated cyanides [158] (evolution of HCN!). For these reasons Caro's acid should only be employed for the detoxification of highly dilute cyanide wastes.

Inorganic and Organic Sulphur Compounds

Most inorganic sulphur compounds — from elementary sulphur to sulphides, sulphur oxyacids and hydrogen sulphide — can be oxidized to sulphate by H_2O_2. Even sulphur dioxide emissions, which are causing pollution of the biosphere on a worldwide scale, can be quantitatively and rapidly removed by hydrogen peroxide in special scrubbers. The sulphuric acid so formed can be recycled to the production process. In spite of this the process is too expensive for industrial application. Table 3.13 presents a review of the reactions of inorganic sulphur compounds with H_2O_2. Amongst the technically important oxidations of sulphur compounds employing H_2O_2 is the detoxification of photographic chemical wastes (fixing, developing and

Table 3.13. The oxidation of inorganic sulphur compounds by hydrogen peroxide

Sulphur compound	Conditions	Reaction/product	Remarks
Sulphur S_8	pH > 12	mainly polysulphides and thiosulphates	further oxidation possible catalysable by ferric ion
Sulphides S^{2-}	pH < 7, catalyst (10–20 ppm Fe^{3+}) 50°–60 °C pH > 8	mainly sulphur, but also polysulphides, thiosulphate etc. $S^{2-} + H_2O_2 \rightarrow S + 2\,OH^-$ formation of sulphate $S^{2-} + 4\,H_2O_2 \rightarrow SO_4^{2-} + 4\,H_2O$	reaction in seconds: further oxidation possible reaction in minutes (mol ratio 1:4); precipitation of sulphur occurs at a ratio of 1:1.
Polysulphides S_x^{2-}	pH > 8	formation of sulphate e.g. $S_3^{2-} + 10\,H_2O_2 + 4\,OH^- \rightarrow 3\,SO_4^{2-} + 12\,H_2O$	The quantity of H_2O_2 required depends on the no. of S Atoms in the polysulphide. Below pH 8 polysulphides decompose to sulphur and sulphide.
Sulphite SO_3^{2-}	pH ≦ 8	formation of sulphate $SO_3^{2-} + H_2O_2 - SO_4^{2-} + H_2O$	reaction time: minutes (reaction times are too long for practical purposes above pH 8)
Thiosulphate $S_2O_3^{2-}$	pH > 4 < 7 catalyst (Mo, W, Ti, V or Zr salts) pH > 7	formation of tetrathionate $2\,S_2O_3^{2-} + H_2O_2 + 2\,H^+ \rightarrow S_4O_6^{2-} + 2\,H_2O$ sulphate formation with catalyst formation of sulphate $S_2O_3^{2-} + 3\,H_2O_2 + 2\,OH^- \rightarrow 2\,SO_4^{2-} + 5\,H_2O$	instantaneous reaction in strong alkalis (15–20% NaOH) and at room temperature
Dithionite $S_2O_4^{2-}$	pH > 7	formation of sulphate $S_2O_4^{2-} + 3\,H_2O_2 + 2\,OH^- \rightarrow 2\,SO_4^{2-} + 4\,H_2O$	Dithionate is obtained as a by-product in acid medium
Dithionate $S_2O_6^{2-}$	temperature increase primary hydrolysis	primary hydrolysis to sulphite and sulphate, further oxidation of sulphite to sulphate $S_2O_6^{2-} + 2\,OH^- \rightarrow SO_3^{2-} + SO_4^{2-} + H_2O$ $SO_3^{2-} \xrightarrow[-H_2O]{+H_2O_2} SO_4^{2-}$	Dithionate is difficult to oxidize, hence a primary hydrolysis is necessary.

Table 3.13. (continued)

Sulphur compound	Conditions	Reaction/product	Remarks
Poly-thionates $S_xO_6^{2-}$	pH > 7	formation of sulphate $S_3O_6^{2-} + 4\,H_2O_2 + 4\,OH^- \rightarrow 3\,SO_4^{2-} + 6\,H_2O$ $S_5O_6^{2-} + 10\,H_2O_2 + 8\,OH^- \rightarrow 5\,SO_4^{2-} + 14\,H_2O$	The H_2O_2 requirement depends on the polythionate. Polythionates are relatively stable at pH = 7
Hydrogen sulphide	wash solution containing NaOH, H_2O_2, catalyst (0.1–0.5 g green vitriol); finally neutralize	formation of sulphate $H_2S + 4\,H_2O_2 \rightarrow H_2SO_4 + 4\,H_2O$ $\downarrow + 2\,NaOH$ $Na_2SO_4 + 2\,H_2O$	A trickle tower (packed) can be employed for the reaction chamber. The gas and scrubbing liquid are contacted in countercurrent, the scrubbing liquid is recycled. Each mol of H_2S requires 4 mol H_2O_2 (100%) and 2 mol NaOH.

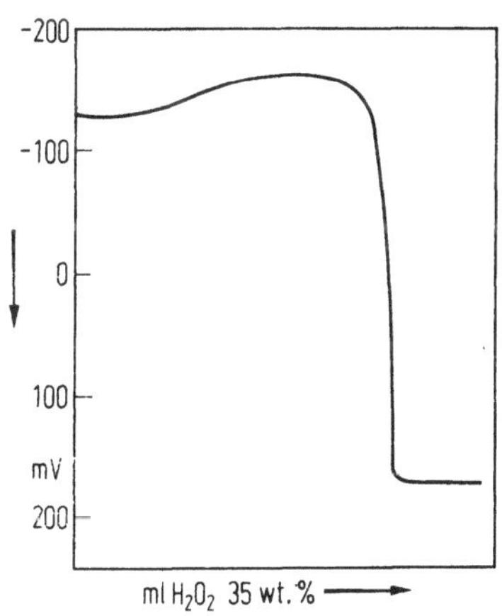

Fig. 3.21. Titration (detoxification) of a fixing bath with hydrogen peroxide [154]

clearing baths) [154]. Fixing baths contain up to 200 g/l thiosulphate as their major component, together with varying quantities of sulphite and bisulphite as well as silver at the soluble thiosulphate complex. The bath solutions that may have been depleted of silver by electrolysis are treated under cooling (exothermic reaction) with hydrogen peroxide in an alkaline medium. The reaction is monitored potentiometrically with a platinum or gold electrode against calomel (see Fig. 3.21). The thiosulphate is oxidized via tetrathionate with the liberation of hydrogen ions which have to be neutralized by alkali. Since the reaction is exothermic cooling must be employed.

Table 3.14. The oxidation of organic sulphur compounds with hydrogen peroxide

Compound class	Conditions	Reaction/product	Remarks
Thiols (mercaptans)	alkaline, ca. 50 acid, ca 50 °C catalyst (iron salt or molybdate)	$2\,RSH + H_2O_2 - R{-}S{-}S{-}R$ (disulphide) $RSH + 3\,H_2O_2 + OH^- \rightarrow RSO_3^- + 4\,H_2O$ (sulphonic acids)	via the intermediate disulphide, reaction time: a few seconds (neutral to weakly alkaline ca. 1 minute)
Thioethers (sulphides)	pH 2–3 ca 50 °C catalyst (iron salt) Temp. 50 °C catalyst (molybdates, zirconium salts, H^+ or OH^- ions)	$R{-}S{-}R + H_2O_2 \rightarrow R{-}S(=O){-}R + H_2O$ (sulphoxides) $R{-}S{-}R + H_2O_2 \rightarrow R{-}S(=O)_2{-}R + H_2O$ (sulphones)	reaction time: a few minutes reaction time: minutes to hours
Disulphide	acid, ca. 50 °C catalyst (iron salt or molybdate)	$R{-}S{-}S{-}R + 5\,H_2O_2 + 2\,OH^- \rightarrow 2\,RSO_3^- + 6\,H_2O$	reaction time: a few seconds

H_2O_2 also destroys the silver thiosulphate complex, causing Ag_2O, AgCl, AgBr and Ag_2S to precipitate. The sulphite in developing and clearing baths is destroyed in an analogous manner.

The reaction of hydrogen peroxide with organosulphur compounds not only brings about the desired detoxification but also offers the possibility of the production of a range of products for secondary application [162]. Table 3.14 reviews the basic reactions of organosulphur waste water constituents. By far the most important reactions of organic compounds to be considered in this context are the oxidations of aliphatic and cyclic thioethers. Reaction to form sulphoxides is generally performed in acetone or glacial acetic acid or in alcohols or in the absence of solvent at about 50 °C and can be accelerated by the addition of acids or iron salts — which is not usually necessary. When higher temperatures are employed or if catalysts are added (molybdenum compounds, zirconium salts, hydrogen ions, hydroxyl ions) the oxidation proceeds via the sulphoxides to the sulphones. When performing detoxification reactions it should be borne in mind that not all organosulphur compounds are transformed into less toxic substances by oxidation. The formation of mustard oils on the hypochlorite or hydrogen peroxide oxidation of dithiocarbamates should be mentioned:

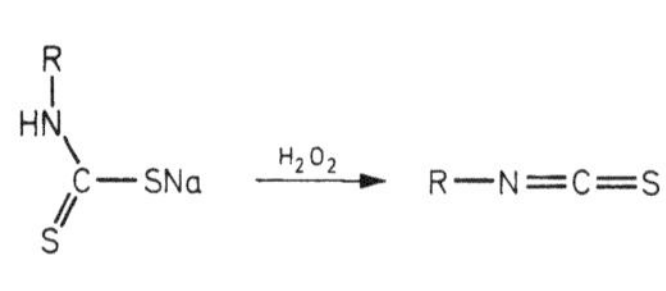

The multiplicity of detoxifications that can be accomplished employing H_2O_2 must not be allowed to lead to its unjustified employment for every group of substances. Before H_2O_2 is employed it should be checked what products can be produced and with what properties. For example, mixtures of alcohols, ketones, aldehydes and carbohydrates can explode, particularly with concentrated hydrogen peroxide solutions. Alkali and alkaline earth metals and their compounds form peroxides, which would then have to be destroyed. Toxic products can be formed under certain circumstances; thus the reaction of H_2O_2 with chloroform in sulphuric acid — particularly in the presence of iron salts — leads not only to the formation of HCl but also of the highly toxic phosgene. In the same way the reaction of organonitrogen compounds with H_2O_2 and derivatives does not invariably result in detoxification [162].

Active Chlorine Compounds

Many chemical processes produce hypochlorite-containing effluents, that may not be led into the drains or biological purification processes without pretreatment. The decomposition disadvantage of increasing the salt load of the waste water with sulphates. Additionally the pH is altered considerably; overdosage with SO_2, can, in the acid range, lead to the release of gaseous chlorine. Oxidation with hydrogen peroxide, which takes place very rapidly in neutral to strongly alkaline solution even in the presence of large quantities of salt, is to be preferred, (the acidic range should be avoided because of the danger of liberating free chlorine).

$$NaOCl + H_2O_2 \rightarrow H_2O + O_2 + NaCl$$

The reaction can be monitored and controlled by following the redox potential [163].

Formaldehyde-containing Waters

The exothermic oxidation of formaldehyde by H_2O_2 in alkaline medium at room temperature leads, with the formation of hydrogen, to sodium formate and water [164].

$$2\,CH_2O + H_2O_2 + 2\,NaOH \rightarrow 2\,HCOONa + 2\,H_2O + H_2$$

It forms the basis of one process for the detoxification of formaldehyde-containing effluents. The effluent is practically free from formaldehyde after only 30 minutes of reaction. In order to reduce the H_2O_2 requirement to about one third, the effluent is first treated at elevated temperature with milk of lime or caustic soda.

Phenols

As in the oxidation of sulphur compounds with H_2O_2, it has been shown that the oxidation of phenols proceeds at an accetable rate when catalysed (by iron salts such as green vitriol, copper or bismuth salts, aluminosilicates doped with noble metals or iron itself) and under the influence of UV radiation. Mixed catalysts consisting of iron and aluminium salts are advantageous for highly dilute phenol solutions. If residual phenol concentrations of <1 ppm are required, then it is necessary to employ 2–3 mol H_2O_2 (100%) per mol phenol on both the laboratory and the industrial scale. The optimal pH in agueous solution is 3–4. In practice, however, an initial pH of 5–6 is chosen, since acids are formed during the reaction which lower the pH to 2–3. The reaction is only slightly temperature-dependent, a range between 20° and 50 °C is recommended [165]. At higher temperatures the H_2O_2 itself begins to decompose.

The first stage of the reaction is the formation of multihydroxylated phenol bodies, which are then oxidized to the corresponding quinones. If excess H_2O_2 is employed, then ring cleavage occurs with the formation of carboxylic acid and possibly even decarboxylation via the peracid. Phenol is unaffected after one hour at a molar ratio of 1:3, while 100% degradation occurs in the same time in the presence of iron salts. Similar results are obtained with substituted phenols, e.g. 2-chlorophenol without catalyst 9% (with catalyst 100%); 2,4-dichlorophenol 0% (100%); pentachlorophenol 0% (100%); p-cresol 6% (100%); 2-nitrophenol 36% (100%); 2,5-dinitrophenol 10% (73%); β-naphthol 9.5% (100%) and 2,4-dimethylphenol 10% (100%) [166].

3.3.5.6 *Oxidation with Ozone*

Ozone is a powerful oxidizing agent, that can react with many inorganic and organic compounds and also possesses a sterilizing action. With a redox potential of +2.07 volts (25 °C) it is one of the most positive substances in the electrochemical series and is only surpassed by fluorine with 2.85 volts (25 °C) (cf. chlorine 1.36 volts).

$$O_3 + 2\,H^+ + 2\,e^- \rightarrow H_2O + O_2$$

Particularly in the presence of moisture, it oxidizes all metals, with the exception of gold, platinum and iridium, to their highest oxidation state oxides. Hydrogen halides, with the exception of hydrogen fluoride, are also easily oxidized. The oxidation of cyanides and complex cyanides is once again of practical significance. Ozone attacks very many organic compounds, of which the reactions with phenols, sulphur compounds and dyestuffs are of particular importance in the destruction of chemicals and the purification of waste waters.

As far as we know there are no reviews concerning the reaction mechanism and the degradation products of pollutants in aqueous solution, even though the degradation of a range of substances has been studied in detail [167, 168].

- Phenol: Is first converted to catechol, which is then oxidized to o-quinone; further degradation yields primarily glyoxalic acid, CO_2, formic acid and oxalic acid. The elimination of phenol and its oxidation products requires 4–6 mol ozone per mol of phenol [169].
- Chlorinated hydrocarbon pesticides: A quantitative reaction was observed for Aldrin and heptachlor. However, the primary oxidation products — Dieldrin and heptachlor epoxide are not harmless. The other pesticides react to a small extent or not at all; this is attributable to the diminution of the double-bond character by substitution with chlorine.

— Cyclohexanol: Reacts with the formation of malonic, glutaric, adipic and oxalic acids.
— Chlorophenols: The aromatic ring is destroyed and the organically bound chlorine transformed into chloride. The most important products are formic acid, oxalic acid, mesoxalic acid semialdehyde, H_2O_2 and CO_2. From 3.5 to 5 mol of ozone are necessary for the complete degradation of one mol chlorophenol. The products of oxidation are biologically degradable.
— p-Toluenesulphonic acid: Here methyl glyoxal, pyrotartaric acid and acetic acid are formed.
— Lignin: The aromatic ring is the main site of attack here, resulting in the formation of carboxylic acid residues; ring cleavage occurs mainly at C atoms bearing oxygen functions.
— Quinoline: Catalytic ozone treatment yields organic acids.
— Urea (in the treatment of swimming pool water): Is converted to nitric acid, CO_2 and water.
— Most carboxylic acids and alkanes and also pyridine are oxidized only with difficulty or not at all [169].
— Parathion: 3 ppm ozone remove 80 ppb, but with the formation of the more toxic Paraoxon, which requires 5 ppm ozone for its degradation; among the other oxidation products are 2,4-dinitrophenol, sulphuric and phosphoric acids.
— Ozone can destroy the mutagenicity of polycyclic aromatics and aromatic amines, whereas it has no effect on that of alkylating agents or nitro and nitroso compounds. The ozonization of hydrazines yielded mutagenic intermediates, which, however, are very sensitive to base-catalysed hydrolysis [170].

The rate of reaction of the degradation of substances by ozone ranges from a few seconds (e.g. olefinic double bonds, dimethyl and trimethylamines) to minutes (e.g. formate, amino acids, methylamine, phenol, chlorophenols, anisole, highly methylated benzenes) up to a few hours (e.g. trichloroethylene, toluene, aldehydes). Compounds such as chloroform, tetrachloroethylene, benzene, chlorobenzenes, aliphatic alcohols, carboxylic acids and NH_3 react even more slowly [169]. In general reaction with ozone first lowers the COD demand. After the molecule has been oxidized to a certain degree, the cleavage of CO_2 begins and the TOC content of the effluent begins to fall. A complete oxidation to CO_2 would be uneconomical however, since some individual degradation products such as acetic acid, do not react any further with ozone; and again it is sufficient to pre-oxidize compounds, that are difficultly biodegradable (this process must by no means be taken as far as acetic acid), with ozone in order to make them accessible for biological degradation. Fluorochlorocarbons (CCl_3F and CCl_2F_2) and carbon tetrachloride are resistant to oxidation by ozone — because of this they are employed as solvents in organic reactions. When working with chlorinated hydrocarbons (such as chloroform) it should be noted that long-term exposure to ozone can release free halogen and possibly lead to the formation of phosgene. It should also be noted that paraffinic hyd*n*ocarbons (e.g. pentane) can dissolve ozone at —78 °C with formation of a blue coloration, but that these solutions are liable to detonate when concentrated. Ozone can also be adsorbed by silica gel at —78 °C with the formation of a deep blue coloration. The desorption of oxygen-free ozone takes place on gentle warming and passage of argon or nitrogen.

Like H_2O_2 ozone has the advantage as a detoxifying agent that further salinization of the effluent water is avoided. Most oxidation products are nontoxic. Furthermore the reaction product O_2 leads to an enrichment of the waste water with oxygen, which has been found to be advantageous for subsequent biological purification processes. Ozone treatment is not yet one of the standard purification methods, but offers an attractive alternative for some industrial and laboratory problems. At the moment it is mainly used on a large scale for disinfection, (e.g. swimming pool water), for the purification of air and drinking water, for the preparation of process water and for the regeneration of used oxidizing agents (e.g. permanganates). Reviews have been published of the applications and pilot-scale investigations in the field of industrial waste water treament [171, 167]. Some examples are:

- Cyanide-containing effluents, preferably in alkaline medium and under pressure, in the ideal case to CO_2 and N_2.
- Silver-containing photographic effluents [172].
- The removal of residual phenol from biologically treated effluents.
- Sulphide or sulphite-containing waste waters.
- Effluents from nitration processes (nitro compounds).
- Acetone-containing waste waters.
- Effluents from the lacquer, paint and plastic processing industries and from the rubber industry.
- Effluents containing detergents [173].

The high reactivity and oxidizing power of ozone limit its employment to particular situations, because — since almost everything is oxidized — highly polluted waters require too great a quantity of ozone and hence of energy. Treatment with ozone is economic for the final polishing purification of mechanically, chemically and biologically pretreated waste waters; that is for the final oxidation of trace quantities, in combination with precipitation, flocculation, adsorption or flotation for example [174]. It is an additional bonus that, as well as its oxidative effect, it also promotes flocculation (destabilization of colloids by reducing their negative charge).

Ozone is most economically produced by the effect of a silent electrical discharge on oxygen, when normal commercial ozonizers produce ozone concentrations of up to 10%. Alongside the production of ozone in an apparatus of appropriate dimensions, which is unlikely to produce difficulties, care must be taken to ensure adequate absorption of the ozone by the aqueous solution being treated. Various types of equipment are employed to ensure that the ozone-air mixture is distributed throughout the solution in the form of the very smallest bubbles. Ozone absorption takes place significantly better at elevated pressures [175]. Research facilities and high pressure processes have been described [176, 177].

Ozone is effective at various pH values of the solution being treated [176]. The mechanism of ozone attack at low pH is, however, not adequately understood [174]. In acid solution it is presumably the ozone molecule itself that is responsible for the oxidation. At alkaline pH the attack takes place via a mechanism where the actual oxidation is performed by primarily produced hydroxyl radicals [178, 179]. Sometimes, therefore, $Ca(OH)_2$ is employed as a catalyst [167]. Treatment with ozone in the presence of $Ca(OH)_2$ is reported to make a particularly efficient removal of organic matter possible. It was possible, for example, to reduce the O_3 consumption per gram TOC by almost a half in the aqueous effluent from a paper and cellulose factory by the addition of lime [169]. Sometimes it is possible to improve the oxidation with ozone by combining it with UV irradiation. Presumably the UV radiation excites the material photochemically so that it is more easily oxidized by ozone. Thus, for example, the oxidation of acetic acid with ozone is rapid under UV irradiation, while in the normal course of events oxidation scarcely occurs [169]. It would only be possible to make CO_2 the major product of ozone treatment by employing a totally uneconomic excess of O_3, also the complete reaction of unsaturated or otherwise activated carbon compounds would slow down tremendously when the stage of low molecular weight carboxylic acids was reached.

Because of its toxicity all work with ozone must be carried out under proper ventilation. The analytical control of ozone oxidations is performed by monitoring the ozone consumption, i.e. a comparison is made between the ozone content of the gas before and after reaction. Modern instruments determine the ozone content of the gas flows electrochemically. To do this an electrolyte (H_2SO_4) in a measuring cell is saturated with ozone by a constant flow of gas and the limiting diffusion current for the cathodic reduction of O_3 at constant electrode potential is measured at a platinum electrode. This parameter is a direct measure of the partial pressure of ozone in the gas mixture. This measurement technique can be employed for the automatic control of ozone purification.

Not every reaction of ozone leads to a nontoxic product. For example, the reaction of ozonized air with haloalkyl amines (e.g. nitrogen mustards) results in the formation of the N oxides which exhibit considerable toxicity. The ozonization of amines in solution or their adsorption on silica gel yields the corresponding nitro compounds [180]. Highly toxic epoxides and peroxides can also be formed in some cases.

Some of the more important oxidations performed with ozone will now be described:

Inorganic Cyanides

According to Fabjahn and Davies [181] ozone reacts rapidly and quantitatively with simple cyanides in slightly alkaline aqueous solution to yield the cyanate, which is then further oxidized by ozone or can decompose hydrolytically to carbonic acid and ammonia. The nascent oxygen released during the reaction may also participate and contribute to the oxidation of the cyanide anion, so that the amount of ozone consumed falls below the stoichiometric amount for cyanide (the total consumption is significantly greater, however, since ozone attacks all the other impurities present too). There is no uniform agreement in the literature concerning the kinetics of the reaction.

$$CN^- + O_3 \rightarrow OCN^- + O_2$$

$$2\,OCN^- + H_2O + 3\,O_3 \rightarrow 2\,HCO_3^- + 3\,O_2 + N_2$$

Heavy metal cyanide complexes can also be oxidized with ozone, whereby if the pH is kept optimal the central metal ion is simultaneously quantitatively precipitated as the oxide or the hydroxide. However, the economic disadvantage of an increased ozone consumption in comparison to the oxidatibn of simple cyanides must be mentioned; this is partly a consequence of the fact that the metals are oxidized to

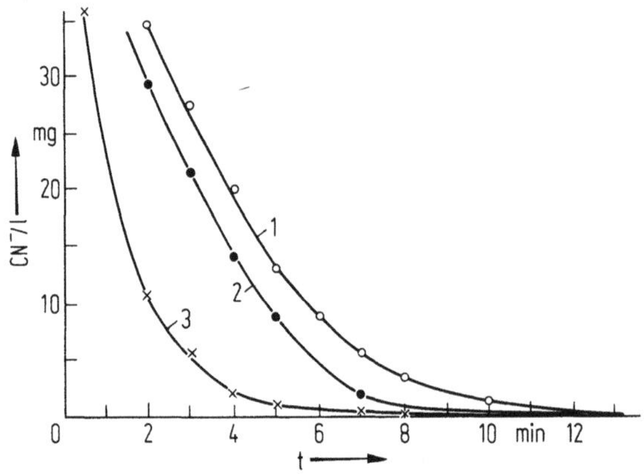

Fig. 3.22. Catalytic acceleration of the oxidation of cyanides by ozone on the addition of copper ions [181]. Total cyanide: 50 mg CN^-/l. T = 20 °C, 3 vol.% O_3 2 l/h. 1 KCN pH = 10. 2 $K_2Cu(CN)_4$ pH = 9. 3 10 KCN + 1 CuCN pH = 9.6

their highest oxidation states and partly because of heterogeneously catalysed decomposition of the ozone on the precipitated metal hydroxide and oxide sludge, which should hence be removed continually from the detoxification bath (metal recovery). The degradation is performed in the pH range 9–11. A rapid homogeneously catalysed decomposition of the ozone begins at or above pH 11. Solutions of simple and complex cyanides of Cu, Cd, Ag, Zn and Ni in concentrations of 50 mg/l are rapidly (5–15 minutes) destroyed at 20 °C by ozone-containing oxygen (3–4 vol. % O_3 gas flow rate 2 l/hour). The very stable Au, Fe and Co complexes are incompletely destroyed under the conditions mentioned (40 %–60 % decomposition). Ferrocyanide is first oxidized to ferricyanide, that is then decomposed with the precipitation of ferric hydroxide. Quantitative degradation of the more stable complexes can also be achieved by employing higher temperatures, UV irradiation or catalysts. The ions of the following heavy metals — in order of their catalytic activity — Cu, Ni, Mn, Sn, Ce, La, V, have proved to be particularly effective catalysts. To take an example, the oxidation process can be considerably accelerated by the addition of catalytic quantities of cupric salts (see Fig. 3.22). It is likely that reaction takes place via complexes with a central univalent copper ion, since in alkaline cyanide solutions cuprous ions are always formed primarily. Although the precise mechanism of this reaction has not been clarified, it has been argued that the direct attack of ozone on the central ion leads to a higher state of oxidation, resulting in an unstable complex which decomposes, whereby the oxidation of the free cyanide to cyanate takes place with simultaneous precipitation of the metal. As in all detoxifications of this type it is recommended that a slight excess of ozone should be employed.

Continuous processes have been developed for employment on the industrial scale, in which the cyanide-containing waste water is contacted with a countercurrent of ozonized oxygen in packed columns. Combined processes, with H_2O_2 for example, should be employed if the cyanide concentrations are $\geqq$0.5 g/l cyanide. Some detoxification processes operate under pressure, when the rate of reaction is increased because of the increase in the solubility of ozone in the aqueous phase in comparison to normal atmospheric pressure. Various industrial plants have been described [169]. The course of the oxidation reaction with ozone can be followed by measuring the residual ozone content of the exhaust gas, or the decrease in the cyanide concentration of the solution treated can be followed continuously by potentiometry (platinum/calomel) or discontinuously by photometry (e.g. pyridine/barbituric acid).

Sulphur compounds

Just as many inorganic sulphur compounds (e.g. hydrogen sulphide, sulphides, SO_2) can be oxidized very rapidly to sulphuric acid or sulphates by sufficient quantities of ozone in the presence of moisture or in aqueous solution, many types of organosulphur compounds — particulatly aliphatic and cyclic thioethers — in waste waters can be oxidized by ozone. For the oxidation of organosulphur-containing chemical wastes it is better to employ tetrachloroethylene or chloroform, when ozone is passed in at room temperature or in the cold until no more ozone uptake is recorded. In some cases it is possible to stop the reaction at the sulphoxide stage; mostly, however, the reaction proceeds as far as the sulphone. The products can be isolated after removal of the solvent.

Phenols

Phenol itself reacts via pyrocatechol with onzonolysis of the bond between the two OH group-bearing C atoms of the aromatic ring to yield muconic acid which is then further degraded by ozonolysis. The organic chlorine of chlorophenols is also liberated as free chloride ion [182, 183].

Combined Processes

Combined precipitation and oxidation techniques unite, for instance, milk of lime precipitation with ozonization, so that the more difficultly oxidizable substances are more easily oxidized and residual amount or reaction products can be removed [184]. The ozone oxidation of phenol is an example. Even at high pH considerable amounts of unwanted products, mainly oxalic acid, remain in solution. In the combined process this is removed as the insoluble calcium oxalate. Conversely the precipitational effect of milk of lime on heavy metals can be significantly improved by combination with ozone treatment at low pH (see Table 3.15) and there is no need for a neutralization step afterwards. A further economic advantage of the process is that the lime can be recovered from the precipitate as CaO (by calcination) and recycled. The organic components are thereby incinerated. The carbonic acid produced in recalcination can be employed for waste water neutralization.

3.3.5.7 Oxidation with Activated Oxygen

The long-recognized bleaching action of sunlight is only possible in the presence of atmospheric oxygen when activated oxygen species and peroxidic by-products are generated photochemically. A proportion of the paramagnetic molecular oxygen which possesses 12 valence electrons, two of which are unpaired, is transformed into electronically excited (singlet) oxygen [185]. Two forms of the latter have been demonstrated spectroscopically. One of them contains both of the valence electrons paired in one

Table 3.15. The improvement of heavy metal precipitation by milk of lime using ozone at pH 9 (%) [184]

Metal	with $Ca(OH)_2$	with $Ca(OH)_2/O_3$
Aluminium	85.1	87.2
Cadmium	96.5	97.0
Chromium	99.8	100.0
Zinc	99.4	100.0
Manganese	99.7	100.0
Nickel	13.3	93.1
Silver	6.5	60.2
Cobalt	65.1	99.6
Copper	100.0	100.0
Iron	100.0	100.0
Lead	44.6	100.0

of the two π^* orbitals, in the other form spin inversion of an electron takes place in the half-occupied π^* orbital of the triplet oxygen. The energy requirements are 22 and 37 kcal/mol respectively (92 and 155 kJ/mol).

The direct generation of excited (singlet) oxygen for practical detoxification purposes is extremely expensive. Helium-neon lasers, the photolysis of ozone or a gas discharge apparatus similar to that used for the generation of ozone are employed. However, the last of these methods only yields concentrations of active oxygen up to 2%. Another method consists of the thermal decomposition of certain peroxides or ozonides, e.g. the cleavage of the ozone-triethyl phosphite adduct, which takes place even at −15 °C. Singlet oxygen has been reported as being produced during the reaction of hypochlorite with hydrogen peroxide.

$$H_2O_2 + OCl^- \xrightarrow{(H_2O)} H_2O + {}^1O_2 + Cl^-$$

However this equation must not be taken as representing a quantitative reaction, since only a proportion of the oxygen formed is in the singlet state. The most elegant method, but again one that is too expensive for detoxification purposes, is the generation of singlet oxygen by the transfer of energy from other photochemically excited molecules (sensitizers, mainly dye molecules such as methylene blue, eosin or fluorescein) to triplet oxygen. Foote et al. established that the product distribution during olefin oxidation with singlet oxygen, generated from H_2O_2 and NaOCl or in a discharge tube, was qualitatively and quantitatively the same as that obtained when employing sensitized photooxidation, from which he concluded that the photogenerated reactive species was singlet oxygen [188, 186]. This can then be employed in sensitized photooxidation (Type II photooxidation [187]) of organic "acceptor" molecules. ("Type I" photooxidation [187] takes place when the primarily excited sensitizer reacts with the substrate to form free radicles, which then add the oxygen, i.e. the reaction proceeds as a substitution analogously to thermal autoxidation via peroxide or hydroperoxide radicals except for the fact that the substrate molecule is photochemically excited. Most photooxidations take place by sensitization according to both these basic mechanisms). Unsensitized reactions occur when the directly photoexcited substrate molecule reacts with oxygen. It seems likely that catalytic oxidation with molecular oxygen also occurs via a similar mechanism [189], since sensitization is necessary even for Type II photooxidation in order to transform oxygen into a more reactive species, i.e. the sensitizer acts as a carrier of the radiant energy, transferring it to the oxygen and ending the reaction in an unaltered state. The catalysts include some transition metals that activate oxygen reversibly or irreversibly, whereby reversibly bound oxygen corresponds to singlet oxygen, the irreversibly bound oxygen to peroxide oxygen [189] of the Type I photooxidation.

It is not possible at the present time to employ activated (singlet) oxygen as a purifying agent, so that these problems are only of theoretical interest. On the other hand, they have been of practical use in the solution of certain problems in the field of natural product synthesis, e.g. the transformation of cyclic dienes with allylic hydrogens (Enreaction) and the synthesis of dioxetanes from electron-rich olefins. Nowadays, singlet oxygen is employed on the tonne scale in industrial synthesis (e.g. for the synthesis of aromas). Recent investigations have established an extremely high effectivity in the destruction of bacteria and viruses [190, 191].

The effect of oxidizing agents (chlorine, hypochlorite, ozone, H_2O_2) is combined with UV irradiation in the Lightox process. The most economic process employs chlorine gas. It is probable that the UV radiation effects the introduction of active oxygen into the reaction scheme [192]. The hydroxide readical also participates [146]. An activation of molecular oxygen for certain oxidations of organic compounds has also been detected during the irradiation (π = 300–400 nm) of $ZnO—TiO_2$ suspensions [193].

3.3.5.8 Electrochemical (Anodic) Oxidation

The oxidative anodic destruction of inorganic and organic substances is the second basic electrochemical purification process alongside the cathodic removal of heavy metals already described. If chloride ion is present in the waste water then there is also the possibility of treatment by oxidation with hypochlorite anion produced in situ at the anode [194]. Such processes have been considered or already tried out for specific decontamination problems, including the treatment of dyestuff effluents. However, the process is not one of selective oxidation but a nonspecific destruction of all organic constituents, which increases the current consumption considerably. Attempts to increase the volume-time yields by the employment of catalysts have not yet led to any significant success. However, the nature of the electrode material employed is of importance [195].

The employment of the process for the removal of free and complex cyanaides has been investigated on the industrial scale. A consideration of the redox potentials for the oxidation of cyanide by atmospheric oxygen reveals that the process is certainly thermodynamically possible. However, it takes place far too slowly to be of practical utility (E_0 CN^-/CNO^- — 0.9 volt; E_0 OH^-/O_2 +0.4 volt). Nascent oxygen produced at the anode is, however, considerably more reactive than the molecular form. The oxidation of cyanide at the anode leads in part to the formation of nontoxic cyanate and in part to the formation of CO_2 and nitrogen. The cyanate is further degraded by hydrolysis. The achievable current yields lie in the range 90%–95%. Both batch and continuous processes are possible [98]. The costs are reported as being only about 1/5 of those for NaOCl or Cl_2 oxidation. The current yield is dependent on the cyanide concentration. The costs are favourable when cyanide concentrations are electrolysed down to about 0.4–0.5 g/l, which are then treated chemically [196]. Anodic oxidation can be employed to destroy 90% of the cyanide in a concentrate and the remainder can be dealt with conventionally. When metal cyanides or complex heavy metal cyanides are electrolysed, the metals are deposited at the cathode; they may, thus, be recovered. There is no unanimous view presented in the literature of the mechanism of anodic oxidation [195, 197], which is surely partly because of the use of differing electrode materials. The best materials for anodes and cathodes are reported to be graphite and iron. One desirable side effect of electrochemical processes is the elctroflotation effect because of the in situ formation of tiny bubbles of electrolysis gases [199]. Laboratory and semitechnical scale investigations have also been reported of the anodic oxidation of phenols, carboxylic acids, nitriles [199], cresols, thiocyanates [195] and sulphur-containing and organophorsphorus compounds [197].

The process of anodic hypochlorite synthesis at neutral pH, already mentioned, constitutes another possibility for the electrolytic oxidation of pollutants in homogeneous phase. The addition of a chloride electrolyte source is not usually necessary (and because of the increase in salinity not desirable), since the chloride content of waste waters is usually sufficient for the anodic synthesis of hypochlorite:

$$Cl^- + 2\,OH^- \rightarrow OCl^- + H_2O + 2\,e^-$$

It is scarcely possible to make a firm distinction between anodic oxidation processes and oxidations by a hypochlorite produced in a primary reaction. Virtually complete destruction of cyanide constituents is reported to be possible with in situ synthesized hypochlorite i.e. anodic oxidation in the presence of chloride; the intermediate products are cyanogen chloride and cyanate. The treatment of cyanide-containing effluents is reported to cost half as much as treatment with chlorine [197]. In the presence of chloride it is also possible to destroy phenol, as well as mercaptans, naphtenates, dyestuffs and traces of heavy oils [199]. The cost of destroying phenol by this method is reported as being half that of ozonization and one fifth that of adsorption on carbon [197].

3.3.5.9 Radiochemical Oxidation

Although it is scarcely employed in practice, substances resistant to biogical degradation, such as phenol [200], chlorophenols or anthraquinone dyestuff wastes, can, in principle, be attacked by radiochemical oxidation [201]. The primary products of irradiation of aqueous phenol solutions are hydroquinone, catechol, p-benzoquinone, 0,0′-diphenyl, 1,2,4-trihydroxybenzene, dihydroxymuconic dialdehyde, maleic acid and CO_2.

The employment of Co^{60} or possibly, in the future, waste materials from the reprocessing of used nuclear fuel rods, as γ-ray sources, makes such methods conceivable as a pretreatment process to ensure a complete biological purification. The problems involved have been discussed in terms of experimental data for the chlorophenol oxidation [201]. The aerated waste waters containing, for example,

3.3×10^{-3} mol/l 2,4-dichlorophenol and 3.7×10^{-3} mol/l 4-chloro-o-cresol (in all more than 90% of the dissolved organics) were irradiated by a ^{60}Co γ-source (dose rate 1.84×10^5 rad/h). After 6 mrad of radiation had been absorbed, the organic chlorine had been completely degraded to inorganic chloride. The phenol groups disappear and H_2O_2 is produced. The pH of the solution drops from 7 to 2. Since the TOC content does not change on irradiation the production of CO_2 and water as oxidation products can be excluded. Organic products, such as oxalic acid, containing highly oxidized carbon functions are formed. The products formed are more easily biologically degraded. It may be said concerning the mechanism that the effect of the ionizing radiation on waste waters containing less than 0.05% dissolved organics, is not attained by primary attack of the organic components. The water is first radiolytically decomposed to radicals and solvated electrons, which then react with the organic components in subsequent reactions.

$$H_2O \rightsquigarrow \cdot OH, \cdot H, e^{\ominus}_{aq}$$

Subsequent reactions
e.g.

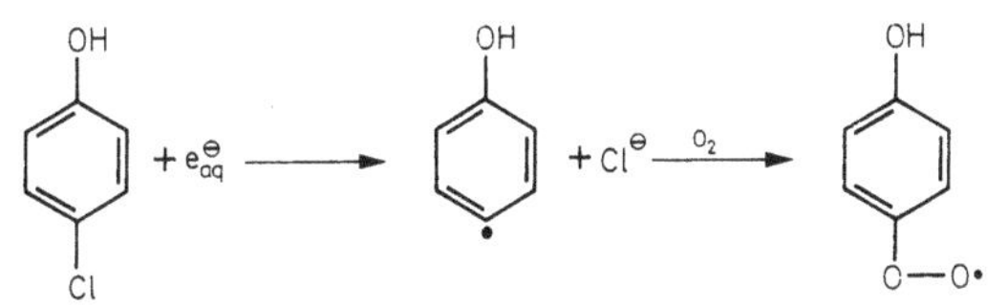

(dissociative electron capture with the formation of a radical and chloride ion)

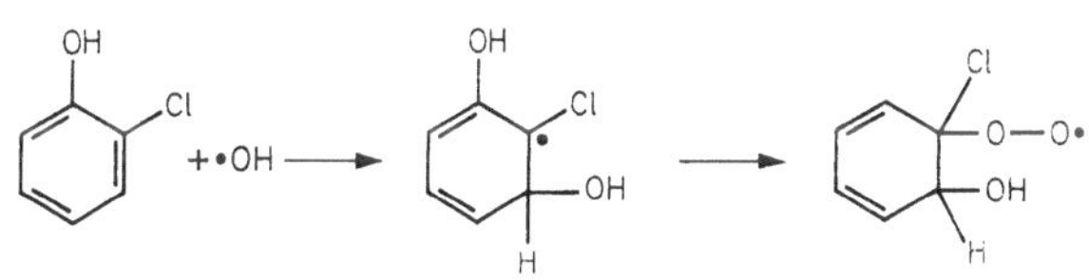

($\cdot$OH addition to a double bond)

The peroxide radicals produced react to open the aromatic rings and oxidize to form smaller molecules such as oxalic and muconic acids.

The oxidation is in fact more complicated; for instance, the oxygen which is led in reacts with solvated electrons to produce radicals (in acidic solution hydroperoxide radicals), which can also enter into the oxidative reaction process. The oxidation process can be inhibited if the primary hydroxide radicals take up electrons while oxidizing organic or inorganic anions.
e.g.

$$\cdot OH + Cl^- \rightarrow OH^- + \cdot Cl$$

Hence, foreign salts (NaCl, NaOH, KNO_3 etc.) increase the radiation dosage necessary for the complete degradation of chlorophenol. Radiochemical degradation falls linearly with increasing pollutant (in this case chlorophenol) concentration.

The degradation of the chlorophenol was determined gas chromatographically. The formation of chloride ion was determined with an ion-specific electrode and directly in the chloride measuring cell of a microcoulometer. The phenolic groups were detected by reaction with 4-aminoantipyrine and the H_2O_2 produced during irradiation was determined spectrophotometrically with $TiOSO_4/H_2SO_4$ reagent.

A new reliable and economic method has been developed at the Massachusetts Institute of Technology for the degradation of polychlorinated biphenyls in waste waters. The irradiation of flowing water with high energy electrons leads to a primary cleavage of the water molecules; the reaction with the PCBs of the hydroxyl radicals liberated leads to biologically degradable compounds, such as aromatic alcohols [224].

3.3.6 Catalytic Processes

The effect of catalysts on a reaction process can direct a reaction via a route involving a lower activation energy and hence increase its velocity; the equilibrium remains unchanged. Many types of chemical reaction can be influenced by catalysts, so that they will generally have been discussed already in the appropriate section. Some important — mechanistically very varied — catalytically active substances are:

— Hydrogen ions (e.g. in hydration and dehydration).
— Hydroxyl ions (e.g. in hydrolysis, condensation and polymerization).
— Metals (particularly the noble metals) and ionic or complexed (e.g. or oxidations, reductions and hydrogenations).
— Specific catalysts (e.g. active carbon for HCl cleavage from β-chloropropionic acid; Lewis acids in halogenation; ion exchangers for the detoxification of organophosphorus esters; hypochlorite for destroying organophosphorus esters; tetracalcium aluminate hydrate for the destruction of many chemical warfare agents.)

Catalysts can, in principle, be either heterogeneous or homogeneous in their action. For instance, activate carbon and various metal oxide hydrates find application as heterogeneous catalysts in waste water purification. The principle of their action is that "compounds" are produced on the catalyst surface by the chemisorption of one or more reactants, whereby the necessary activation energy for attack is reduced. Homogeneously catalysed detoxification reactions of practical importance employed until now have consisted, in the main, of various hydratation and oxidation processes. However catalytic processes have not attained anything like the importance in detoxification that they possess in the heterogeneous catalysed combustion of gases and vapours. A particular case of catalysis, that may be regarded as a mixed heterogeneous-homogeneous catalysis, is the adsorption and catalytic degradation of soluble toxic substances by synthetic, expandable, layered crystals dealt with on p. 191.

The phase transfer catalysts constitute quite another type of catalytic possibility. Liquid-liquid phase transfer catalysed two phase reactions are made possible by the employment of ammonium and phosphonium salts and also crown ethers; in these, inorganic reactants are transferred from the aqueous to the organic phase in the form of relatively loosely bound ion pairs. This makes many reactions possible in heterogeneous aqueous media that would normally require a nonaqueous solvent.

The employment of carrier-bound catalysts, such as oxidation catalysts, can, besides making possible a considerable acceleration of the rate of reaction, also lead to considerable savings in the consumption of oxidizing agent or in the use of air. Thus investigations have been described where textile industry effluents containing 500 ppm sulphides were reduced to 1 ppm by atmospheric oxidation in the presence of a pearl-form catalyst (Mn on Al_2O_3). This was possible with air at a temperature of 20 °C in just 40 minutes, while no significant sulphide elimination occurred within 40 minutes in the absence of the catalyst. When ozone was employed as the oxidizing agent the time required was reduced from 26 minutes without catalyst to 7 minutes with catalyst, which led to a saving of 36% in the amount of ozone [222]. The catalytic decomposition of hypochlorite by Co_3O_4 spinel on an inert carrier is another example [223]. In order for them it be economic in use the working lives of carrier-bound catalysts must be of sufficient length; the factors to be taken into consideration here are not just mechanical attrition and questions of regeneration but also irreversible poisoning, which can be a significant factor. This means that each particular application must be preceded by extensive investigations in order to allow an estimate to be made of its economic viability.

The purpose of this section is to complement the "scattered" possibilities of influencing detoxification reactions by catalysis, that have been described, by including several more practical examples performed in aqueous solution. One significant example of a heterogeneously catalysed wet detoxification process is the oxidation of cyanide with air; this has already been discussed in detail employing other oxidizing agents. In the process to be described here the cyanide-containing water is led through a column or trickle tower, packed with activate carbon, against a countercurrent of air. In this the cyanide is first adsorptively bound and then catalytically oxidatively degraded, with emphasis on the partial oxidation to cyanate [147]. The cyanate complexes of zinc, cadmium and copper can also be degraded by this process; on the other hand, nickel complexes can only be incompletely degraded and iron complexes not at all. The best results are obtained with rinse waters containing small concentrations of cyanides, when the cyanide concentrations can be reduced by 99%. The detoxifying capacity of the active carbon amounts to several kilograms of cyanide per day and per cubic metre. The catalyst that is eventually rendered inactive by adsorbed carbonates and hydroxides can be regenerated by first rinsing with acid and then neutralizing. The metallic impurities are thereby precipitated as hydroxides. Adsorbed complex cyanides are cleaved and spontaneously oxidized during the regeneration process. This method will certainly gain in importance in the future for the purification of cyanide-containing effluents. Even today wet catalytic oxidation processes are being investigated and employed in combination with other processes such as subsequent precipitation. Hypochlorite-containing solutions can be economically destroyed by highly catalytically active, finely divided, high-valency metal oxide hydrates (particularly Co/Ni mixtures). For this purpose the preprecipitated Co/Ni oxides and/or carbonates are added to the solution to be detoxified. Recovery and recycling are possible [202].

Processes involving homogeneous catalysis are rarely employed for detoxification. Acid and base catalysed reactions find wider application in organic chemistry. Here catalysis occurs because the acid or base transforms one of the reactants to a reactive intermediate by either donating to it or accepting from it a proton. This proton as then later to be removed or returned. A distinction is made between generalized acid catalysis, where the concentration of the acid catalyst is decisive, and specific acid catalysis, where only the concentration of the conjugate acid of the solvent appears in the rate equation, e.g. [C] = [H_3O^+]. The situation is analogous for base catalysis.

The catalysed hydrolysis of nitriles has been investigated very extensively. Alongside such specific catalysts as copper oxide and copper-containing polymers with basic nitrogen-containing residues [204], the hydrolysis of a range of nitriles can be accelerated by bases, e.g. barium hydroxide [205] or acids [206].

One reaction that is very familiar in the practice of detoxification is the catalytically influenced hydrolysis of esters of phosphoric and of phosphonic acids. Detailed investigation has revealed that their hydrolysis can be catalysed by OH^- ions, but hydrogen ion also accelerates the hydrolysis process in some cases (e.g. DFP, TEPP). Because of the differing reaction mechanisms various catalysts can lead to the formation of different products. For instance the hydrolysis of Tabun and other organophosphorus esters containing dimethylamino groups proceeds by the cleavage of cyanide in the presence of OH^- ions and in the presence of HCl by the cleavage of dimethylamino groups [207]. The rate of hydrolysis can be significantly increased by raising the temperature. Synthetic ion exchange resins can also be employed to affect the rate of hydrolytic cleavage as in small portable water detoxification apparatus.

"Cation catalysis" is a further possibility for influencing the hydrolysis. It was known, even before the Second World War, that the hydroxides of the rare earths possessed considerable catalytic activity. During the 1950s it was discovered in the United States, in the context of a large-scale research programme, that the copper chelate complexes of tetramethylethylenediamine, of dipyridyl and of diaminoethane, affect the hydrolysis of DFP the most, which also applied to other organophosphorus. esters. But the cupritetrammine complexes are also excellent. For a time such copper complex catalysts found application for the gentle detoxification of fluorophosphonate and fluorophosphate esters of the Sarin/Soman type [208].

The copper-catalysed phosphate ester cleavage is an unwanted reaction of organophosphate pesticides, which has been the cause of the loss of activity observed in certain formulations. As well as copper chelates a range of other metal chelates has been investigated, but only the ZrO(II), UO_2(II) and MoO_2(IV) chelates gave anything like comparable results for the hydrolysis of Sarin. It is believed that the mechanism consists of an electrophilic polarization of the fluorine-phosphorus bond by the Cu(II) ion, which facilitates nucleophilic attack by water. Sulphide-containing effluent waters can also be detoxified by air or oxygen at pH 10–11 under the influence of $MnSO_4$ or $KMnO_4$ [209].

Now to turn to "anionic catalysis" of the hydrolysis of phosphate and phosphonate esters. The important example here is that of the hypochlorite anion, which has already been discussed in terms of its high reactivity and oxidizing ability. However, its action in phosphate ester hydrolysis is primarily catalytic, which does not exclude other reactions as a consequence of its high reactivity. It is believed that the hypochlorite makes a direct befunctional attack on the ester to be detoxified and facilitates the departure of the nucleofuge leaving group by a polarization of the P = 0 bond [210]. The effect of hypochlorite is very pH-dependent, for example, only about 1/8 the concentration of chlorine is necessary at 250 °C and pH 7 as at pH 6 [208]. Chromate, molybdate and tungstate anions also exert a catalytic effect, albeit a weaker one.

The recently described electrocatalysis consists of the employment of electrochemically active substances, without the supply of external electrical current, in conjunction with atmospheric oxygen as an oxidizing agent or hydrogen as a reducing agent. The catalyst is brought into contact with the solution being treated either in the form of a suspension or of a solid bed. In the case of oxidation with atmospheric oxygen, the pollutant/oxidation product electrode potential must be sufficiently negative compared with that of oxygen reduction. A mixed potential is set up at the catalyst, when the pollutant is oxidized and the equivalent amount of oxygen is reduced (the opposite applies to reduction with hydrogen). An adequate reaction rate is ensured by a large catalytic surface and high catalytic activity. High electrical conductivity and good electrocatalytic properties, with respect to the particular reaction, are decisive for the catalyst. Active carbon, for instance, whose electrical conductivity has been increased by thermal treatment and which has been coated with an electrochemical catalyst that depends on the particular application, is a suitable material [211]. It is certainly possible to oxidize cyanide with atmospheric oxygen, but the reaction proceeds too slowly for practical purposes. In the presence of an electrochemical catalyst (in alkaline solution at 20 °C) a mixed potential is set up between − 0.97 volts and +0.40 volts (with respect to the standard hydrogen electrode) [211].

$$CN^- + 2\,OH^- \rightarrow CNO^- + H_2O + 2\,e^- \quad (E_0 = -0.97\ V)$$
$$1/2\,O_2 + H_2O + 2\,e^- \rightarrow 2\,OH^- \quad (E_0 = +0.40\ V)$$

The large potential difference means that dilute solutions of cyanide can be oxidatively removed down to very low residual concentrations. An active carbon, which has been ignited to 1100 °C in an inert atmosphere (nitrogen, hydrogen or vacuum) and coated with Pt, Ag or Ni, is employed as the catalyst [211]. The same catalysts can also oxidize SO_2 which is dissolved in alkaline solution as the sulphite. Electrocatalytic reductions have been investigated on the laboratory scale, for example the reduction of chromate by hydrogen (and other reducing agents) at a tungsten carbide-coated active carbon in acidic solution [211]. Such processes have not yet been employed on an industrial scale for destruction of pollutants.

Alongside the further development of physical, physicochemical, chemical and nonspecific processes for the detoxification or immobilization of biologically active and otherwise dangerous chemicals, it is quite certain that, in the future, the development and realization of various catalytic processes will be of increasing importance. Possibilities for degradation and detoxification can be expected by the application of various types of catalysts, from synthetic hydrolysis catalysts (e.g. polymeric imidazoles) via polymeric phase transfer catalysts (e.g. polystyrene-anchored onium functions) and polymeric metal and ion carriers (i.e. the immobilization of homogeneous catalysts), to enzymes covalently bound to polymers [212]. The aim is to try to imitate the high specificity and catalytic activity of the natural polymeric catalysts — the enzymes. The great advantage of insoluble polymeric catalysts is in their easy separation and the possibility of their reuse; the, at present, insoluble problem of the separation from metallic salt catalysts (e.g. often Cu^{2+}) could be solved in this manner. The problems involved and the directions of development are discussed in [212].

References

1. Bradke, H. J.: Galvanotechnik 64 (1973) 556.
2. Weber, H. H., Mann, Th.: Naturwissenschaften 64 (1977) 82.
3. Wuhrmann, K.: Vortragsveröffentlichungen des Hauses der Technik, Essen. Heft 28: Tagung vom 15. 10. 1964.
4. Sixt, H.: Chemie-Ing.-Tech. 53 (1981) 844.
5. Schefer, W., Wälchli, O.: Chimia 34 (1980) 349.
6. Mangold, K.-H., et al.: Abwasserreinigung, Leipzig: VEB Deztscher Verlag für Grundstoffindustrie 1975.
7. Martin, P.: Abwassertechnik 1978 (3) 30.
8. Kämpf, H. J.: Chem. Industrie 30 (1978) 37.
9. Becker, K. P.: Chem.-Ing.-Tech. 51 (1979) 549.
10. Diesterweg, G., et al.: Industrieabwässer. Juni 1978, 7.
11. Braun, R.: Österr. Abwasserrundschau 1980, 126.
12. Ansorge, D. W.: Das Umweltmagazin. Dezember 1980, 50.
13. Howe, R. H. L.: Process Biochemistry 1969 (2), 18.
14. Hogan, P., Kuhn, A. T.: Oberfläche-Surface 18 (1977) 255.
15. Zlobarnik, M.: Chem.-Ing.-Tech. 53 (1981) 600.
16. Martinetz, D.: Immobilisation, Entgiftung u. Zerstörung von Chemikalien. Leipzig: VEB Deutscher Verlag für Grundstoffindustrie 1980.
17. Hopfe, D.: Fortschr. d. Wasserchem. u. ihrer Grenzgebiete. Heft 3. Berlin: Akademie-Verlag 1965, S. 95.
18. Kerl, M.: Münchener Beitr. Abwasser-, Fisch-, Flußbiol. 22 (1972) 198.
19. Verfahrensberichte zur physikalisch-chemischen Behandlung von Abwässern. 2. Bericht: Abwassereindampfung. Hrsg. vom Verband der Chemischen Industrie, Köln 1975.
20. Dimitirou, M. H.: Chem. Industrie 27 (1975) 315.
21. Dimitriou, M. H.: Chemie-Ing.-Tech. 45 (1973) 1417.
22. Verfahrensberichte zur physikalisch-chemischen Behandlung von Abwässern. 5. Bericht: Abwasserreinigung durch Extraktion. Hrsg. vom Verband der Chemischen Industrie, Köln 1976.
23. Schalk, W., Gmelin, J.: Verfahrenstechnik 7 (1973) 227.
24. Marvel, C. S., Richards, J. C.: Analytic. Chem. 21 (1949) 1480.
25. Kiezyk, P. R., Mackay, D.: Canad. J. Chem. Engng. 49 (1971) 747.
26. Wurm, H.-J.: Chemie-Ing.-Tech. 48 (1976) 840.
27. Hemmann, Ch., et al.: Verfahrensberichte. Chemische und physikalisch-chemische Verfahren der Abwasserbehandlung in der chemischen und artverwandten Industrie. Berlin: Wissenschaftliches Informationszentrum der AdW der DDR 1976.
28. Stenger, K., et al.: Chemiker-Ztg. 99 (1975) 220.
29. Saier, H.-D.: Chem. Labor-Betrieb *28* (1977) 6; Rautenbach, R., Rauch, K.: Chemie-Ing.-Tech. 49 (1977) 223.
30. Dytnertsky: Membranprozesse zur Trennung flüssiger Gemische. Leipzig: VEB Deutscher Verlag für Grundstoffindustrie 1977.
31. Hardwick, W. H.: Chem. and Ind. (London) 1970 (9) 297.
32. Staude, E.: Chemiker-Ztg. 96 (1972) 27.
33. Verfahrensberichte zur physikalisch-chemischen Behandlung von Abwässern. 4. Bericht: Abwasserreinigung mittels Reversosmose und Ultrafiltration. Hrsg. vom Verband der Chemischen Industrie, Köln 1976.
34. Saier, H.-D., et al.: Dechema-Monographien 80 (1976) 1, 211.
35. Strathmann, H., Saier, H.-D.: CZ-Chemie-Technik 3 (1974) 174.
36. Oswald, E.: Metalloberfläche 28 (1974) 165.

37. Marquardt, K.: Betriebstechnik 16 (1975) 5, 11.
38. Knobloch, H.: Dechema-Monographien 80 (1976) 1, 259.
39. Marquardt, K. in: Hartinger, L.: Taschenbuch der Abwasserbehandlung. München/Wien: Carl Hanser Verlag 1976. Kapitel: Umgekehrte Osmose, S. 149.
40. Staude, E.: Umschau Wiss. Tech. 74 (1974) 747.
41. Desai, S., et al.: An economically attractive application of reverse osmosis to refinement of a petrochemical effluent stream. AIChE Dymos. Ser. (New York) 68 (1972) 120. 379.
42. Walch, A.: Chemie-Ing.-Tech. 48 (1976) 307.
43. Schmidt, A.: Angewandte Elektrochemie. Weinheim: Verlag Chemie 1976, S. 267ff.
44. Weiner, R.: Galvanotechnik 63 (1972) 614.
45. Marquardt, K.: Metalloberfläche 27 (1973) 169.
46. Verfahrensberichte zur physikalisch-chemischen Behandlung von Abwässern. 3. Bericht: Adsorptive Abwasserreinigung. Hrsg. vom Verband der Chemischen Industrie, Köln 1975.
47. Huber, L.: Münchener Beitr. Abwasser-, Fisch.-, Flußbiol. 22 (1972) 180.
48. Mann, Th.: VDI-Berichte 218 (1974) 151.
49. Fritz, W., et al.: vt-Verfahrenstechnik 13 (1979) 536.
50. Grandjaques, B. L., et al.: CZ-Chemie-Technik 1 (1972) 477.
51. Port, E.: Gas- u. Wasserfach, Ausg. Wasser/Abwasser 120 (1979) 527; Umwelt 1979 (3), 175.
52. Haberl, R.: Entfernung organischer abbaubarer Stoffe bei der Abwasserreinigung. Vortrag auf dem 16. Seminar des Österr. Wasserwirtschaftsverbandes. 26.–30. 1. 1981, Raach.
53. Neff, I., et al.: Chem. Industrie 1975 (6) 309.
54. Oberländer, G., Funke, W.: Energietechnik 17 (1976) 560.
55. Rosenzweig, M. D.: Chem. Engng. 82 (1975) 2, 60.
56. Bonnet, J., et al.: Tech. Eau Assainissement 397 (1980) 15.
57. Fox, C. R.: Hydrocarbon Processing. November 1978, 269.
58. Bieling, H., Kreutzenberger, G.: Möglichkeiten der Abtrennung und Rückgewinnung von Quecksilber aus Abwässern und Alkalilaugen der Chloralkalielektrolyse. Vortrag auf der Hauptjahrestagung der Chemischen Gesellschaft der DDR, Dresden 1977.
59. Dosch, W.: Zivilverteidigung (Bad Honnef) 1970 (7/8) 72; 1970 (10) 35; 1970 (12) 39; 1971 (1) 39.
60. Dosch, W., Keller, H.: Zivilverteidigung (Bad Honnef) 1972 (2) 68.
61. Stumm, W.: Gas, Abwasser, Wasser 57 (1977) 134.
62. Harmsen, H.: Privatmitteilung.
63. Klute, R.: Umwelt 1979 (4) 303.
64. Winkler, F.: Wissenschaft u. Fortschritt 24 (1974) 503.
65. Schmidt, K., Schladebach, K.: Zellstoff u. Papier 22 (1973) 2, 45.
66. Reuter, J.: Chemiker-Ztg. 98 (1974) 222.
67. Reuter, J.: Umwelt 1981 (1) 27 J.
68. Neukirchen, B.: VDI-Berichte 260 (1976) 395.
69. Schmitz, W.: Chem. Industrie 1973 (8) 499.
70. Weil, L., et al.: Dechema-Monographien 80 (1976) 1, 105.
71. Contesse, E.: Umweltschutz 1981 (12) 251.
72. Martin, P.: Metalloberfläche 28 (1974) 161.
73. Bujard, W.: Abwassertechnik 22 (1971 Heft 12-1972 Heft 2) S. V.
74. Das techn. Umweltmagazin, Oktober 1976, 28.
75. Rommenhöller GmbH (BRD): Abwasserneutralisation. Firmenschrift.
76. Pöpel, F.: Textilveredlung 14 (1979) 61.
77. Janke und Kunkel KG: Neutralisation mit Rauchgas. Formenmitteilung in: Verfahrenstechnik 10 (1976) Nr. 1.
78. Verfahrensberichte zur physikalisch-chemischen Behandlung von Abwässern. 7. Bericht: Flockung und Fällung. Hrsg. vom Verband der Chemischen Industrie, Köln 1977.
79. Bischofsberger, W.: Österr. Wasserwirtschaft 28 (1976) 177.
80. Schlegel, H.: Galvanotechnik 63 (1972) 514.
81. Götzelmann, W.: Galvanotechnik 67 (1976) Nr. 5.

82. Hartinger, L.: Die Metalle im Abwasser — ihre Toxikologie und die Chemie ihrer Ausfällung. IWL-Forum 66/V, S. 1–43 (Druck: Schäfer, Köln 1968).
83. Allisson, S.: Chem. Rundsch. 27 (1974) 28, 17.
84. Jola, M.: Fachber. für Oberflächentechn. 1974 (10) 211.
85. Hartinger, L.: Taschenbuch der Abwasserbehandlung. München/Wien: Carl Hanser Verlag 1 (1976), 2 (1977).
86. Yano, T., et al.: Dechema-Monographien 80 (1976) 1, 179.
87. Aratani, T., et al.: Bull. chem. Soc. (Japan) 51 (1978) 2705; 52 (1979) 218.
88. Dittrich, V.: Wasser, Luft u. Betrieb 15 (1971) 15.
89. Zietz, U.: Gas- u. Wasserfach, Ausg. Wasser/Abwasser 120 (1979) 259; Berichte zur Abwasser- u. Abfalltechnik. Int. Workshop 8./9. 6. 1978 anläßlich der IFAT München.
90. Mudrack, K., Stobbe, G.: Wasser, Luft u. Betrieb 18 (1974) 289.
91. Hartkorn, K. H.: Städtehygiene 1973 (9), 213.
92. Leumann, P.: Gas- u. Wasserfach, Ausg. Wasser/Abwasser 114 (1973) 272.
93. Roth, C., Mühlhäuser, M.: Gas- u. Wasserfach, Ausg. Wasser/Abwasser 114 (1973) 585.
94. Leumann, P., Lutz, W.: Gas, Wasser, Abwasser 57 (1977) 370.
95. Leumann, P.: Galvanotechnik 68 (1977) 715.
96. Stumm, W., Sigg, L.: Z. Wasser Abwasser Forsch. 12 (1979) 73.
97. Cervenka, L., Timmermann, F.: Umwelt 1976 (3) 190.
98. Gleisberg, D., et al.: Angew. Chem. 88 (1976) 354.
99. Götzelmann, W., Spanier, G.: Galvanotechnik 54 (1963) 265.
100. Oehme, F., et al.: Galvanotechnik und Oberflächenschutz 7 (1966) 75.
101. Verfahrensberichte zur physikalisch-chemischen Behandlung von Abwässern. 9. Bericht: Abwasserreinigung mit Ionenaustauschern und Adsorberharzen. Hrsg. vom Verband der Chemischen Industrie, Köln 1978.
102. Schlegel, H.: Galvanotechnik 56 (1965) 73.
103. Weiner, R.: Galvanotechnik 64 (1973) 99.
104. Marquardt, K.: Metalloberfläche 23 (1969) 231.
105. Arnold, K.-H.: Chemie-Ing.-Tech. 47 (1975) 583.
106. Hartinger, L.: Galvanotechnik 68 (1977) 721.
107. Kühne, G.: Chemiker-Ztg. 96 (1972) 239.
108. Götzelmann, W.: Wasser, Luft u. Betrieb 20 (1976) 473.
109. Martinola, F.: Chemie-Ing.-Tech. 51 (1979) 728.
110. Bergk, K. H., et al.: Z. Chem. 17 (1977) 85.
111. Oehme, Ch.: Jahrbuch vom Wasser 38 (1971) 345.
112. De Jong, G. J., Rekers, C. N. J.: 3. Symposium über Ionenaustauscher. Symposiumsbericht. Balatonfüred (Ungarn), Mai 1974.
113. Akzo Zount Chemie (Niederlande): The Akzo Imac TMR process for the removal of mercury from waste water. Firmenschrift 1974; De Jong, G. R., Rekers, C. N. J.: J. Chromatogr. 102 (1974) 443.
114. Wiedmann, H.: Wasser, Luft u. Betrieb 15 (1971) 87.
115. Verfahrensberichte zur physikalisch-chemischen Behandlung von Abwässern. 7. Bericht: Flockung und Fällung. Hrsg. vom Verband der Chemischen Industrie, Köln 1977.
116. Faensen, W.: Zentrale Entgiftungsanlage Bielefeld. Konzeption-Leistung-Grenzen. Tagung Nr. 351-76 (1976). Haus der Technik, Essen (BRD).
117. Fabjahn, Ch., Bauer, P.: Galvanotechnik 67 (1976) 307; Peroxid-Chemie GmbH (Höllriegelskreuth, BRD): Entgiftung von Chromsäure mit Wasserstoffperoxid. Firmenschrift.
118. Marquardt, K.: Galvanotechnik 69 (1978) Nr. 1, 2, 4.
119. Martin, P.: Abwassertechnik 1967 (3) 9.
120. Kreysa, G.: Chemie-Ing.-Tech. 50 (1978) 332.
121. Verfahrensberichte zur physikalisch-chemischen Behandlung von Abwässern. 1. Bericht: Abwasserverbrennung. Hrsg. vom Verband der Chemischen Industrie, Köln 1975.
122. Verfahrensberichte zur physikalisch-chemischen Behandlung von Abwässern. 5. Bericht: Abwasserreinigung durch Naßoxydation. Hrsg. vom Verband der Chemischen Industrie, Köln 1976.
123. Randolf, R.: Wasserwirtschaft 18 (1968) 342.

124. Randall, T. L.: Ind. Waste. Proceedings of the Thirteenth Mid-Atlantic Conferenc, June 29–30, 1981, S. 501.
125. Braden, R., Schulz-Walz, A.: Naßoxydation von halogenhaltigen Industrieabwässern. Bericht der Bayer AG; Chemie-Ing.-Tech. 54 (1982) 692.
126. Bernadiner, M. N., et al.: Z. Vses. Chimičeskogo Obščestva (Moskva) 17 (1972) 2, 162.
127. Lohmann, U., Tilly, A.: Chemie-Ing.-Tech. 37 (1965) 913.
128. Machu, W.: Das Wasserstoffperoxid und die Perverbindungen. Wien: Springer-Verlag 1951, S. 73.
129. Teletzke, G. H.: Chem. Engng. Progr. 60 (1964) 1, 33.
130. Schoeffel, E. W., Seegert, N.: Wasser, Luft u. Betrieb 10 (1966) 541.
131. Schulz-Walz, A., et al.: Chemie-Ing.-Tech. 53 (1981) 295.
132. Huesler, H.: Abwassertechnik 22 (1971) 1, III.
133. Oehme, Ch.: Das Umweltmagazin. März 1980, 20.
134. Wysocki, G., Höke, B.: Wasser, Luft u. Betrieb 18 (1974) 311.
135. Höke, B., Wittbold, H. A.: Wasser, Luft u. Betrieb 13 (1969) 250.
136. Valente, J.: Gas, Wasser, Abwasser 57 (1977) 1, 70.
137. Rauh, J. S.: Disinfection and oxidation of wastes by chlorine dioxide. 21. Annual Meeting Mississippi Water Pollution Control Association. 17./18. 4. 1978, Jackson.
138. Reinigung und Entgiftung phenol- und phenol/formaldehydhaltiger Abwässer. Firmenschrift Degussa.
139. Kojima, Y.: Chem. Industrie 26 (1974) 447.
140. Bucksteeg, W., Thiele, H.: Gas- u. Wasserfach. Ausg. Wasser/Abwasser 98 (1957) 36, 909.
141. Jola, M.: Fachber. für Oberflächentechn. 10 (1972) 5, 170.
142. Stumm, W., et al.: Z. Hydrol. 16 (1954) 1.
143. Von Beckerath, K., et al.: Metalloberfläche 30 (1976) 385.
144. Weiner, R.: Die Abwässer der Galvanotechnik und Metallindustrie. Saulgau: Leuze Verlag 1965, S. 176.
145. Rehn, F., Czimber, J.: Galvanotechnik 56 (1965) 86.
146. Meiners, A. F., et al.: An investigation of lightcatalyzed chlorine oxidation for treatment of wastewater. R. A. Taft Water Research Center, Rep. TWRC-3. U.S.-Departm. Interiors, FWPCA, Cincinnati, Ohio 1968.
147. Jola, M.: Die Vernichtung von cyanidhaltigen Abfällen. Separatdruck aus Chem. Rundsch. 27 (1974) Nr. 22–24.
148. Krüger, H., et al.: Chemiker-Ztg. 99 (1975) 132.
149. Schwarzer, H.: Umwelt 1981 (6) 482.
150. Hahn, F.: Chemie-Ing.-Tech. 46 (1974) 11, A 396.
151. Hahn, F.: CZ-Chemie-Technik 3 (1974) 5, 197.
152. Pötschke, H.: Wasser, Luft u. Betrieb 19 (1975) 504.
153. Krüger, H., et al.: Chemiker-Ztg. 99 (1975) 3, 12.
154. Hahn, F., Meier, F.: Chemiker-Ztg. 95 (1971) 467.
155. Wagner, R., in Aurand, K., et al.: Organische Verunreinigungen in der Umwelt. E. Schmidt Verlag, Berlin (West) 1978.
156. Feuerstein, W.: Untersuchungen über die Anwendung von Fentons Reagenz zur chemisch-oxydativen Abwasserbehandlung. Dissertation. Fakultät für Chemie der Universität (TH) Karlsruhe, 1982.
157. Lewis, S. N., Augustine, R. L.: Oxidation, Techniques and Applications in Organic Synthesis. Vol. 1. New York: M. Dekker Inc. 1969, S. 215.
158. Schwarzer, H.: Galvanotechnik 66 (1975) 22.
159. Entgiftung cyanidischer Abwässer mit Persauerstoffverbindungen. Firmenschrift Degussa.
160. Knorre, H.: Entgiftung von Abwässern aus Härtereien mit H_2O_2. Vortrag auf dem 34. Härterei-Kolloquium. 4.–6. 10. 1978, Wiesbaden.
161. Oehme, F., Disam, J.: Galvanotechnik 58 (1967) 236.
162. Weigert, W. M., et al.: Chemiker-Ztg. 99 (1975) 106.
163. Beseitigung von Natriumhypochlorit mit Wasserstoffperoxid. Firmenschrift Degussa.
164. Entgiftung formaldehydhaltiger Abwässer mit Wasserstoffperoxid. Firmenschrift Degussa.

165. Oxydation von Phenolen mit Wasserstoffperoxid. Firmenschrift Peroxid-Chemie (Höllriegelskreuth, BRD).
166. Keating, E. J.: Industrial Water Engineering. Dezember 1978.
167. Verfahrensberichte zur physikalisch-chemischen Behandlung von Abwässern. 8. Bericht: Abwasserreinigung durch Ozonisierung. Hrsg. vom Verband der Chemischen Industrie, Köln 1978; Kurz, R.: Untersuchungen zur Wirkung von Ozon auf Flockungsvorgänge. Dissertation. Fakultät für Chemieingenieurwesen der Universität (TH) Karlsruhe, 1977.
168. Gilbert, E.: Proceedings der Konferenz „Oxydationsverfahren in der Trinkwasseraufbereitung. 11.–13. 9. 1978, Karlsruhe.
169. Verfahrensberichte zur physikalisch-chemischen Behandlung von Abwässern. 8. Bericht: Abwasserreinigung durch Ozonisierung. Hrsg. vom Verband der Chemischen Industrie. Köln 1978.
170. Burleson, G. R., et al.: Environm. Mutagenesis 1982 (4) 469.
171. Kandzas, P. F., et al.: Z. Ves. Chimičeskogo Obščestva (Moskva) 17 (1972) 2, 169.
172. Gorbenko-Germanov, D. S., et al.: Chim. Prom. (Moskva) 1975 (2) 101; Hendrickson, T. N., Daignault, L. G.: J. Soc. Mot. Pict. Telev. Eng. (Easton) 82 (1963) 727.
173. Grossmann, A., et al.: Zesz. Nauk. Politechn. Slask., Inz. Sanit 1970 (16) 27.
174. Dietrich, K. R.: Chemiker-Ztg. 100 (1976) 68.
175. Greiner, G., Grünbein, W.: Dechema-Monographien 75. (1974) Nr. 1452–1485, 399.
176. Grünbein, W.: Chemie-Ing.-Tech. 46 (1974) 339.
177. Rüb, F.: Wasser, Luft u. Betrieb 19 (1975) 4, 147.
178. Sandmann, H.: Reinigung industrieller Abwässer. Vortrag auf der BBC-Abwassertagung, Hemmenhofen am Bodensee, 14./15. 10. 1976.
179. Hoigne, J., Bader, H.: Water Research 10 (1976) 377.
180. Keinan, E., Mazur, Y.: J. org. Chemistry 42 (1977) 844.
181. Fabjahn, Ch., Davies, R.: Wasser, Luft u. Betrieb 20 (1976) 175.
182. Gould, J. P., et al.: Wat. Pollut. Control. Fed. 48 (1976) 47.
183. Gilbert, E.: Vom Wasser 43 (1974) 275.
184. Abwasserreinigung mit dem BBC-Kaloz-Verfahren. Firmenschrift Nr. 102 der Brown, Boveri & Cie.
185. Lechtken, P.: Chemie in unserer Zeit 8 (1974) 1, 11.
186. Foote, Ch., et al.: J. Am. chem. Soc. 90 (1968) 975.
187. Autorenkollektiv: Einführung in die Photochemie. Berlin: VEB Deutscher Verlag der Wissenschaften 1976, S. 383.
188. Foote, Ch., et al.: J. Am. chem. Soc. 86 (1964) 3879.
189. Scheve, J., Scheve, E.: Z. Chem. 14 (1974) 172.
190. Adam, W.: Chemie in unserer Zeit 15 (1981) 190.
191. Wallis, C., et al.: in Berg, G., et al.: Viruses in Water. Amer. Public Health Association, Washington 1976.
192. Sontheimer, H.: Jahrbuch vom Wasser 37 (1970) 171.
193. Kinney, L. C., et al.: Photolysis mechanism for pollution abatement. R. A. Taft Water Research Center, Rep. TWRC-13. U.S. Departm. Interior, FWPCA, Cincinnati, Ohio 1969.
194. Kuhn, A. T., Lartey, R. B.: Chemie-Ing.-Tech. 47 (1975) 129.
195. Kuhn, A. T.: J. Appl. Chem. Biotechnol. (London) 21 (1971) 2, 29.
196. Conrad, J., Jola, M.: Chem. Rundsch. 24 (1971) 23, 529.
197. Tomilov, A. P., et al.: Chim. Prom. (Moskva) 48 (1972) 4, 267.
198. Jola, M.: Fachber. für Oberflächentechn. 11 (1973) 151.
199. Beck, F.: Elektroorganische Chemie. Berlin: Akademie-Verlag 1974, S. 342.
200. Sato, K., et al.: Envir. Science & Technology 12 (1978) 79, 1043.
201. Gilbert, E., Güsten, H.: Chemiker-Ztg. 101 (1977) 22.
202. BRD-OS 2 624 642 (1977), Zarth, W., Zimmermann, K. (Krupp).
203. U.S.-Pat. 3 846 495 (1974), Svarz, J. J., Nalco Chem. Co.
204. BRD-OS 2 429 269 (1975), Watanabe, Y.
205. U.S.-Pat. 3 876 691 (1975), Lincoln, R. M., Atlantic Richfield Co.
206. BRD-OS 2 438 263 (1975), Norton, R. V., Sun Ventures Inc.
207. Larsson, L.: Acta chem. scand. 6 (1952) 1470.

208. Lohs, Kh.: Synthetische Gifte. Berlin: Militärverlag der DDR 1974.
209. Martin, J. L., Rubin, A. J.: Vortrag. 33. Annual Ind. Waste Conference, Lafayette 1978.
210. Epstein, J., et al.: J. Am. chem. Soc. 78 (1956) 4068.
211. Faul, W., Kastening, B.: Galvanotechnik 68 (1977) 699; Chemie-Ing.-Tech. 50 (1978) 533; Technische Informationen der Kernforschungsanlage Jülich 1977, Nr. 14; 1978, Nr. 17.
212. Manecke, G., Stork, W.: Angew. Chem. 90 (1978) 691.
213. Schwartz, W.: Bild der Wissenschaft 1976 (2), 60.
214. Gräf, H.-D.: Pharmazie in unserer Zeit 6 (1977) 43.
215. Manecke, G.: Chimia 28 (1974) 467.
216. Bailin, L. J., Hertzler, B. L.: Environm. Science & Technology 12 (1978) 673.
217. Munnecke, D. M.: Process Biochemistry 13 (1978) Nr. 2.
218. Tengerdy, R. P., et al.: Appl. Biochem. and Biotechnol. 6 (1981) 3.
219. Trommer, W., et al.: Umwelt 1981 (5), 410.
220. Ward, W. J.: Chloride dioxide, a new selective oxidant/disinfectant for wastewater. Int. Ozone Institute Forum on Ozone Disinfection. 2.–4. 6. 1976, Chikago.
221. Rauh, J. S.: Disinfection and oxidation of wastes by chlorine dioxide. 21. Annual Meeting Mississippi Water Pollution Control Association. 17./18. 4. 1978, Jackson.
222. Gnieser, J.: Chem. Rundsch. 1977, Nr. 34.
223. U.S.-Pat. 4 073 873 (1978), Caldwell, D. L., et al., Dow Chemical.
224. Das techn. Umweltmagazin 1978 (4), 46.
225. Hogan, P., Kuhn, A. T.: Oberfläche-Surface 18 (1977) 255.
226. Zlokarnik, M.: Chem.-Ing.-Tech. 53 (1981) 600.
227. Marquardt, K., in Hartinger, L.: Taschenbuch der Abwasserbehandlung. München/Wien: Carl Hanser Verlag 1976, Kapitel: Umgekehrte Osmose, S. 160.
228. Verfahrensberichte zur physikalisch-chemischen Behandlung von Abwässern. 9. Bericht: Abwasserreinigung mit Ionenaustauschern und Adsorberharzen. Hrsg. vom Verband der Chemischen Industrie, Köln 1978.
229. Miller, R. A., et al.: Evaluation of catalyzed wet oxidation for treating hazardous waste. 7th. Annual Research Symposium, U.S. Environm. Protect. Agency, Philadelphia, Pennsylvania, March 18, 1981; Destruction of toxic chemicals by catalyzed wet oxidation. Purdue Ind. Waste Conference, West Lafayette, Indiana May 1980.
230. Alberti, B. N., Klibanov, A. M.: Biotechnol. and Bioengng. Symp. No. 11, 373 (1981).
231. Carr, J. D.: Ferrate Ion: Potential uses in advanced wastewater treatment. Nebraska Water Resources Center. Project Completion Report A-053-NEB, 1982.
232. Stefanou, E., Giger, W.: Environm. Sci. Technol. 16 (1982) 800.

Incineration of Chlorinated Hydrocarbons

A. Robin

Industrial complexes manufacturing chlorinated organic compounds must provide a satisfactory and lasting solution to dispose of chlorinated residues.

In land incineration plants a very serious problem is the need to dispose of large volumes of hydrochloric acid or inorganic chlorides, after neutralisation of the acid.

The process described hereafter satisfies the environmental protection requirements by an ample margin. It allows the recovering of high quality commercial hydrochloric acid solution at desired concentration or, after distillation, anhydrous HCl.

For this process a solid know-how has been developed and, at present time, these plants give effluents satisfying the most stringent specifications of antipollution regulations, while producing the maximal volume of commercial HCl.

Contents

1 Foreword

Chlorinated hydrocarbons are one of the major product groups manufactured by the chemical industry, both as end products and as intermediates for other products.

In the manufacture of chlorinated organic derivatives, monomers or chlorinated solvents, the yield of waste products is often as much as 5% of the main product. For example, the manufacture of 300000 tons/year of chlorinated solvents may result in 4000 to 15000 tons/year of various wastes containing more than 60% chlorine by weight.

Governments of industrialized countries have prohibited the direct discharge of these wastes into rivers and lakes, but such wastes have been found in landfills.

Thus, industrial complexes must provide a satisfactory and lasting solution to the pollution problems caused by the chlorinated residues they generate.

Large volumes of chlorinated wastes are disposed of by incineration in land-based incineration plants. The most serious problems of these plants are the production of large quantities of inorganic chlorides resulting from the neutralization of hydrochloric acid with alkaline materials, the cost of neutralizing agents and corrosion.

Marine incineration is another technique. However, burning chlorinated wastes aboard incinerator ships at sea does not allow the recovery of any of the waste's valuable components for subsequent recycling, or of any of the considerable calorific value of the waste chemicals.

Finally, a completely satisfactory solution to the problem of chlorinated waste disposal has to

- achieve a recycling of at least part of the chemical material involved,
- have a high operational reliability,
- be economically viable in recovering energy,
- satisfy the present and projected environmental protection regulations.

A completely original process, the V. R. C. process, has been developed in France to fulfil these conditions.

2 The V.R.C. Technology

The V.R.C. process was developed at the Saint-Auban plant, in the South of France, by scientists from several companies: Péchiney, Saint-Gobain, Progil, Rhône-Poulenc, Chloé-Chimie and Atochem.

This process, which is named V.R.C., from the French "valorisation des résidus chlorés" (economic utilization of chlorinated residues), effectively destroys chlorinated hydrocarbons by incineration, while simultaneously producing high-quality hydrochloric acid. It also recovers 70% to 75% of the heating value of the wastes as high-quality steam.

The incineration unit located at Saint-Auban treats about two tonnes/h chlorinated wastes, or 16000 tonnes/year. The output of the V.R.C. plant is 4.3 tonnes/h of high quality hydrogen chloride as 33% aqueous solution, or, after distillation, 1.4 tonnes/h 100% HCl gas.

The furnace operates at a temperature of 1200 °C, with a residence time of 5–6 s. It burns low viscosity liquid wastes and viscous fluids as well as mixtures containing gases and solid particles.

This plant has run continuously since start-up, in 1974, and is operating at capacity to-day. Atochem itself operates two other plants located in Saint-Auban and Fos. Similar plants have been licensed and built in Spain (one), Morocco (one), Russia (four), France (one) and in the U.S. (one). Negotiations are under way for two new plants in the U.S. and three or four in Europe.

3 Discussion of the Process

The general equation for the combustion of a chlorinated hydrocarbon in the presence of an excess of air can by written as:

$$C_xH_yCl_z + \frac{4x + y - z}{4} O_2 \rightarrow x\, CO_2 + z\, HCl + \frac{y - z}{2} H_2O + Q \text{ cal} \quad (1)$$

But, at the optimal incineration temperature (about 1200 °C) hydrochloric acid is in equilibrium with chlorine, according to Deacon's equilibrium equation:

$$4\, HCl + O_2 \rightleftharpoons 2\, H_2O + 2\, Cl_2 - 20.8 \text{ kcal}.$$

This reaction is exothermic from left to right.

The equilibrium in the core of the furnace is arrived at from the kinetics of each partial reaction, and is different from that of the actual combustion reaction.

The equilibrium constant is:

$$K_p = N \frac{[H_2O]^2 \cdot [Cl_2]^2}{[O_2] \cdot [HCl]^4} \quad (3)$$

$[H_2O]$, $[Cl_2]$, $[O_2]$ and $[HCl]$ being the partial concentrations of water, chlorine, oxygen and hydrochloric acid and N the total number of molecules in the combustion gases.

Calculation reveals that the value of K_p is related to the temperature by the formula:

$$\log K_p = \frac{6034}{T} - 6.97 \quad (4)$$

Chlorine is an impurity in the gases coming from combustion and can be eliminated by caustic soda treatment after the hydrochloric acid has been separated by water absorption.

From equilibrium (2) and relations (3) and (4), it can be seen that, in order to decrease the chlorine's partial pressure in the gases after combustion, the following conditions have to be fulfilled:

- a high temperature of combustion,
- a water vapor partial pressure as large as possible taking into consideration the thermal and water balances of the system,

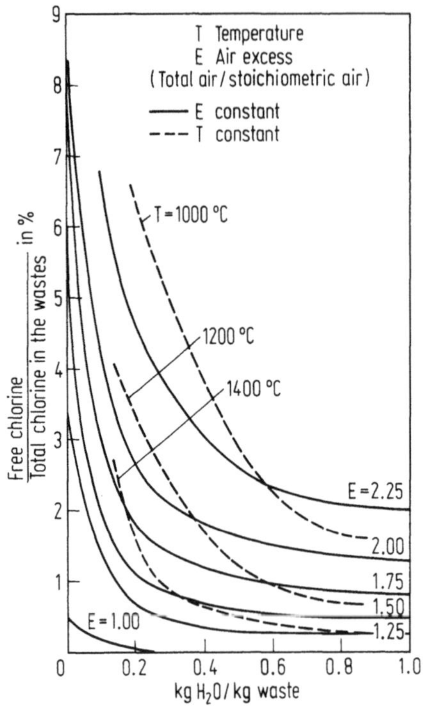

Fig. 3.1. Incineration of wastes from vinyl chloride manufacture average formula: $C_2H_2Cl_{1.8}$

$$Cl = 71\,\% \text{ by weight}$$

Relationship between free chlorine, temperature, excess of air and water content.

— reduction of the oxygen excess to the minimum consistent with complete combustion
— a furnace designed in such a way that the residence time of the gases at high temperature be long enough to enable complete combustion and establishment of Deacon's equilibrium
— a very efficient "quencher" to "freeze" Deacon's equilibrium state.

All the process operating data can be determined using a computer program taking into account the waste's composition and desired capacity (combustion temperature, material and heat balances, equipment sizing).

As an example, the diagram above shows the relationship between various parameters in the combustion of wastes from vinyl chloride manufacture.

For different temperatures and values of air excess, the graphs indicate the relationship between free chlorine and water vapor concentrations.

4 Operating Conditions

The result of the calculations mentioned above, plus the experience accumulated in numerous trials carried out in industrial units, led to the following conclusions:

4.1 Combustion of Wastes

The use of patented burners of a static type [5] provides complete combustion of chlorinated wastes so long as the heating values of the residues are over 2200 to 2500 kcal/kg (about 3700 Btu/lb).

Wastes containing up to 74% chlorine can be burnt in this way. For example, wastes from vinyl chloride manufacture, with an average formula $C_2H_2Cl_{1.8}$ which contain about 71% chlorine by weight, are burnt without the need for a supplementary fuel.

For compounds with deficient calorific values, the addition of fuel oil, natural gas, hydrogen, etc. leads to stable conditions with combustion gas temperature about 1200 °C, without any modification of the burners.

4.2 Combustion Temperature and Residence Time of Gases

In this process, the exothermicity is important and the optimum incineration temperature is around 1200 °C. It has often been observed, especially if chlorine content is high, that the combustion is stabilized only above 900 °C. The first condition is total combustion, the combustion fuel being usually the waste itself.

The burner must be installed in a combustion chamber properly dimensioned to handle the large size flame and to make the maximum use of high temperature radiation.

4.3 Excess of Air and the Water Balance

A large excess of air would reduce the combustion temperature to 800–900 °C, which is too low to guarantee the burning of certain stable organic compounds (polychlorobiphenyls, for instance) and would lead to the formation of a larger amount of chlorine.

In addition, air in excess introduces an additional volume of nitrogen into the process which lowers the HCl pressure in the combustion gases, leading in turn to a more difficult absorption in water.

Consequently, the incineration temperature should not be adjusted by using an excess of air, which ought to be reduced to a minimum.

Direct water injection into the furnace to control the exothermicity would be problematical because small quantities of water quickly disturb the system's water balance, leading to more dilute HCl solutions.

The technique used here to permit a satisfactory temperature adjustment (1200 °C) and a H_2O/HCl equilibrium corresponding to the best recovery rate (33% HCl solutions) consists of recycling directly into the furnace an aqueous HCl solution, preferably withdrawn from the quench loop.

In this way, partial pressure of water in the hot combustion gases is increased while the partial pressure of oxygen is reduced.

4.4 Free Chlorine and Quenching

The manipulations described above, aimed at achieving the best conditions for Deacon's equilibrium, lower the free chlorine output in the core of the furnace. The quench freezes the concentrations achieved.

Chlorine is only trapped in the final caustic soda absorption.

5 Description of Plant

The description and flow-sheet below correspond to an industrial plant able to incinerate 2000 kg/h of chlorinated wastes.

The main sections of the plant are:
- incineration,
- waste heat boiler,
- quench and washing tower,
- HCl absorption and storage,
- HCl distillation and compression,
- final processing of tail gases.

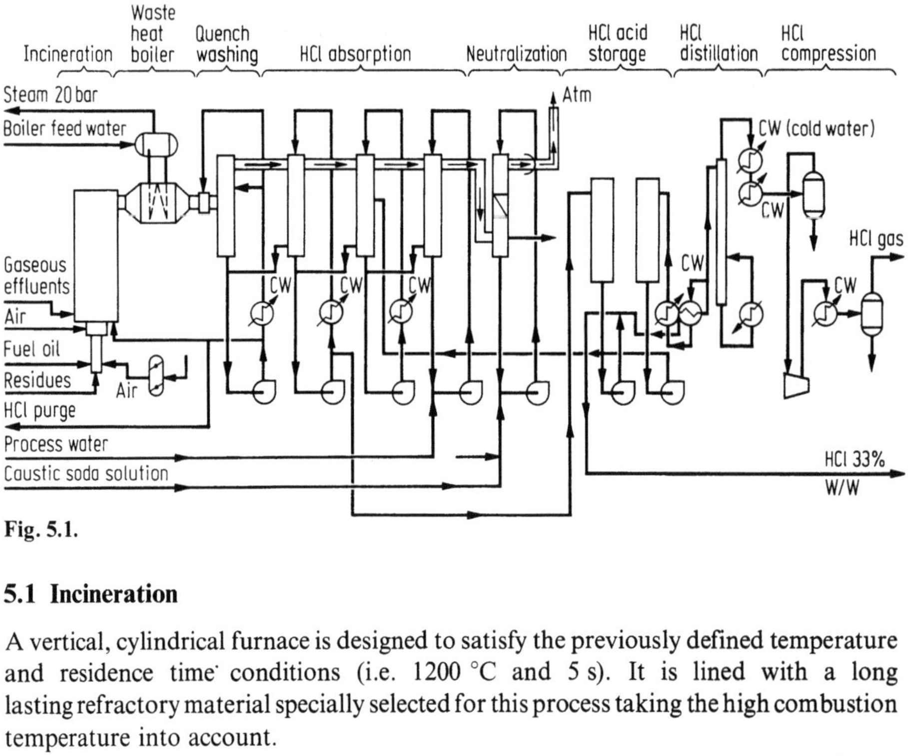

Fig. 5.1.

5.1 Incineration

A vertical, cylindrical furnace is designed to satisfy the previously defined temperature and residence time conditions (i.e. 1200 °C and 5 s). It is lined with a long lasting refractory material specially selected for this process taking the high combustion temperature into account.

The furnace burner, of static type, was developed by Rhône-Poulenc and Atochem. It is covered by a patent.

The system provides an excellent dispersion of the liquids and insures a complete combustion. It also handles viscous fluids, liquids containing small solid particles and gases. The system requires no pressurization of the liquid and is gravity fed by a very small head. The air pressure requirement is about 0.4 bar.

5.2 Waste Heat Boiler

In the V.R.C. process, the calorific value of the waste can be used to generate steam. In such a case, a boiler is positioned between the point where the combustion gases leave the furnace and the quench. Up to 70–75% of the calorific value is recovered and the combustion gases are cooled to 350 °C before entering the quench.

The steam generated has an average pressure of 16 bar (280 psi) and a temperature of 220 °C.

5.3 Quench and Washing Tower

The quencher, which is patented [4] is made of graphite. It was developed in cooperation with the French Company VICARB (Grenoble — France). This very light, extremely reliable piece of equipment, reduces the temperature of the combustion gases to 50–80 °C, depending on the differences in the chemical constituants of the waste and the temperature of the cooling water.

The temperature of the quench is controlled by recycling cooled HCl solution. The same solution is also sprayed directly into the washing tower.

A cooling loop connected to the washing tower is designed to operate in such a way that no HCl absorption takes place. A purge, withdrawn from the loop, eliminates ashes, metallic impurities and other materials. This purge (approximately 25% w/w HCl solution) can be diverted to a tank where it decomposes the stored hypochlorites resulting from the final neutralisation of effluent gases (see final processing of tail gases).

5.4 HCl Absorption and Storage Equipment

A series of absorption units ensures the gas circulation and maintains a slight depression in the furnace, because of a venturi system. High concentration of the solution is reached by means of countercurrent mixing with the combustion gases.

One, two or three isothermal absorption units are required, depending upon such factors as the chemical composition of the residues, the temperature of the cooling water and the desired concentration of the HCl solution.

The HCl is usually collected in the form of commercial 33% w/w solution.

5.5 Distillation and Compression of HCl

If desired, HCl can be obtained as an anhydrous gas (less than 100 ppm of water).

In this case an azeotropic distillation is conducted under 8 bar in a special column. The resulting gaseous HCl is compressed and the 21% HCl azeotropic solution is recycled from the bottom of the column to the absorption equipment.

5.6 Final Processing of Tail Gases

A dosed amount of caustic soda is injected into the neutralization stage in order to react with all the chlorine and that HCl which has escaped absorption in previous stages.

This step does not cause major changes in the pH and output conditions.

After this treatment, the gas is virtually free of chlorine and HCl (less than 30 ppm). Moreover neither phosgene nor nitrogen oxides have been detected in it. Consequently the gas can be discharged into the atmosphere.

The solution from the caustic soda treatment contains previously small amounts of sodium chloride and sodium hypochlorite. As mentioned, the acid purge withdrawn from the washing tower can be used to decompose sodium hypochlorite into sodium chloride and chlorine.

The gases produced on decomposition contain besides chlorine CO_2, from the partial decomposition of the sodium carbonate and inerts. They are returned to the combustion furnace to participate in Deacon's equilibrium.

The resulting brine is led into the sewage system.

5.7 Requirements

For one ton of wastes with the general formula $C_2H_2Cl_{1.8}$ (waste from V.C.M. plant), the average requirements are:

	33% w/w HCl solution	anhydrous HCl
purified water (m^3)	1.6	0.1
consumed well water (m^3)	2.0	2.5
recycled cooling water at 25 °C (m^3)	425.0	475.0
NaOH 100% (kg)	12.5	17.3
electric power (kWh)	150.0	200.0
steam (ton)	—	1.0
manpower	1 man per shift[a]	
maintenance	5 to 6% of capital investment i.e.: spare parts 2% working hours 4%	

[a] however no additional manpower is needed for an integrated unit

6 Range of Application of the V.R.C. Process

Generally speaking, the chlorinated wastes are quite different from each other. The V.R.C. process can be adapted to most requirements, by adjusting each section of the plant to cope with the particular physical and chemical characteristics of the waste and with the local conditions.

The following unit, (fig. 6.1) which is much less sophisticated than the standard one described above, has been operating for 15 years at Saint-Auban.

The V.R.C. process was basically designed to incinerate the wastes produced, as by-products, by the following plants:

- vinyl chloride monomer (VCM)
- trichlorethylene (Tri)
- perchlorethylene (Per)
- carbon tetrachloride (Tetra)
- 1,1,1, trichloro-ethane (T 111)

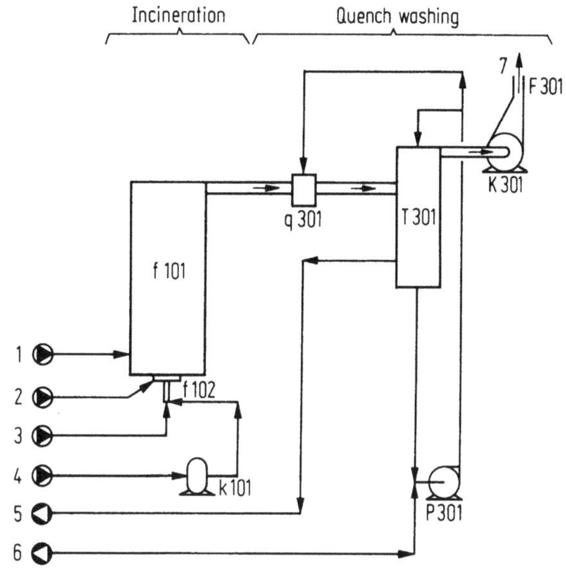

Fig. 6.1. f 101 incinerator
f 102 waste burner
k 101 atomizing air blower
q 301 quench
T 301 washing drum
P 301 washing pump
K 301 exhaust fan
F 301 stack

1 vent gas (possibly)
2 combustion air
3 liquid wastes
4 atomizing air
5 product acid
6 plant water
7 flue gas.

Nevertheless it has other applications. Its efficiency has been demonstrated for the combustion of polychlorobiphenyls (PCBs).

The necessary tests were carried out over 10 days in the Saint-Auban plant with careful monitoring, using pure commercial spent PCBs returned by customers, and without additional fuel.

The sampling procedures and the analytical methods, which were the most accurate available, were selected to comply with the recommendations of the U.S. E.P.A., namely the "train five method" [2] concerning isokinetism in the sampling of gas.

The main results of gaseous effluent analysis were as follows:

HCl	30 ppm	
Cl_2	1.5 ppm	
$COCl_2$	0.5 ppm	(detection limit: 0.5 ppm)
NO_x	20 to 60 ppm	
unburnt particles	from 0 to 2 ppm	(detection limit: 0.2 ppm)
PCBs	13 ppb = 60 mg/hour	
$\frac{\text{PCBs out}}{\text{PCBs in}}$	0.12 ppm	
yield of destruction:	$100\left(1-\frac{0.12}{100}\right) = 99.9999\,\%$	

Semi-oxidized products of PCB combustion, such as polychlorodibenzofuran or polychlorodibenzodioxin were not present at the detection level of 1 ppb.

These figures clearly indicate that the V.R.C. process achieves complete combustion of PCBs. It complies with the recommendations of the E.P.A. in the Federal Register [1] and with the requirements of a French decree [3] concerning the destruction of spent PCBs.

7 Conclusions

Chemical producers faced with the problem of disposing of chlorinated hydrocarbon wastes have few options. The V.R.C. process is of particular interest for wastes having high chlorine content. It is approved by the French authorities for the incineration of PCBs. Environmental authorities in Holland and Germany are also interested as an alternative to incineration at sea.

The process looks quite simple. However to build and operate such units properly, requires a great of know-how about equipment, materials and details of construction.

References

1. U.S. Federal Register *44*, no. 106, p. 31551, May 31, 1979.
2. U.S. Federal Register *42*, no. 160, Part II, August 18, 1977.
3. Conditions d'emploi des polychlorobiphényles. Arrêté du 8 juillet 1975. Journal officiel du 26 juillet 1975, p. 7600.
4. French Patent no. 2086574, 1970.
5. French Patent. Provisional no. 81 13081, 3 july 1981.

Sludge Treatment

M. Chambon and A. Navarro

Residual sludges containing toxic products present a particular risk for the environment in the sense that they constitute dispersed media in contact with the external environment.

This favors the passing in solution of theses products by physical dissolution, hydrolysis, biodegradation, etc... At this risk must be added the one constituted by the more or less active, mobile interstitial liquid phase.

Two types of solutions may be proposed:

The first consists of separating the liquid and solid phases by the most efficient manner possible. Mechanical dewatering, for example, can lead to a solid which, properly washed, can be directly landfilled.

Among the many solutions of this type, we describe a flat rotating vacuum filter.

The second approach consists of solidification or fixation sludge treatment processes. Many such processes have been commercialized within the past years under various trademarks [3] using different reagents and different reacting conditions. These are, for instance:

- Processes using hydraulic binders (lime, cements, ...)
- Processes using pozzolanic type reactions
- Use of other inorganic binders
- Polymerization of thermoplastic or thermosetting organic reagents.
- Coating processes (asphalt).

The purpose followed is double-headed. It is to confer mechanical resistance to the fixed sludge so that firstly it can be landfilled in a physical state compatible with the working conditions on site, and secondly to make it more resistant to the action of solubilisation mechanisms (in particular by rain water). We will develop here only recent examples of applications of these techniques.

All industrial sludges do not consist of solid and liquid phases only, but also of non-miscible emulsified substances (hydrocarbon sludge, oil slick, paint sludge, ...).

The treatment of these sludges cannot employ neither traditional liquid-solid separation techniques nor standard fixation techniques.

That is why we present here an interesting treatment appropriate for these sludges as well as for polluted liquids generated during filtration or solidification operations.

Contents

Chemical Waste
Managing Editor: K. R. Müller

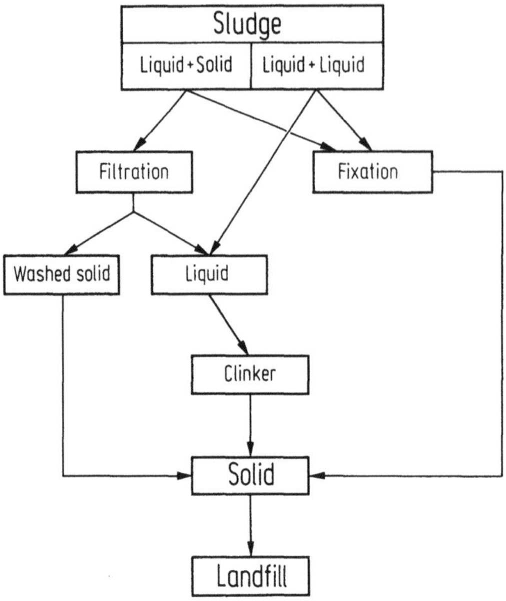

1 Sludge Filtration

Sludge filtration mostly requires an efficient filter which must realize both a good retention of solid waste and a good washing for a complete elimination of liquid pollutants.

The main function of the UCEGO filter (patented by Rhône-Poulenc) is to separate liquid from solid in the slurry generated during industrial processes. The filter separates any size of solid particles from a few tens to several hundred microns, depending on the operating conditions.

The retention of liquid of the slurry cake is generally considerable (25 to 35% by weight) representing an extra pollution if the cake cannot be washed efficiently.

This washing on UCEGO filters is carried out in a countercurrent system through the feed boxes distributing the washing liquid (generally water) radially over the entire surface of the solid cake.

The UCEGO filter is a flat, rotating vacuum filter available in 14 different sizes with the surface area under vacuum ranging from 8.8 to 307 m^2.

Figures 1.1 and 1.2 show the essential characteristics of these filters which are as follow:

- Great rigidity of the table, ensuring complete flatness and hence a regular thickness of the filtered solids, leading to efficient washing and drying.
- Original design of the unit holding the solids, reducing to the minimum the perimeter of the air and washing by-pass.
- Compact construction which still further reduces this perimeter and the draining time of the filtrates.

Simplicity of general design which limits the number of moving parts and the transmitted forces, which hence simplifies the carrying structures. Rotation speeds are higher than those of concurrent filters and, for a comparable effective area, the filtration capacities are also higher.
All these caracteristics result in a considerable reduction in the maintenance costs of the UCEGO filter.

According to the nature of the slurry being treated, this filter can be arranged to provide either two or three countercurrent washing stages.

The UCEGO filter has been specifically designed for the difficult cases of phosphoric acid production. Besides, it may be used for filtering of other products, and particularly in cases where the advantage of a short retention time of the solid matter on the filter is looked for, as is the case of sludge filtration.

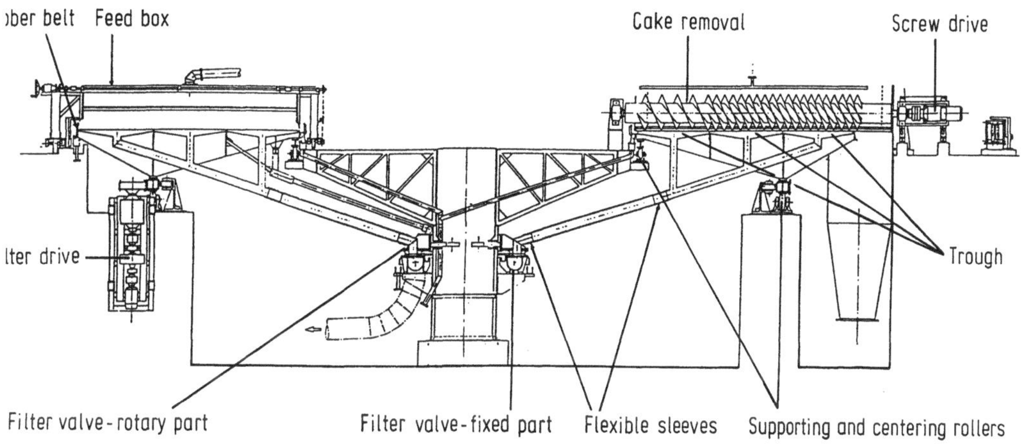

Fig. 1.1. UCEGO Filter — Cross Section (RHONE POULENC)

2 Sludge Treatment by Mixing with Clay Materials

Traditional solidification techniques show poor results when the wastes to be treated contain important quantities of organic substances.

To mitigate this disadvantage, we have studied the use of the specific properties of clay materials which are, in theory, readily available in sufficient quantity on landfill sites and whose adsorbant properties towards organic molecules are well known.

An experiment has been set up in this sense since 1979 at Menneville (France), a landfill site owned by the Company FRANCE DECHETS, and supported by the French Ministry of Environment and A.N.R.E.D. (National Agency for Recovery and Disposal of Waste) [1].

15 tons of petroleum sludge (50–70% water and 30–50% organic matter), mixed with clay, resulted in the production of a very firm solid with the following composition:

- clay: 45%
- water: 32%
- hydrocarbons: 22.5%
- cement: 0.5%

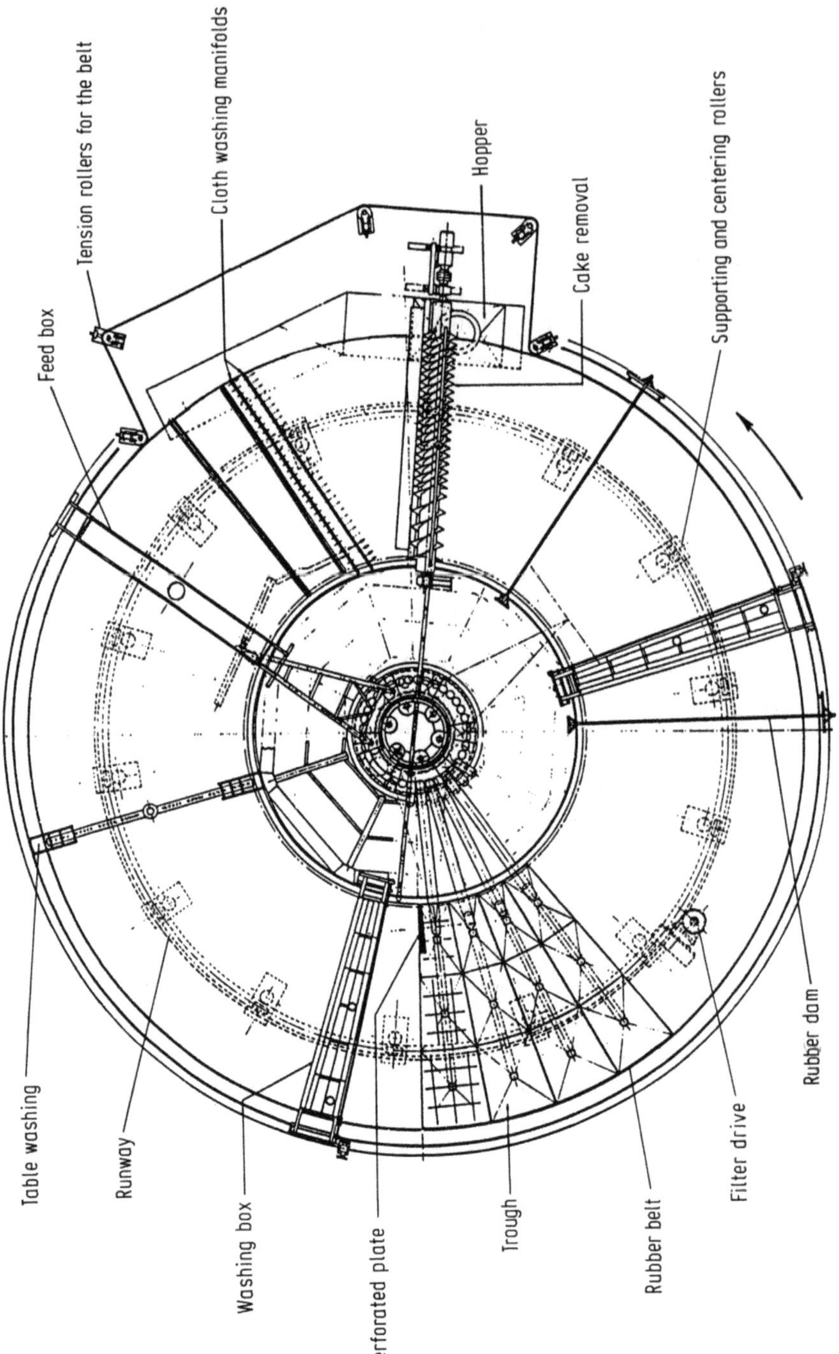

Fig. 1.2. UCEGO Filter — TOP view (RHONE POULENC)

Independently of the fact that the treatment conferred a very good mechanical frimness to the initial sludge, leaching tests with water in the laboratory have shown a very important attenuation of the initial pollutant load with respect to the organic pollution and toxicity.

The COD decreased from 13000 mg/l for the supernatant liquid of the sludge to 140 mg/l for the leachate of the solidified product, whereas the initial Daphnia toxicity of 2500 equitox/m^3 became less than one.

After 2 years of field experiments, the analysis of leachate resulting from the percolation by rain water showed that the extracted pollution (essentially during the first 12 months) was quantitatively very much weaker than could be foreseen form the laboratory tests results (Table 2.1).

This study [2] has been continued in the laboratory on other industrial wastes representative of sludges treatable by this technique (Table 2.2).

A mass ratio $\frac{\text{clay + total residue}}{\text{water}}$ of about 2 (varying from 1.5 to 2.5) is suitable to produce a paste which is neither too fluid nor too viscous.

For certain suldges, the water content is large enough, whereas for others we had to add water to enable mixing.

The added water can besides be replaced by a mother liquor weakly loaded with organic matter.

Table 2.1.

	Laboratory		Field	
	Extrated quantity mg/l	Hold-up by the solid	Extrated quantity mg/l	Hold-up by the solid
COD	1400	98%	25	99.96%
BOD	360	94%	9	99.85%

Table 2.2.

	Sludge	Principal Caracteristics
A	Production residue of nitrophenols and aminophenols	Aqueous solution with: 4.5% inorganic matter and 4% organic matter COD:6300 mg/l
B	Aqueous effluents containing chlorophenols	Aqueous solution: pH = 12.9 COD = 3450 mg/l degree of mineralization:45000 mg/l
C	Treatment residues of hydrocarbon wastes	Very dense black sludge Water:37% Heavy hydrocarbons:39% Total residue dried at 103–105 °C = 24%
D	A.P.I. sludge (American Petroleum Institute)	Heterogeneous mixture Water:72% Total residue dried at 103–105 °C: ~14% Hydrocarbons:14%

Table 2.3.

		%				Mass ratio		
		Organ. Comp.	Total residue (T.R.)	water	clay	Clay / waste	clay + T.R / water	Org. matter / clay + T.R.
A	Nitrophenols + Aminophenols	1.6	1.7	33.7	63	1.7	1.9	2.5
B	Chlorophenols	0.2	1.2	30.6	68	2.1	2.3	0.3
C	Hydrocarbon sludge	5.9	3.5	31.4	59.2	4.0	2.0	9.4
D	A.P.I. sludge	5.8	5.9	31.5	56.8	1.3	2.0	9.2

Table 2.3 shows the composition of the obtained pastes.

These pastes lose about 90% of their water content to the air in about 15 days. Those containing organic substances soluble in water tend to exude.

The other pastes shrink after drying without any splitting.

In all cases there appears a certain efflorescence (inorganic substances but also nitrophenol, for example).

Leaching tests in an automatic percolimeter have shown that C and D resist very well both from a mechanical and chemical point of view.

The pastes A and B show a poor mechanical resistance to water action which results in a more important extraction of undesirable substances (Table 2.4).

Mixing studies have also been carried out by the C.E.R.I.P.E.C. (paint industry technical center) on steam distillation residues of paint cleaning solvents [4, 5].

The following results have been observed, for example, for a one-phased waste with a total residue content of about 26%:

— With ratio sludge/clay of 1/1 (by volume) a solid containing 53.9% total residue was obtained.

Table 2.4. Results of leaching of pastes

	Parameter	% Extrated pollutant	Resistance to water
A	COD Phenol Nitrates	9.3% 5.9% 3.4%	Swelling followed by a slow disintegration
B	COD Chloride	100% 95%	rapid disintegration
C	COD	very weak	good
D	COD Hydrocarbons	very weak very weak	good good

— A leaching test of 18 hours (10% solids and 90% water) resulted in a leachate with the following characteristics:

pH:	7.6	SO_4^-:	74 mg/l
COD:	675 mg/l	phenols:	0.84 mg/l
Cl^-:	not detected	Pb:	0.16 mg/l

A comparative study of different simplified commercial processes has shown that mixing with clay allows a good holding back of hydrocarbons, detergents, and phenol, but this technique is disadvantaged by the poor physical resistance shown by the resulting pastes.

However, important improvements can be made, at short term, in particular in relation with the addition of other materials.

The undertaking of these processes on an industrial scale is acutally in the initial stage and it is not illusory to forecast important progress in this field if more fundamental studies on the fixing mechanisms and on the criteria of selection of wastes and clays are not short-circuited and if the technical economical conditions of working operations are fully mastered.

3 Use of Domestic Waste Incineration Residues

The treatment of municipal solid waste by incineration, which is undergoing an important development, sets up, because of this fact, the problem of the incineration residue [6].

Many techniques exist to valorize the whole or a part of this residue by:

— metal extraction
— production of materials
— use in civil engineering, etc. . . .

Landfill disposal of this waste proceeds according to precise rules which take into account the specific properties of this residue. The study of these properties, in last few years, has been an occasion to set up new techniques using this residue in the conditioning of industrial organic sludge and in the treatment of specific effluents like landfill leachates, for example.

3.1 Composition and Properties of the Residue

Generally, the residues are black solids, more or less divided and very heterogeneous, consisting of bits of glass and highly oxidized metallic objects.

The other constituting solids are in the form of powder and fritted aggregates, either fused or vitrified.

The exact composition depends very much on that of the original wastes, on preliminary treatments, and on the performances of the incinerator.

For example, we can give the following values (Table 3.1).

This residue is an insulating substance of very low heat capacity.

Appropriately arranged in large heaps, the residue is the center of an important thermic effect and physical and physico-chemical phenomena which constitute many

Table 3.1. Example of the composition of a domestic waste incineration residue

Apparent density	1 g/cm³	Specific area (B.E.T.)	30–70 m^2g^{-1}
Solid density	2.53 g/cm³	lime (CaO) content	4–15% (by weight)
Unburnt organic matter	4–14% (by weight)	Scrap-iron	12–30% (by weight)
		Silica (SiO_2)	40–50% (by weight)
Hold-up capacity for water	20–35% (by weight)	Fe_2O_3, Al_2O_3, MgO, MnO_2, ...	Na_2O, K_2O, Zn, Pb, ...

barriers to the washing away and the dispersion of pollutants under the leaching action of rain water.

Those properties have been used for treatment of industrial sludges and aqueous effluents.

3.2 Treatment of Paint Sludge

The production of paints generate specific wastes:

— production errors,
— unsold charges,
— cleaning solvents,
— distillation residues resulting from solvent recovery operations.

Generally, the incineration of these wastes does not set up any unsolvable problems, but nevertheless it runs into certain obstacles:

— the risk of solubilization of metallic elements contained in the ashes,
— the weakening of the refractory lining materials of the incinerator,
— air pollution,
— eventual addition of fuel.

All efforts have principally been centered on solvent recovery, in particular, by evaporation or steam distillation and in all cases there remains a sludge to be disposed of.

In the case of steam distillation, this sludge consists of:

— water,
— non-metallic elements (Cl_2, S, P),
— metallic elements (Co, Mn, Pb, Zn, Cr, Ti),
— and organic constituents (polymers, alcohols, ketones, esters, etc ...).

The C.E.R.I.P.E.C. has obtained encouraging results by mixing this sludge with domestic waste incineration residue [4, 5].

The preliminary studies permit the forecasting of the possibility of developing this technique and extending it to other suldge types.

Indeed, the COD, the phenol, and the metallic contents are strongly attenuated in the leachates, and the products obtained are mechanically compatible with landfill disposal.

For example, in the case of a cleaning paint solvent containing 14% total residue (dried at 103–105 °C), the mixing of two volumes of domestic waste incineration residue with one volume of sludge results in a solid and a released aqueous phase.

The COD of the supernatant of the sludge is 42000 mg/l and the concentrations in Cr and Zn are 85 and 52 mg/l, respectively.

The leaching of the solidified product generates a leachate having a COD of 340 mg/l, containing no Zn and 1.3 mg/l Cr.

3.3 Landfill Leachate Treatment

It is conceivable to use domestic waste incineration residues for treatment of waste water because of the properties of incineration residues, which are:

- thermic effect
- basicity of the material
- adsorption capacity
- porosity
- filtrating properties.

The spraying of landfill leachates on pyramidal heaps of incineration residue may constitute an interesting stage in the treatment line.

This can, furthermore, be extented to other polluted effluents as, for example, those generated by the conditioning or filtration of industrial sludges.

Apart from the major decrease in final volume of the effluent by evaporation and water fixing, there is also an important attenuation of the organic pollutant load (BOD, COD, TOC, phenols) by oxidation and adsorption, and of the metallic pollution (precipitation).

This system, which is the object of a patent, is functioning on industrial waste landfill sites.

Table 3.2 gives the first results obtained by percolating 150 m^3 of leachate on a heap of 4300 tons of incineration residues during 500 hours. The attenuation of pollution observed in the 75 m^3 collected after percolation is spectacular. The purifying capacities of the heap were not yet exhausted.

Numerous improvements can still be made on this process, in particular with regard to the addition of judiciously chosen chemicals, intending to improve greatly the efficiency and especially the purifiying capacity.

It remains also to determine the place of this system in a complete treatment unit (flocculation, biological and tertiary stages) with respect to the nature and volume of effluents to be treated and, above all, the quality goals fixed by the natural environment.

Table 3.2. Composition of landfill leachate percolating on incineration residue heap

	Influent leachate	Abatement for effluent
Color	Black	Colorless
Odor	Fecal	Odorless
COD	6422 mg/l	85%
Water hardness	122 °F	91%
Fe	13.5 mg/l	99%
Zn	0.25 mg/l	80%
$N(NH_3)$	211.5 mg/l	63%
N-Organic	67.5 mg/l	71%
Phenols	14.5 mg/l	99.6%

4 Conclusion

We did not wish to carry out a complete review of all sludge treatment processes.

Instead, we have tried, through some examples, to demonstrate the progress realized in techniques that have shown their proofs (e.g., filtration), and also to inform the reader on new, improvable techniques, which, however, already have promising industrial applications (use of clay materials and domestic waste incineration residues).

Without minimizing the merits of other techniques, our choice has been quided by our personal experience and by the studies which we have followed on the field.

References

1. A. Navarro, J. Nicole, P. Revin, "Recherche en laboratoire et dans des casiers expérimentaux sur le comportement de résidus industriels polluants en décharge étanche." Symposium protection des sols et devenir des dechets La Rochelle, 22–24 novembre 1983, Secrétariat d'état à l'Environnement et à la Qualité de la Vie, Agence Nationale pour la Récupération et l'Elimination des Déchets (A.N.R.E.D.), pp. 363–386, Conventions: 7847–7853 81138 to 81149.
2. J. Veron, J. M. Blanchard, A. Navarro, "Attenuation du caractère polluant des déchets industriels par malaxage à l'argile." in: Symposium La Rochelle (1), pp. 209–222.
3. F. Colin, "Evaluation des performances des procédés de fixation des boues utilisés en France." in: Symposium La Rochelle (1) pp. 199–208, Ministère de l'environnement: Convention 8001465 00223.
4. Bulletin du C.E.R.I.P.E.C., No. 54, Novembre 1980 Z. I. Petite Montagne B.P. 1416 91019 — EVRY CEDEX F.
5. J. M. Blanchard, A. Navarro, M. Charreton, "Contraintes et Perspectives dans le traitement des déchets de fabrication des peintures." XIVème Congrès A.F.T.P.V., Aix Les Bains 1981, pp. 25–29.

6. P. Revin, J. M. Blanchard, "The impact of landfilling on a porous geological site with residue resulting from the inceneration of municipal waste." I.S.W.A. — 5ème Symposium Européen des eaux usées et des résidus urbains E.A.S. Munich, 22–26 juin 1981.
7. J. Simond et I.N.S.A. de Lyon (J. M. Blanchard, A. Navarro, P. Revin), "Procédé de traitement d'effluents liquides pollués." Brevet France no. 82.19.718., 1982.
8. J. M. Blanchard, P. Revin, J. Simond, "Le traitement des lixiviats de décharge." In: Symposium La Rochelle (1), pp. 253–266.

Societe France Dechets, 71, rue Bretonnet — 78970 — Mezieres/Seine F.
C.E.R.I.P.E.C., 181, avenue Jean Jaurès — 69353 — Lyon Cedex 2 F.
Filtres U.G.E.C.O. — Rh ne Poulenc, 18, avenue d'Alsace — La Défense 3 — 92400 — Courbevoie adresse postale: Cedex 29, 92097 — Paris La Defense

Waste Disposal Site Sanitation

H. Schirmer

At the beginning of the 1970s, Boehringer Ingelheim KG set up a waste disposal site at a former gravel pit near the company's premises. To investigate subsoil conditions beneath the pit floor, bore holes were sunk. Beneath the pit floor a very uneven clay-marl layer was found and, above this, a ground-water body in a sand-gravel mixture. In accordance with the level of technology at that time, the site was sealed off by being cemented over; the recently developed special cement Pectacrete was used. Then, laboratory and production residues, sometimes common salt and water-insoluble, bitumen-type distillation residues were deposited at the waste disposal site. Infiltrating water crossed the sealing layer into a drainage ditch before reaching the sewage plant via a pump installation.

After some years, very high chloride concentrations occurred in the garden wells of houses in the immediate vicinity. There was a strong suspicion that the cement layer was not meeting requirements and that contaminated infiltration water was leaking out.

On the basis of the particular geological situation, i.e. that there is a water-impermeable clay-marl layer at an economic depth, the sanitation proposal to construct a sealing wall around the site was implemented. The wall is fixed into the clay-marl layer to a depth of 2.5m and is intended to deflect oncoming water around the site. Sink wells were created at deep points of the site to collect infiltrating water and convey it to the sewage plant. This ensures that the water level within the site is always lower than that outside. Therefore, the hydraulic gradient thus obtained prevents infiltrated water diffusing outwards through the wall. Following extensive investigations, a sealing-wall compound was developed which has a low degree of permeability, high pressure resistance, high resistance against substances corrosive to concrete and is easily dealt with. Construction of the sealing slit wall was carried out in one stage. Ground water already rose by more than 1 m in the S part of the site during construction.

4 years after construction of the sealing wall, a rise in ground water of almost 3.0 m has been observed at times at some points.

Infiltrated water consists of precipitation water and ground water diffusing through the wall. According to theoretical calculations, precipitation accounts for around 0.55 l/s, diffusing ground water 0.77 l/s. The monthly mean for water actually pumped off the site is between 1.1 and 1.6 l/s.

Six-monthly investigations of the ground water upstream and downstream of the site as well as inside reveal that conductivity, chloride concentration and COD values have fallen sharply following construction of the sealing wall downstream. To date, however, values measured upstream of the site have not been attained. A further reduction can be expected as a result of dilution. At the observation marker inside the site, however, there was a considerable increase in individual parameters because of elution from the deposited waste substances.

It can, therefore, be established that the sealing slit wall fulfils impermeability requirements and that the actual aim of sanitation, prevention of continuous contamination of the ground water downstream of the waste disposal site, is achieved.

Contents

Chemical Waste
Managing Editor: K. R. Müller

1 Foreword

In the following report, the sanitation of an existing Boehringer Ingelheim KG waste disposal site is described. The Boehringer company is a chemical concern with around 22,000 employees worldwide and chiefly operating in the pharmaceutical, but also in the veterinary, medical, laboratory diagnosis, chemical, baking agents and pesticide sectors.

The location of the town of Ingelheim and the waste disposal site, in what is termed the Mainz Basin, necessitate a brief explanation of the geological features of this basin, since these factors decisively influenced the implemented sanitation proposal.

2 The Ingelheim Waste Disposal Site

2.1 Geology of the Mainz Basin

The valley and plateau landscape between the Odenwald, Spessart, Taunus, Hunsrück and Pfälzer Bergland (Palatinate Uplands) consisting of Tertiary and Pleistocene (ice age) rocks is termed the Mainz Basin.

The first tectonic sedimentation in the Tertiary valley area took place around 250–300 million years ago at the point where the Saar-Saale valley running NE–SW crosses the Rhine rift valley running N–S. From the Eocene to the Tertiary period, the thus originating Mainz Basin formed a flat, roundish sea inlet which flooded the northern Rhine valley depression. Here, Rupelian clays, creeping marl and Cyrenian marl were deposited in the course of time. These water-impermeable layers at an economically attainable depth are the essential requirements for later site sanitation.

As a result of elevations the sea retreated by the end of the Oligocene period and marl deposits were formed. In the course of further geological development the sea advanced; this was followed by the Basin being largely cut off and salt water being replaced by freshwater. During the lower Pliocene period, "dinotherium sands" were deposited. These were the sediments of the primeval Rhine river. Today's Rhine only took its present course in the Upper Pliocene period after the watershed was breached at the level of the Kaiserstuhl. The Mainz Basin attained its present form in the last 3 million years of alternating ice ages and interglacial periods. As a result of high volumes of water conveyed by rivers and streams all the year round, substantially greater quantities of sludge and rubble were transported away than deposited and, in connection with an up-lifting of the Rhine-Hesse (Rheinhessen) area, the waters formed a deep cleft.

2.2 Location, Geological, and Hydrological Characteristics of the Waste Disposal Site

The Ingelheim waste disposal site, a former gravel pit, is located at the northern edge of the Mainz Basin in the upper region of the Lower Terrace of the Rhine (Rhein-Niederterrasse) on the left bank of the Rhine between Mainz and Bingen. The site lies about 3 km from the river. For many years, the normal level of the Rhine at this point has been NN +79.30 m. The edges of the waste disposal site are around NN +92.00 to NN +99.00 m and those of the floor NN +86.00 to NN +83.00 m.

The existing and planned waste disposal site extends about 200 m from north to south and around 360 m from east to west, with a total area of around 6.4 ha (see Fig. 2.1).

Profiles used for pit supporting walls and numerous soil probes essentially showed the following subsoil structure:

(a) 0–1.5 m under GOK[1]: slightly argillaceous, humus sand; 1.5–6.0 m under GOK[1]: medium sand
These sands are termed "drift sand" which, geologically, collected into dunes in the Würm glacial and postglacial periods, respectively. The Pleistocene lower terrace of the Rhine is regarded as a supply area.

(b) 6.0–15.0 m under GOK[1]: sand and gravel
The sand and gravel layer, also subjected to former degradation, is 9–15 m thick in the area of the waste disposal site and was deposited about 100,000 years ago. These deposits can be described as the middle terrace of the Rhine river.

[1] Grade ground level.

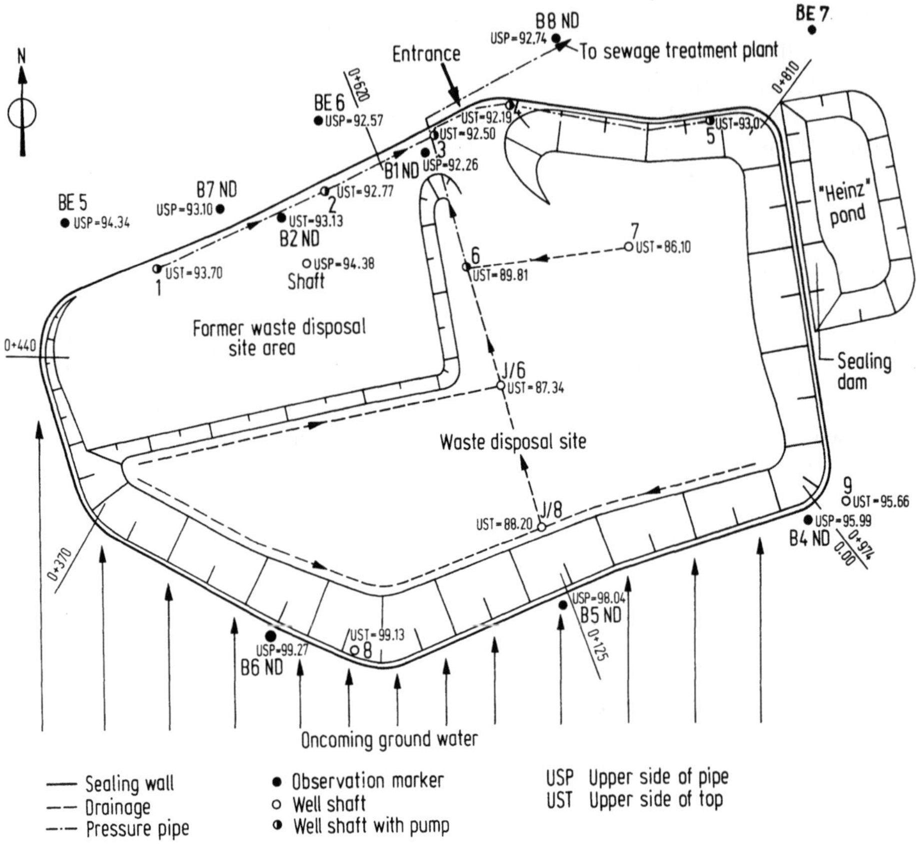

Fig. 2.1. Ground plan

(c) Beneath the gravels and sands is a layer of clay marl, "Cyrenian marl", from grey-green to grey-yellow in colour. In the area of the pit it has a proven thickness of more than 6.0–8.0 m;
individual investigations even yielded thicknesses of over 25 m.

(d) Beneath the clay marl is a layer of "creeping sands" 50–100 m thick. This layer of creeping sand consists majorly of argillaceous, highly water-permeable soil materials.
Therefore, a total of 100 m of sealing complex layers can be relied upon.

The hydrological characteristics of the gravel-pit area were determined by considering the location at the foot of the extended slope terrain declining to the north, the layer structure of the subsoil and the submorphology of the clay-marl surface. Moreover, ground water flow is determined by precipitation water which, because of an impervious marl layer, cannot seep through at the summit of the hill and usually flows downhill as surface water or just below the surface to seep away in the area of the creeping sand covering the Middle Terrace of the Rhine.

In the area of the waste disposal site, the clay-marl surface has a very uneven morphology. There are differences in depth of up to 7.0 m.

The ground water generally flows downhill in a NNW direction. Ground water depth varies greatly because of the morphology of the clay marl. Essentially, the surface of the ground water follows the same gradient as the clay-marl base.

2.3 Origin of the Waste Disposal Site

Before the FRG Waste Disposal Law *(Abfallbeseitigungsgesetz)* came into force on 7 June, 1972, the disposal of wastes was not uniformly regulated. It was the norm to store accumulating wastes near production sites. Precautionary measures regarding the permeability of a waste disposal site and the problem of infiltrating water must be regarded in the light of the level of knowledge at that time. It cannot be overlooked that in recent years a special technology concerning the construction and operation of waste disposal sites had firstly to be developed.

Since there was a partially worked sand and gravel pit available directly outside the Boehringer company gates, it was an obvious thought to use this pit for disposing of accumulating wastes. Quarrying of sand and gravel in this pit began before World War II and was carried out by local, small companies. After the War the requirement for sands and gravels increased considerably, so that the pit had already undergone considerable development. The vertical quarrying limit extended as far as the exposed ground water zone. The stipulation not to cut into this was imposed on the companies.

To ascertain subsoil conditions in the NW area of the former gravel pit, 16 drill holes were sunk to a maximum depth of 5 m. Beneath the pit floor, a sandy gravel between 0.8–4.2 m thick was generally found. The Tertiary clay-marl layer beneath this (Cyrenian marl) was described as heavily argillaceous with a stiff consistency.

The irregularly undulating surface of the clay marl, which inclines generally from north to south, was recognised. Soil probes found ground water inside the pervious, sandy gravel. The depth of the ground water depended on the shape and elevation of the marl surface and fluctuated between 0–3.3 m.

This made it clear to clients and the technical authorities that appropriate measures had to be taken against contamination of the ground water.

Given the hydrogeological conditions encountered, there were two obvious possibilities for protecting the ground water from contamination:

(1) Sealing the floor of the waste disposal site.
(2) Blocking ground water flow in the area of the pit by means of a "watertight" sheet pile wall or injection zone.

re (1)
At that time, there were still no proven sheets available for sealing off waste disposal sites. Therefore, sealing off meant reinforcing the outcropping sandy gravel with the aim of achieving a high level of imperviousness.

re (2)
Blocking the ground water flow using a steel-sheet pile wall was also left out of consideration because of expected corrosion and the fact that the sheet-wall fasteners would not be watertight, particularly during banking-up of ground water.

Extensive injections were known from dam construction where maximum watertightness involving a considerable outlay had a K_f value of 3×10^{-7} m/s. This did not seem quite sufficient for sealing off the waste disposal site. Moreover, there was no appropriate company available nearby to carry out the construction measures inevitably entailing a major technical outlay.

Following the decision to cement the floor of the waste disposal site, the Dyckerhoff Zementwerke offered a recently developed cement especially for such projects. This was the Pectacrete product. The advantage of this special cement was the coating of the individual cement particles with a wax layer in a special process. Only when the cement is introduced into the outcropping ground material is the wax destroyed so that, from this point, the cement is reached by the mixing water and hydration can begin. This had the major advantage that the material could be handled even during rainy weather.

In the laboratory, a sample from the outcropping ground material was mixed with the special cement. At an optimum water content of 7 % this yielded pressure resistances of around 9.2 N/mm^2 after 28 days. A cement layer at least 20 cm thick with a Pectacrete content of 20 kg/m^2 was recommended. This layer was to be laid down in one operation, its thickness corresponded to that used in road-building at that time. Attention was also drawn to the possibility of cracks. To eliminate this weak point the cement manufacturer proposed an additional bitumen layer.

The permeability of the cement layer was also of interest. An investigation report dated 31 March, 1971 ascertained a K_f value of 2.97×10^{-9} m/s at a Pectacrete content of 5 %. Apart from the possibility of crack-forming, this is a respectable value still demanded even today for substructure sealing.

Following classification of these requirements, the construction of an initial section with an area of around 16,000 m^2 was planned for the NW region of the site.

Soil elevation conditions were adapted to the ground-water levels ascertained by drill probes. These probes revealed a site floor having to be manufactured in 22 m multipitch roof-type strips. The lateral gradient selected was 3.5 % from south to north, 0.5 % from north to south. The longitudinal gradient selected was 0.25 %. The slope on the north and west sides was to be cemented to a depth of approximately 1.5 m above the subsoil, also using Pectacrete. At the deepest point, the upper surface thus constructed was about 1 m above the ground-water level found in this area. Many years of observations of ground-water ponds in the pit area showed reasonable fluctuations of a maximum of 0.3 m. The entire cemented surface drained at deep points on the east side into a recipient trench inlaid with concrete semi-shells, terminating with a submerged pump at a pump station to be constructed. The intention was to construct a DN 100 pressure pipe made of PE from this point up to the sewerage system on company premises. The manufacturer's suggestion to provide the finished, cemented layer with a bitumen layer was taken up — a cast asphalt layer was incorporated into planning.

Construction then took place in summer 1971 during favourable weather conditions. The client deemed it possible to dispense with the additional sealing of the concrete layer recommended in the planning stage. Following consultation with technical experts in the competent ministries and after taking into account experience at other sites, cementing in connection with proper drainage of infiltration water appeared sufficient.

On completion of this measure, laboratory and production residues were deposited at the site and covered over with building rubble and excavated earth to a depth of around 2 m. The accumulation was around 4 m thick in spring 1975. Also among the deposits were common salt and water-insoluble, bitumen-type distillation residues. Over the following years, high chloride concentrations occurred in the garden wells of the houses in the immediate vicinity, around 100 m north of the waste disposal site. It was suspected that the high level of salt in the garden wells was attributable to site deposits. Therefore, an exploratory excavation pit about 4.5 m deep was sunk inside the site on 19 Sept. 1974. This breached the cement layer. The exploratory pit terminated in the sand and gravel layer beneath the cement. Samples were taken at intervals of around 1 m from the wall profiles and substructure cement including the gravel-sands outcropping beneath it. The samples were hermetically packed and their water content investigated by an institute. To summarise, it was ascertained that an infiltration water zone was forming in the substructure of the waste disposal site. The sample from the cement layer had a water content of only 5.78 weight % and, in the open area of the exploratory pit, no cracks were visible to the naked eye. In contrast, the sandy gravel soil material found in the substructure above the layer was saturated with water at 18.5 weight % and contained free infiltration water. Despite these positive findings, the possibility that the infiltration water had migrated into the ground water-bearing material around cracks in the cemented area could not be ruled out.

3 Possible Sanitation Measures for the Waste Disposal Site

Because of continuous contamination of the ground water downstream of the waste disposal site, the authorities demanded measures aimed at its sanitation. Various proposals were put forward, some of which should be mentioned briefly:

- One proposal was based on covering a section of the planned site area (see ground plan) with extensive polyethylene sheets onto which the entire quantity of waste (80,000 m^3) would then be redeposited. In an extension of this proposal, a clay-marl covering was also to be added to the layer of sheets since they were not regarded as non-ageing.
- A second proposal provided for two layers of sheets with additional drainage between the layers. However, since the sheets of both layers are manufactured at the same time they age at the same rate, thus providing no additional reliability.
- The particular geological situation in the Mainz Basin with the naturally occurring sealing in the substructure resulted in the proposal to enclose the site with a sealing wall; this would have to bond deeply enough into the clay marl and deflect oncoming ground water around the waste disposal site. At deep points of the site, ground waterlowering wells were to be sunk to collect infiltrated ground water and conduct it to the sewerage system. Despite initial resistance by the authorities, this idea prevailed.

On the basis of, relatively, only slightly higher costs, the decision also allowed for additional site capacity since the pit available was large enough.

4 Sealing Wall Construction

4.1 Planning Requirements

Before the sealing wall was planned, an investigation of the site floor was carried out. Bore-holes were sunk at intervals of 20 m in the entire area of the future site. These bore-holes were then precisely measured and charted with regard to layer and elevation, which therefore provided a relatively exact picture of the clay-marl surface. This zone proved to be very irregular with deep grooves and steep elevations; it was possible to ascertain a general difference in elevation of the clay-marl surface of around 7 m from the southern to the northern edge. The thickness of this blocking layer as ascertained by boring was 8 m. However, deep bore-holes sunk in the vicinity of the waste disposal site indicated a thickness of up to 100 m (see also Sect. 2.1.). Laboratory investigations of the clay marl yielded a lime content of between 15.5 and 29.5 %. The permeability coefficients for the clay marl were $K_f = 2.1 \times 10^{-11} - 9.2 \times 10^{-11}$ m/s.

After it had been determined that the substructure seal is sufficiently thick and, furthermore, largely impervious and regular in granular composition, it was possible to implement the idea of pressing a water-impermeable pipe into the clay to divert the ground water.

4.2 Function

The function of the sealing wall is to seal off the planned area of the waste disposal site from the external ground water flow. The quantity of ground water infiltrating the site is considerably reduced, and is pumped off using specially placed pumps.

Besides this sealing function, the sealing wall has an even more important purpose. As a result of very low permeability, ground water banks up outside the sealing wall and therefore the ground-water level is higher than previously. An appropriate drainage system within the site ensures that the water level is always lower inside than outside. The hydraulic gradient thus obtained thereby keeps the infiltration water away from the sealing wall and only allows external ground water to penetrate the sealing wall.

4.3 Sealing Compound Technology

Following numerous investigations regarding composition of the suspension, the following mixture was used per m³ for the slit: 32.7 kg tixoton 15 (bentonite)
200.0 kg blast furnace slag cement HOZ 35 L in accordance with DIN 1164
924.0 kg water

During excavation of the slit trench, this bentonite-cement compound supports the earth walls. After reaching the final depth, the compound slit trench in the remains, hardens via the hydration process of the cement and ensures sealing.

A sealing wall compound used in the single-stage process differs in some of its properties from bentonite suspensions and concrete which, in the traditional slit-wall construction method, assume the functions of support fluid and sealing.

As opposed to bentonite suspensions used in slit-wall construction, sealing-wall compounds have a:

- higher proportion of solid substances and therefore higher density,
- higher filtrate water loss,
- higher viscosity,
- higher pH value.

Sealing-wall compounds differ from usual concrete technology regarding:

- substantially high water/cement factors (usually water/cement factor ~ 5),
- presence of cohesive fine constituents (bentonite),
- pressure resistances which increase very slowly and are clearly measurable for several years,
- long working times following pumping into the slit trench.

Essentially, the mixture should fulfil the following stipulations:

(a) The required permeability of the suspension taken at the building site must reach a value of $K_f = 1 \times 10^{-7}$ m/s.
(b) Pressure resistance must be 200 kN/m^2.
(c) Erosion stability must be ensured.
(d) The mixture ratio must yield a stable suspension, i.e. the finished mixture must not display sedimentation behaviour.

4.4 Sealing Wall Components and Interaction

4.4.1 Bentonite: General Aspects

Bentonite is a clay discovered by a geologist in 1890 near Fort Benton, Wyoming (USA), from which the name is derived.

The chief constituent of bentonite is the clay mineral montmorillonite, with an average content of 70–90%. In addition, bentonite also contains accompanying minerals such as illite (10–19%), quartz (3–5%) and other minerals (3%) (mean values).

Owing to its many favourable properties, bentonite is used today in special operational areas in various branches of industry, e.g. ceramics, casting and founding, the chemical and pharmaceutical industry and drilling and civil engineering.

The properties of bentonite are determined by its chief constituent, montmorillonite, an aluminium hydrosilicate. Its structure and capacity for internal crystalline swelling enables it to absorb liquid by changing the intervals between silicate layers. When proper activation takes place, the platelet-shaped montmorillonite crystals swell during dispersal in water to the extent that individual crystal lattice layers can be prised apart by shear force almost like the pages of a book. This yields a very large number of extremely fine particles as large as colloids. Moreover, cation-exchange capacity enables bound calcium ions to be replaced with sodium ions, which have much greater swelling capacity.

4.4.2 Properties and Interaction of Sealing wall Components: Bentonite and Cement

The tixotone CV 15 used for the sealing wall is a special sodium bentonite particularly suitable for constructing sealing walls in single-stage processes. It is characterised by an especially high swelling capacity and an equally high water-binding strength.

On the basis of colloid size, montmorillonite particles have an extremely large specific surface of up to 1000 m^2/g. Given the composition of the sealing wall compound, 32.7 kg tixotone CV yields a bentonite surface of 3.27×10^7 m^2 per m^3 sealing compound. This large surface area is sufficient to bind all the mixing water. The mixture ratio yields a stable suspension showing no sedimentation behaviour.

Cements, however, have a low water-retaining capacity; they can only bind approximately 40% of their own weight, 25% of this chemically, 15% physically. Therefore, 200 kg cement stabilise only around 80 l water. This is attributable to the essentially smaller specific surface compared with bentonite. The specific surface of the HOZ 35 L blast-furnace slag cement used is around 3,300 cm^2/g. The total surface of the cement particles at 200 kg given 1 m^3 sealing compound is only around 66,000 m^2. A comparison of the specific surfaces yields a bentonite/HOZ ratio of 495:1.

To prepare the sealing compound on site or in the laboratory, a bentonite suspension is first of all prepared (bentonite + water) and then mixed with the cement in a second mixing operation. A stable sealing compound is thereby obtained, i.e. a cement-stabilising bentonite with two additional reliability components.

The bentonite provides stability, water binding capacity and, therefore, resistance against contaminated infiltrating water, as well as imperviousness. The cement provides not only erosion resistance but also an improved sealing effect.

Compared with another commercially available type of cement, the blast-furnace slag cement used has the advantage that hydration begins later. This facilitates longer processing times for the compound during excavation activity without disturbing the setting process of the sealing wall compound.

4.5 Sealing Wall Stability and Quality Controls

The long-term stability of the sealing materials against sometimes very aggressive infiltration water had also to be discussed. To investigate the influence of this water on the permeability of the sealing-wall compound and to determine its stability extensive tests were carried out. In a long-term investigation using contaminated infiltration water a decrease in the permeability coefficient (K_f value) was ascertained during the period under investigation, i.e. the sealing-wall sample became less permeable. Other tests demonstrated the stability of the sealing-wall compound against substances which attack concrete.

To enable analysis of the behaviour of the sealing wall, two special cylindrical containers made of PE were constructed, each capable of holding 25 individual test samples with 10 cm diameter and a height of 10 cm (see Fig. 4.1). The circular base of the container is divided into five segments and, vertically, it has five decks. A cylindrical, perforated hollow body is placed on the container, which is open all the way round. Using a wire cable, the entire container can be lowered into a well shaft. One container was deposited inside the waste disposal site (in infiltrated water) and the other outside it (in the ground water); care had to be taken that the samples were standing in water. After 2 years, two samples were withdrawn from the wells and their permeability to water laboratory-tested. Permeability of the sample left in ground water was 1.3×10^{-8} m/s and for that in infiltrated water 3.1×10^{-8} m/s. Compared with the mixture used in the suitability tests, permeability had declined in both samples.

The samples from the ground water had a typically blue colour while those deposited in infiltrated water were black. Structural changes or dissolution could not be ascertained.

Fig. 4.1. Containers for storing test bodies

The conclusion is that the sealing compound fulfils stability and impermeability requirements.

5 Implementation of Construction Measures

Construction work was begun in May 1979. Firstly, the sometimes vertical slopes were regulated and brought to the statically required gradient of 1:2.

Most of the soil substances required for this could be obtained from the floor of the future waste disposal site. Concerning the necessary drainage operations, the costs of researching the clay surface were more than recouped. By specifically locating drainage channels it was possible to obtain valuable backfill material. Simultaneously, capacity could be increased by around 20,000 m^3. During the initial earth moving work in clay as well as the subsequent construction of the drainage system, the irregularity of the clay surface was confirmed. Sudden deep channels and elevations of up to a meter later justified selection of the relatively great fixing-in depth of 2.5 m.

The sealing wall was constructed in a single-stage procedure in which a self-hardening filling compound was used. Construction was according to the "pilgrim step procedure". Firstly, primary lamellas were driven into the ground at fixed distances (see Fig. 5.1) and, as soon as the sealing-wall compound had attained a compact consistency in these, the secondary sheets were constructed with a 30 cm overlap. This created a homogeneous seeling wall fixed into the clay zone to a depth of 2.5 m. This achieved the aim of extending the impervious clay soil (substructure seal) upwards into the outcropping pervious sand-gravel material above (Fig. 5.2).

Owing to the high degree of vertical precision demanded, the client decided on a residual reinforced concrete guide wall for guiding the excavator. The guide wall, the base of which was covered with concrete on completion of sealing wall works as a sealing wall protection, now serves as a drainage ditch for eventually approaching surface water.

A 50 cm-thick layer of gravel remained on the site floor above the clay zone. This

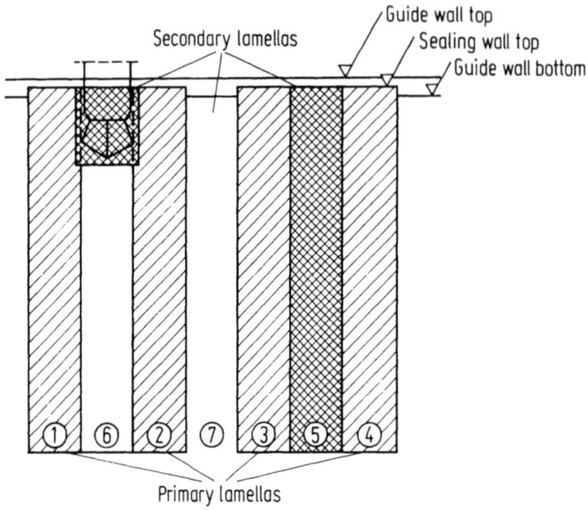

Fig. 5.1. Pilgrim step process

gravel layer was linked to a ring drainage system. The ground water-lowering wells mentioned above were sunk at deep points of the clay surface inside the waste disposal site. These wells have a depth of up to 16 m beneath the subsequent site. Internal diameter is 1.5 m. Submerged pumps were installed to deal with infiltrated water. Normal concrete pipes perforated in the lowermost section were used as well-pipes.

Since a banking-up of approaching ground water had been expected from the outset following construction of the sealing wall in the S area, 3 DN 100 observation pipes were sunk down to the clay floor south of the sealing wall before building commenced. Continuous observation of this marker showed a still uncompleted rise in the ground water of up to 1.2 m (Fig. 5.3) following construction of the sealing wall.

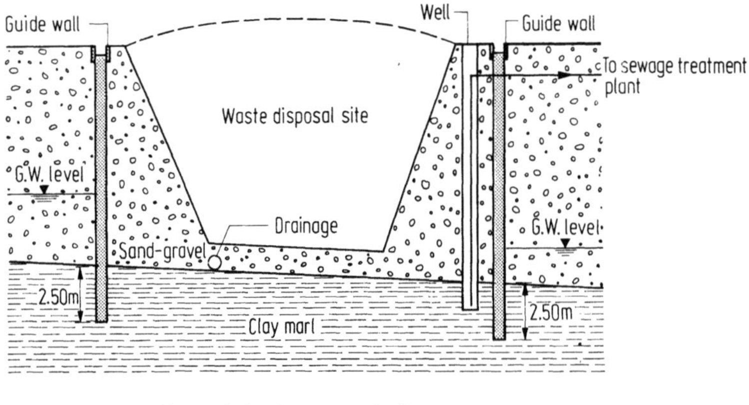

Fig. 5.2. Waste disposal site (cross section)

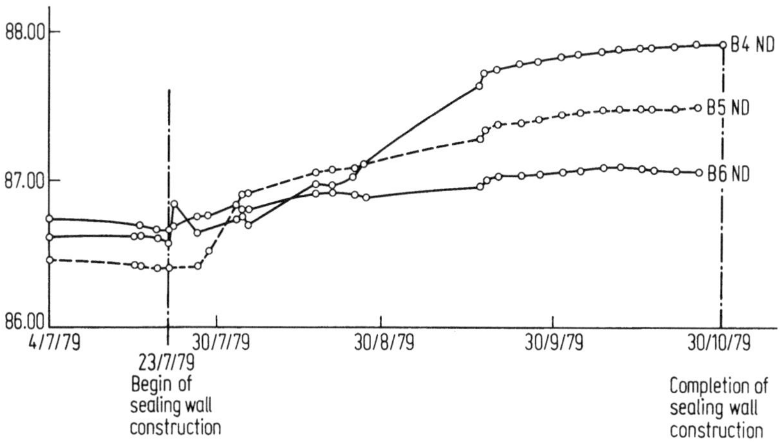

Fig. 5.3. Increase in WFLL-water levels during construction period

6 Infiltration Water Quantities

Infiltration water consists of precipation on the waste disposal site and ground water penetrating the sealing wall.

The amount of water infiltrating as precipitation depends on the infiltration rate for the outcropping subsoil. The rate inside the planned waste disposal site is around 50 %, for the existing site around 40 %. An existing site area of around 69,000 m^2 and mean yearly precipitation of 530 mm yields an annual mean of 0.55 l/s.

The determining parameters for calculating the amount of infiltration from the ground water are the permeability of the wall and the hydraulic gradient between the inside ınd outside wall. Permeability of the wall is calculated at the required value of $K_f = 1 \times 10^{-7}$ m/s. In initial calculations made shortly after completion of the wall, the difference in water level between the inside and outside wall was put at 3 m in the southern section of the site. On the basis of oncoming ground water considerable banking-up was expected here while, along the N, W and E sides of the site, increased mean levels of only 1 m were assumed. However, well observations conducted over several years revealed that banking-up has in fact occurred not at the S but E side. At times, levels were up to 3.0 m and existing gravel pits sometimes filled with water. On the S side, ground water banked up to a maximum level of 1.5 m and levels on the W side fell in line with the overall declining ground-water level. At the N side of the site, the ground-water level fell slightly. When the subsequent gradient is taken into account, calculation of the amount of infiltrated ground water yields Q = 0.77 l/s. Therefore, the total amount of water is calculated at 1.33 l/s. The actual mean monthly amount of water pumped from the site, measured by inductive flow-meter and read off continuously, is between 1.1 and 1.6 l/s.

Therefore, it can be stated that the sealing wall complies with required permeability standards, and levels of infiltrating water are as expected.

7 Ground-Water Investigations

To monitor permeability of the sealing wall, ground-water investigations have been conducted in the observation wells inside and outside the waste disposal site at six-monthly intervals ever since the wall was constructed. The filtrate's colour, odour, cloudiness, conductivity, pH value, temperature and oxygen content are monitored. In addition, the filtrate is analysed. If conductivity, chloride concentration and COD values are given diagrammatically over the period under investigation to represent wells BE 5 and BE 7 downstream of the site, B 5 ND upstream of the site as well as B 2 ND inside it, the positive effects on the quality of the ground water can be seen very clearly.

Figure 7.1 shows the investigation results yielded by marker B5 ND. Here, conductivity is between 1000–1500 μS/cm and the chloride concentration around 60–100 mg/l. COD values are minimal; contamination of the ground water is not indicated.

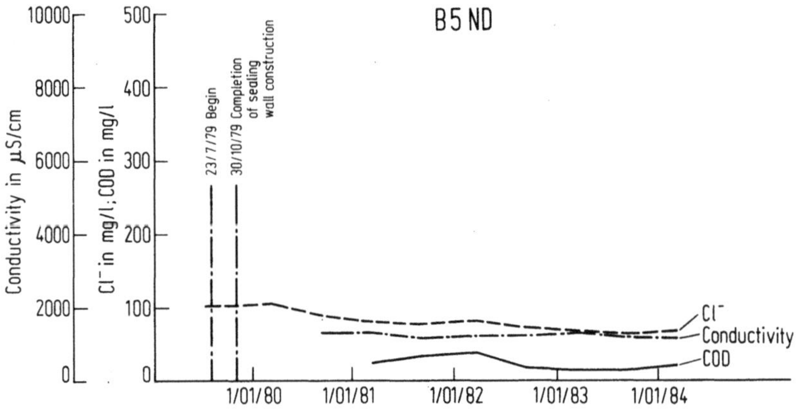

Fig. 7.1. Ground water investigation B5ND

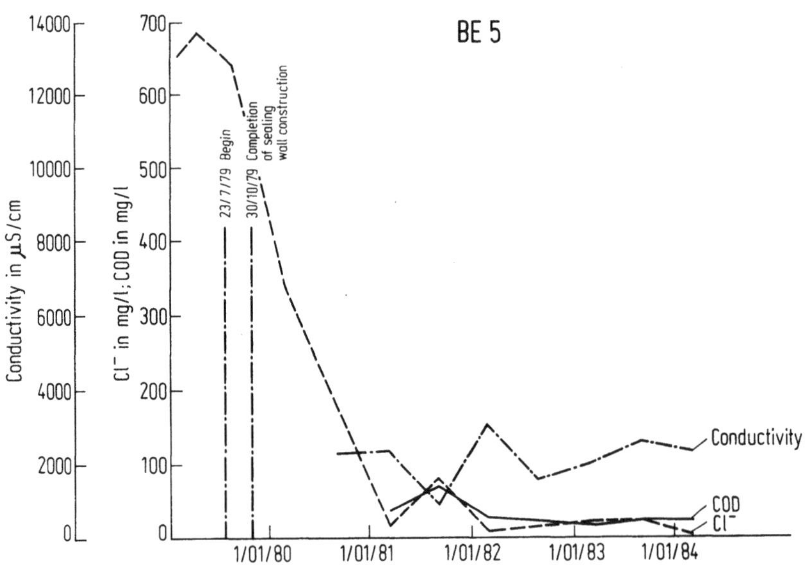

Fig. 7.2. Ground water investigation BE5

Downstream of the site, however, observation marker BE 5 and BE 7 (Figs. 7.2 and 7.3) reveal a major decrease in conductivity and chloride ions. Original concentrations of 1000 mg/l in BE 7 have declined to around 250 mg/l. Conductivity has fallen from 12000–17000 μS/cm to around 3000 μS/cm. COD values are also lower.

Investigation results yielded by observation marker B2 ND (inside the site) show a major increase in conductivity, chloride concentrations and $KMnO_4$ (COD was not measured) (Fig. 7.4). Soluble substances are continuously eluted from the existing site area via infiltrating precipitation water, causing the increase in analytical data.

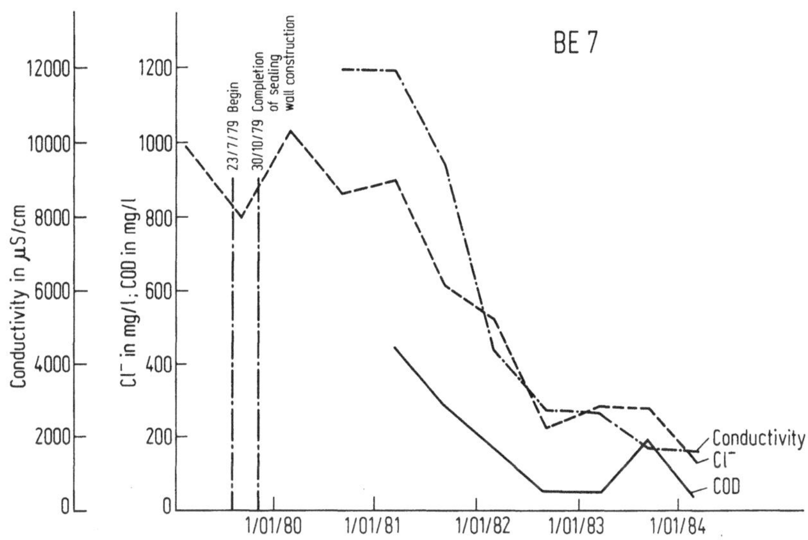

Fig. 7.3. Ground water investigation BE 7

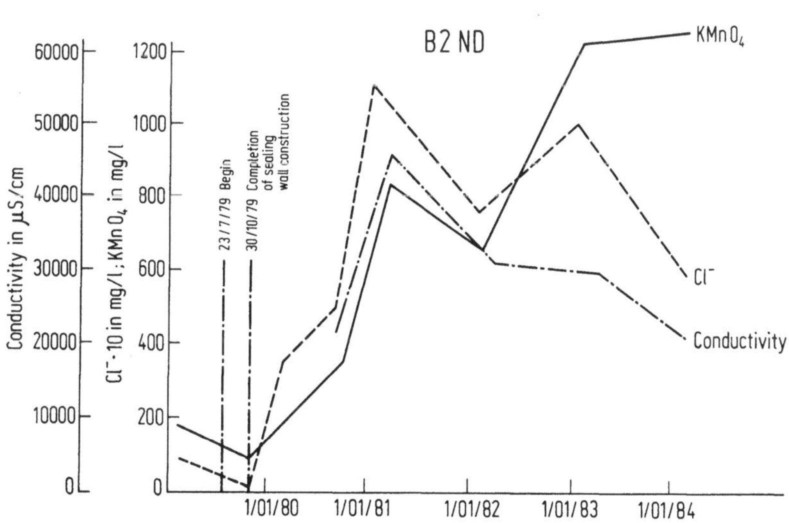

Fig. 7.4. Ground water investigation B2ND

In summary, the results show that contamination of the ground water downstream of the waste disposal site has greatly declined following construction of the wall. However, it has still not been possible to attain the low values determined upstream of the site. On the basis of the dilution effect, a further slow reduction of the values is to be expected. Therefore, it can be assumed that contaminated infiltration water no longer reaches the ground water from the existing waste disposal site.

8 New Aspects Concerning Sanitation of Contaminated Sites

The environment is impaired by diverse existent, formerly originated pollution and by contaminated sites. There are, thus, many locations/areas where the groundwater is either polluted or, at least, threatened to become polluted. The increasing awareness regarding environmental problems, but, in particular, the fact that people are concerned about safeguarding the supply of drinkable water have led to the development of continuously novel and improved sanitation methods in regard of existent, formerly originated pollution. With that, obstructing and interrupting the paths of contamination has been a major consideration aiming at sanitation. This is tantamount to saying that the pollutant area should in some way become encapsulated. The elements used to bring about encapsulation may consist of horizontal grouting curtains, vertical grouting curtains, sealing walls, subterraneous curtains, foils, etc. The task in this respect is to find the individual modules, or building blocks, to make up a multi-barrier system that will comply with the requirement to prevent mass transfer (i.e., permeation of substances from one given area to the next). Along these lines, it is important that the materials dispose of a special resistance against attack from contaminated waters as well as against attack from pure pollutants. Besides, it will be of great importance that operations on the "construction site 'dump' . . ." can be controlled and that repairs also can be carried out. Only then will it be possible to guarantee that, in the long run, the danger area will be kept under control.

8.1 Vertical Dump Sealing Wall Systems

With a great many cases, only subterraneous curtains will qualify as an enclosing vertical sealing measure that, with natural clay layers or artificial horizontal sealing layers, will provide for a "fixing-in" underneath the substance of the disposal site. As a rule, these subterraneous curtains will be produced either by using a one-phase, or a two-phase system. With the one-phase system, hardening agents will be added to the supporting liquid consisting of bentonite and water — the supporting suspension will then serve as sealing substance within the subterraneous curtain. With the two-phase system, the supporting liquid will be exchanged for a permanent and hardening sealing substance.

The company ED. ZÜBLIN AG developed two sealing systems providing for the additional installation of plastic membranes (consisting of either one or several layers) thus achieving a considerable reduction of permeability vis-à-vis emergent contaminated water. With System 1, a single-layer sealing membrane will be lowered section-by-section using ballast substances into the still soft suspension (see Fig. 8.1).

Fig. 8.1. Lowering of the foil into the sealing wall

The individual sections will be seperated from one another with the aid of sinking tubes. Connecting the foil sections with one another is effected with the aid of a special plug-in connection. The tightness of this connection can be checked by means of an internally placed tube. The sealing membranes will be manufactured using high density polyethylene foils having a thickness between 2 to 5 mm.

System 2 represents a multi-layered drainage plate. When using this system, two high density polyethylene foils, together with an intermediate drainage layer, are sunk into the sealing mass. The leakage waters diffusing through the sealing mass and one of the foils will then flow through the drainage layer having a thickness that is greater than 10 mm. The connection of the individual sections is effected via a plug-in type connection consisting of slit tubes by means of which the leakage water having penetrated into the drainage layer can then be carried away in a manner that will be controllable.

With unadulterated (pure) water, the foils will, to a large extent, be tight. With attacks resulting from leakage water and their subject matter there will, however, be increased permeability values. Tests have shown that, with foils, permeation processes will take place. However, by comparison with a mineral wall, permeability will be extremely low. With the multi-layered drainage plate, one may even assume a zero leakage rate since, due to the drainage layer at the second sealing barrier, there will to all intents and purposes be no more water containing pollutants.

We may thus summarize the essential advantages as follows:

- by comparison with the usual sealing walls, the system will be tighter by several powers of ten;

— possibly defective spots will be excluded because of the installation of foils;
— the sealing foil will prevent flow and leakage processes within the wall; merely small quantities of chemical substances will diffuse through the barrier (blocking) membrane;
— because of the use of drainage plates, the system can be controlled and repaired section-wise.

8.2 Protection by Means of Soil Injections

Soil injections constitute an important construction element regarding the protection against existent, formerly originated pollution. Vertically applied sealing measures may in this connection either consist of sealing walls or grouting curtains. Horizontally applied sealing measures will, however, be brought about by means of an injection floor. This floor will be brought about at an appropriate level below the layer of pollutants by the grouting of injection liquids. When using this method, injection lances (rods) will be pressed downward (from the surface) either vertically or in an inclined (tilted) position until they reach the requisite depth, or, alternatively, this may be achieved by flushing, resp. drilling, and the material to be injected will then be pressed down into the ground in sections; alternatively, the injection lances (rods) will be pushed forward either horizontally or in an inclined (tilted) position with the personnel operating from galleries specially constructed for this purpose.

Depending on the particular purpose one may use different materials as injection material. Cement is used with medium-sized gravel, clay cement is used with fine gravel, bentonite with coarse sand, and chemical injections are used with fine sand. According to grain size, the materials are either used as solutions, emulsions, suspensions or else as pastes or mortars. The chemical injections predominantly consist of hard or soft gels on a silicate basis. For quite a while, this method has already proved to be successful with non-contaminated soils. The reaction (conduct) of the injection materials vis-à-vis contaminated waters is still largely unknown.

In regard of this area, the Dynamit Nobel company has done research work with the aim of developing sealing media showing increased resistance and lesser permeability against chemical attack.

The gel-generating substances developed on a silicate basis are, in this connection, subdivided into two groups. One group comprises injection mixtures leading to compaction gels, while the other group comprises those that generate pure sealing gels. In due time, compaction gels are expected to form a "SiO_2" structure improving the mechanical properties of the permeated grain system. Therefore, the use of these gels should only be restricted to applications that call for compaction properties. Moreover, with compaction gels, gel water will be released (syneresis) at the moment of transition from a gel to a crystalline substance. In the course of this process, syneresis values of up to 40 percent will be observed within a time span of 160 days.

On the other hand, sealing gels are not expected to form a "SiO_2" structure, but are rather expected to stay in contact with the groundwater as long as possible thus maintaining the gel structure. This means that such gels, with respect to the silicate content, will be of substantially lower concentration by comparison with compaction gels, and they will therefore enclose, in terms of volume, much more water and they will, in the final analysis, show very low syneresis values.

A typical sealing gel will, for instance, be formed on the basis of water gels and sodium aluminate. This gel will be characterized by a lower SiO_2 content, by a considerable binding capacity, and by a considerable long-term stability. The long-term stability will very much depend upon the degree of polymerization of the basic component used, i.e., of the silicate and the aluminate. The basic components used have the added advantage of not being noxious in terms of environmental protection. The importance of this property lies in the fact that the syntheresis water splitting up would otherwise represent a pollution hazard to the ground (subsoil).

When determining the resistance of this gel vis-à-vis chemical substances, it showed, among other things, an excellent tightness when tested against a solution containing sulfuric acid and phenol; when tested against chlorinated hydrocarbons, the compound did not, however, prove to be stable.

Further developed gels based on silicate that are also stabler against attack by organic chemicals are considered to modify the so-far hydrophilic character of the above-mentioned gel to a hydrophobic one. In this process, a gel was generated whose organic component will be chemically integrated into the silicate gel. The gel thus generated is strongly hydrophobic, it shows improved binding in regard of the sand structure, and the organic binding may neither be dissolved nor washed out. When tested against organic chemicals this gel showed an excellent sealing effect.

With respect to cement-bound materials, corresponding additives likewise have been developed leading to a greater tightness of the sealing wall material also as regards certain chemical solutions. Moreover, the additives are chemically bound to the mineral components of the sealing walls; according to tests so far carried out, they are neither hydrolyzable nor can they be washed out.

In the future, exact investigations will continue to be necessary depending on the presence of pollutants requiring sanitation, informing us about the resistance of the wall system vis-à-vis attacking chemicals. It will be necessary to develop new gels, showing a yet lesser permeability and greater resistance; these investigations should at once establish where and how the sealing system is being attacked.

9 Conclusion

Early in the seventies, the Boehringer Ingelheim KG company installed a dump (disposal site) in a former gravel pit near the site of the company. Impact drilling was carried out for the purpose of investigating ground conditions on the level of the mine floor. On this occasion, one came upon a layer of clay marl of a highly wavy character below the floor and, above this layer, a groundwater mass was found contained in a sand and gravel mixture. According to the state of the art prevailing at the time, the dump body was sealed by means of a layer of mortar; in this connection, the newly developed special cement "Pectacrete" was used. Afterwards, laboratory and production residues, in part also sodium chloride as well as distillation residues that, according to their nature, were similar to bitumen and that were insoluble in water, were deposited. The resulting leakage water flowed across the sealing layer into a drainage ditch and then reached the cleaning plant via a pump station.

After a few years, very high chloride concentrations were found to be present in the garden wells of the nearest houses. It could therefore be assumed that the layer of mortar did not correspond to the requirements and that contaminated leakage water thus was flowing out.

Because of the particular geological situation, i.e., the fact that there was a water-tight clay marl layer below the dump at a depth that could economically be attained, the proposal made with respect to sanitation, i.e., setting up a sealing wall around the dump, was carried out. The sealing wall would bond in for 2.5 m in the clay marl layer and was expected to conduct the groundwater flowing in around the disposal site. Lowered wells were constructed at the low points of the dump that were to collect the resulting leakage water and to conduct this leakage water to the cleaning plant. This would guarantee that the level of water within the dump would always be lower than the exterior level of water. The hydraulic gradient thus obtained would prevent diffusion of the leakage water through the wall to the outside. By means of comprehensive tests, a sealing medium was developed the properties of which comprised low permeability, high compression strength, excellent resistance against substances attacking concrete, as well as excellent processibility.

The construction of the subterraneous sealing curtain was effected using the one-phase method. In the southern part of the dump, the groundwater rose by more than one meter (1.0 m) already during the construction.

When checking the water level situation four years after the construction of the sealing wall, a rise of the groundwater by almost 3.0 m was to be noted at some places.

The resulting quantities of leakage water are composed of precipitated (rain) water and of groundwater diffusing through the wall. According to theoretical computations, about 0.55 l/s result, in this connection, from precipitation, while 0.77 l/s (of groundwater) diffuse through the wall. The quantities of water actually pumped off the disposal site attain average monthly values between 1.1 and 1.6 l/s.

Tests carried out every six months concerning the groundwater found in the up-stream, resp. downstream, as well as within the dump showed that the conductivity, the Cl-concentration, as well as the COD-values strongly decreased following the construction of the sealing wall in the downstream area. The values resulting from analyses carried out in the upstream area of the dump could, however, not be attained. Because of the dilution (thinning out), one will have to seek a further reduction. Because of the elution noted in connection with the residues deposited, a considerable increase of the individual parameters was observed regarding the level observed within the dump.

Thus, it may be stated that the subterraneous sealing wall corresponds to the stipulated tightness requirement and that the real aim of the sanitation, i.e., to prevent a continuous impairment of the groundwater in the downstream area of the dump, has indeed been attained.

The considerable number of contaminated sites as well as the sustained discovery of new areas of pollution originated much earlier and the growing concern about the thus created hazards to the environment leads to the development of continuously novel, improved sanitation methods. The method of encapsulating polluted areas that originated much earlier is in the foreground of respective considerations. With recently developed sealing wall systems, one-layered, resp. multi-layered plastic membranes are being installed into the substance of the subterraneous curtains. This leads to a considerable reduction of the permeability and to greater safety.

With respect to injection mixtures and additives, new developments are also noted that aim at increasing the resistance against the attack resulting from contaminated water.

In regard of the "construction of dumps", measures carried out in the future ought to be controllable and repairs should be possible. Only then will it be possible to guarantee that the danger area will be kept under long-term control.

References

1. Abfallbeseitigungsgesetz-AbfG-in der Fassung vom 05. 01. 1977.
2. Wasserhaushaltsgesetz-WHG-20. 10. 1976.
3. Erbslöh & Co. „Technische Informationen für die Bau- und Bohrtechnik".
4. Merkblatt Nr. 3 „Die geordnete Ablagerung (Deponie) fester u. schlammiger Abfälle aus Siedlung und Industrie, Sonderdruck Bundesgesundheitsblatt 12/1969.
5. Verwaltungsvorschrift über die geordnete Ablagerung von Abfällen in Rheinland-Pfalz 9/1980 Ministerialblatt des Landes Rheinland-Pfalz.
6. Sanierung kontaminierter Standorte Dokumentation eines Arbeitsgesprächs UBA.
7. H. Simons, W. Hänsel, H. Meseck, „Beständigkeit von Deponieabdichtungen aus Ton gegen Sickerwässer", Vortrag am 24. 06. 81 auf dem ISWA-Symposium im Rahmen der IFAT 1981, München.
8. Mitteilung des Lehrstuhls für Grundbau u. Bodenmechanik, 8 (1982), Prof. Dr. Ing. H. Simons, „Dichtungswände und -sohlen".
9. W. Hänsel, „Entwicklung und Eigenschaften von Basisabdichtungen und ihre Überwachung".
10. Deponie-Merkblatt der Länderarbeitsgemeinschaft Abfall (LAGA) vom 1. 9. 1979.
11. W. Brechtle, „Grundlagen für die Abdichtung von Deponien mit Ton, Deponiebasisabdichtungen", herausgegeben von K. Stief, Erich-Schmidt-Verlag 1979.
12. H. Simons, „Tone als Dichtungsmaterial für Deponien", Vortrag im Haus der Technik in Essen am 8. 4. 1981.
13. K. Stief, „Maßnahmen zur Erfassung und Verminderung von Sickerwasser aus Deponien, Fortschritt in der Deponietechnik", Erich Schmidt-Verlag 1978.
14. Falke, H. (1960), Rheinhessen und die Umgebung von Mainz.-Samml. geol. Führer, *38*, 156 S., Borntröger Berlin.
15. Geib, K. W. (1950), Neue Erkenntnisse zur Paläogeographie des westlichen Mainzer Beckens.-Not.-Bl. Hess., Landesamt f. Bodenforschung (VI), *1*, Wiesbaden.
16. „Sanierung u. Erweiterung einer Abfalldeponie durch eine umschließende Dichtungsschlitzwand", H. Schirmer, Wasser u. Boden Heft 11/1980, Verlag Paul Parey.
17. Salomo, K. P., „Technische Möglichkeiten zur Sanierung gefährlicher Altlasten", Müll und Abfall Heft 3/1985, Erich-Schmidt-Verlag.

The Utilization of Vegetable Wastes by Composting

H. Propfe

The postulation of the Club of Rome to handle carefully our limited resoures and to recycle raw materials must be extended to organic materials.

Humus is one of the most important matter we need. No life exists without it. Therefore the interruption of the cycle of organic matter by dumping or disposal of organic refuses can be desastrous. By obeying some fundamental rules vegetable wastes can be converted easily into fertile humus. New aspects in gathering, treating and application of organic raw materials and compost show that industrial composting is not only an oecological but also an economical way for treating vegetable wastes.

Contents

1 Why Composting?

Composting is real waste disposal as compared with tipping [1]. Tipping merely removes the waste away from one's sphere of influence. The tipped materials remain mostly unchanged on the waste disposal site for lengthy periods and this can lead to later contamination of the water table [2] or the air. Because it involves an utilization of the residues composting brings about actual disposal, since the materials are returned to the natural nutritional cycles.

The actual disposal occurs via decomposition, alteration and also via the resynthesis of new raw materials. Composting fulfils Justus von Liebig's admonishment to return to the natural food cycle everything that man has removed from the vegetable world by harvesting. Liebig's admonishment to preserve the mineral nutritional cycle may be expanded today to a preservation of the cycles of "life-giving substances". This aspect certainly ought to be taken into account when analysing the possibilities for the disposal of the wastes produced by industry.

The fertility of agricultural land was maintained for centuries by the application of organic wastes, stable manure in particular. Nowadays we are able to increase the fertility of the land several times over by the application of mineral fertilizers. However, in certain cases, where no further adequate provision of humus is made, there is a tendency for the land to become exhausted, towards decreased fertility, erosion and towards diminished resistance to soil-carried diseases.

Supplying the land with sufficient organic material is thus a necessity, which does not concern just the agricultural industry. Other industries can make the necessary material available. Organic residues should not, for the above-mentioned reasons, be simply disposed off but should be brought to a meaningful re-utilization. The possibilities of the new composting techniques described below can play their part in ensuring the proper utilization of industrial vegetable wastes [3].

2 The Industries Concerned

Medicaments were amongst the oldest types of chemicals manufactured, that were produced exclusively from vegetable sources. Even today the production of pharmaceutical raw materials from vegetable sources plays an important role [4]. But plant tissue is not just processed by the pharmaceutical industry. The range of chemical materials extracted from plants is very large, ranging from edible oils extracted from oil seeds, which are grown on enormous plantations, to raw materials for cosmetic production which are extracted from plants growing in the wild.

All types of plant cultivation and harvesting produce problems of residue utilization, since, in most cases, the whole plant is not used but rather particular raw materials are extracted from parts of the plant. If the amounts processed are small or the processing is carried out at the cultivation site then disposal of the residues is usually no problem. If, however, raw materials are obtained from plants on an industrial scale, then the large capacity of industrial processes necessitates the starting material being collected and transported long distances. Then the residues must once again be transported long distances if they are to be properly utilized or even if they are just to be tipped.

The types and amounts of vegetable wastes can be extraordinarily varied. Nowadays the following alternatives are available for their disposal:

1. Introduction into flowing water.
2. Dumping in the sea.
3. Tipping/dumping.
4. Burning/gasification.
5. Composting.

Composting, as it is discussed here, means the safe disposal or possible utilization of the wastes.

In the long term waste disposal by simple dumping can be neither economically nor ecologically acceptable. Firstly, ever increasing amounts of waste will compete for small amounts of tipping space. Secondly, simply to throw away potentially useful materials and considerable quantities of energy is flying in the face of the principle of retaining the natural cycles. This practically eliminates the first three possibilities. Re-utilization will gain increasingly in importance. Waste paper, glass and, of course, metal recovery has long been economical and is practised in some sectors of the waste disposal industry, e.g. in household and industrial waste disposal.

Apart from the possibilities of recovering useful raw materials and saving tipping space, the possibility of producing energy is also beginning to attract attention. All these aims can be realized, in a simple manner, by means of composting. Combined processes, aimed at the production of useful raw material together with possible energy production, will probably have a real chance of competing with other methods of waste disposal in the future.

Waste disposal is regulated by law in most countries and sometimes the local or regional authority runs the necessary facilities. As a result of this very few industrial plants have bothered to set up their own disposal systems for at least a portion of their wastes. The increasing costs of waste disposal have now motivated many companies to set up their own disposal facilities in which at least part of the waste produced can be processed. The residues which are left are mainly tipped or burned in the communal facilities. Most plant residues are tipped because they do not cause any problems. The amounts produced are also often too small for the operation of a special treatment plant. On the other hand, many facilities exist which accept vegetable wastes from industrial concerns and would be able to process larger quantities.

The following summary lists production sectors which may produce vegetable wastes and is meant to indicate where these materials can be utilized or perhaps are already processed. In the first instance the wastes can be roughly divided into solid waste and sludge or paste-like wastes. Very often materials in both these categories are produced in one industrial facility. No thought is usually given to their combination. It would, for instance, be very practicable to compost a combination of the tree bark and the sludges produced in a pulp or paper mill.

Nevertheless, the distinction between solid and liquid wastes is a valid one because their differing water contents necessitate differing pretreatments. Solid plant waste is mainly produced by industries that only process particular portions of the plant and separate this portion from the rest of the plant mechanically. This is the case, for example, in the milling industry, edible oil production and the conserve and food-manufacturing industries. Thus peanuts and cocoa beans are usually delivered complete with seed cases. The parmaceutical industry, on the other hand, often produces moist or porridge-like residues, since it is usual to extract vegetable matter with liquids. This also applies to the sugar-refining and alcohol-producing or processing industries. The biotechnological synthesis of chemicals, e.g. the production of antibiotics and organic acids, which is so highly developed nowadays, also produces mainly sludge, residues of mycelia for example.

The following industries are possible producers of organic wastes from vegetable residues (see Table 2.1).

As has been mentioned previously, solid residues require different pretreatment than pasty or sludge-like residues. Since sludges are difficult to manipulate and their high water content means that transport and disposal costs are enormous they must be brought into either a pumpable or a compact condition. Further drying would be very expensive. Composting offers enormous cost advantages, since sludges are very suitable for adjusting compost to the correct moisture content. Thus expensive drying can be avoided if suitable dry composting material is available for admixture.

Table 2.1.

Industry	Waste
1. Foodstuffs industry	
a) Edible oil production	— oil seed residues, straw
b) Fat production	— mucilage
c) Milling	— chaff, husks, straw
d) Animal feedstuffs	— chaff, husks, straw
e) Conserves	— pods, skins
f) Confectionary	— coffee and cocoa husks, solid and liquid residues
g) Fermenting industry	— molasses/fermented molasses, press residues
h) Sugar refining	— molasses, extraction residues, liquid wastes
i) Herb and spice manufacture	— extraction residues
2. Textiles/clothing	
a) Spinning	— lanolin, carding and scutching residues
b) Weaving	— weaving wastes
3. Timber industry	
a) Paper making	— bark, sawdust, sludges
b) Furniture industry	— shavings, sawdust
c) Sawmills	— sawdust
4. Chemical Industry	
a) Pharmaceutical industry	— mycelia, herbal residues
b) Cosmetic industry	— herbal residues
c) Dyestuff industry	— extraction residues
d) Manufacture of pesticides	— herbal residues

In order to save costs it is worth investigating other industries for possible complementary composting materials. The use of public sewage and local authority active sludges is not recommended, however, since these contain unknown amounts of harmful materials. Only when sewage works treat an influent from purely domestic sources is a co-operative utilization of its active sludge possible. If on-site composting is not feasible it may be possible for several manufacturers to co-operate in setting up a communal composting facility. Composting wastes does not involve a great deal of of capital cost, which is one advantage over incineration, which is very cost-intensive.

All that is required are suitable starting materials to ensure rapid composting and a qualitatively valuable end-product.

3 The Composting Process

The most important factor in a successful composting process is the careful choice of vegetable waste. If sufficient nutrients and organic materials are available then the decomposing micro-organisms can operate optimally. The major factors are the C:N ratio, the aeration and the moisture content. The C:N ratio is the relation between the nitrogen and carbon contents of the raw materials. This ratio determines the decomposition rate, since the micro-organisms can only oxidize excess carbon if they have sufficient ammonia [5]. The C:N ratio falls during composting, since, as can be seen from scheme 1, carbon dioxide is formed which represents a loss during composting. Both the volume and actual amount of the waste are drastically reduced on composting. The losses when the compost is ripe, i.e. at the end of the transition phase, can be up to 30% of the original amount.

Optimal aeration is important because rotting is an aerobic process. Oxygen from the air is required by the micro-organisms for the oxidation of the carbon. As composting takes place the carbon dioxide content of the air in the compost rises and can reach 10%. However, during the decomposition phase aeration should be sufficient to keep the carbon dioxide content below 6% if possible, to ensure that anaerobic pockets do not form [6]. Anaerobic processes mean putrifaction. Putrifaction may cause unpleasant odours because sulphide compounds can be formed.

Moisture content is a decisive parameter for all biological processes. The micro-organisms become dormant at moisture contents of less than 20%. Poor oxygen supply must be expected with water contents of more than 80%, so that the optimal moisture content lies between approximately 50 and 60%. In nature the decompostion of the organic material takes place on the surface of the ground. Industrial composting, however, is performed in heaps to accelerate the process. This causes thermal insulation which brings about an increase in temperature in the heap during the decomposition phase. Temperatures of more than 70 °C are reached in this way. A lowering of the temperature indicates a slowing-down of the decompostion process (see Fig. 3.1). The raised temperature has, at the same time, a sanitizing effect on the compost [7]. All the readily utilizable organic substances, including cellulose, proteins and lipids, are consumed rapidly during the decomposition phase. More difficultly decomposable substances, such as lignins, are only attacked by micro-organisms during the synthesis phase. Three different phases of composting are recognized according to the temperature course.

Different micro-organisms are active during the three different composting phases. Actinomycetes are particularly active during the decomposition phase. Fungi are particularly active during the transition phase that follows. The addition of appropriate compost starters, containing suitable micro-organisms, greatly accelerates the rotting process and also steers it in the desired direction.

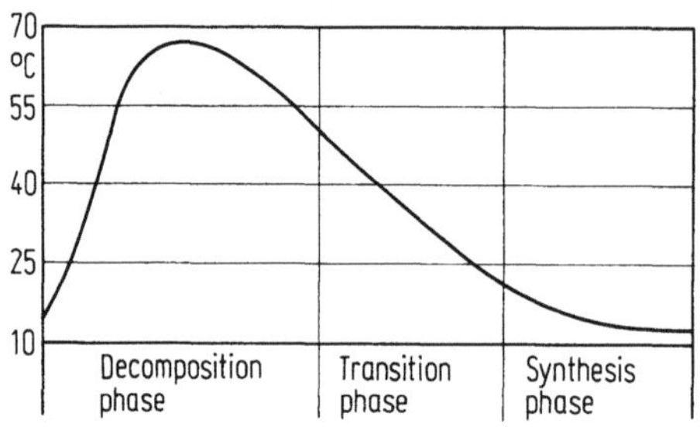

Fig. 3.1. Temperature changes in a compost heap

4 The Composting Area

The traditional form of composting is performed in heaps 1 to 1.3 m high and 1.5 to 1.8 m wide at the base. The composting area should be accessible to vehicles at all times of the year. A large proportion of the composting area and the transport access should be made up. Making up means that the drainage should be good. Hard core is unsuitable, however, since it can too easily become mixed with the humus and would have to be removed subsequently by sieving. A gradient of 1–3% should be maintained in the composting area during making up. This allows the drainage of rainwater out of the heaps and ensures that the standing for the heaps remains dry. The bases of the heaps require special preparation since they play an important role in drainage and aeration processes. The ground should be porous so that excess water can easily escape. Possible contamination of subterranean water supplies by harmful constituents of the leach water should be checked from case to case by appropriate analysis. Particular attention should be paid to moistening the heaps through in regions of abnormally low rainfall. In such cases the heaps should be situated in depressions if possible and the ground made waterproof so that water can collect. If, however, there is a danger of seasons of high precipitation then arrangements must also be made for the drainage of excess water. It is advantageous to align the heap along a north-south axis since this leads to a more uniform heap climate.

Particular attention must be paid to the aeration of the heaps. This can be done in a number of possible ways which have proven themselves in practice. The simplest measure is to make the heaps as small as mentioned above which avoids the formation of anaerobic pockets. The chimney effect [8], for example, uses the natural draught produced in pipes inserted vertically into the heaps to mobilize the air, the increased quantities of air greatly accelerate the rotting process. Better aeration can also be achieved by laying perforated plastic pipes — drainage pipes — at the base of the heap. Air can be sucked out of the heaps through these pipes, thus providing uniform aeration and more rapid decomposition. If the ground on which the heaps are built is not made up then concrete blocks can be laid as a base with spaces between them to act as aeration channels. Excess leach water is also able to escape through these channels.

A fundamental distinction is made between the preliminary and final rotting stages. Figure 6.1 illustrates the composting process as it can be arranged in most cases. The initial intensive high-temperature fermentation takes place during preliminary rotting when the heap must be turned over several times. The oxygen and nitrogen consumption is naturally particularly high during this first decomposition phase. Adequate nutrients should, therefore, be provided right at the beginning of the composting so that the rotting process gets under way quickly. Because of the high oxygen consumption the heaps should be no higher than 1.5 m in the absence of artificial aeration, otherwise anaerobic pockets will be created with all their accompanying disadvantages. The temperature profile of the heap does not only give information concerning the phase of the rotting process, as in Fig. 3.1, but also concerning the timing of the turning-over of the heaps. There is a well-cooled zone of about 10 cm at the surface of the heap. Immediately beneath this is the zone of most rapid decomposition and hence of highest temperature, since it is here that the best oxygen supply is usually to be found. The intense fungal growth within this zone can be recognized from its grey coloration.

The more advanced the rotting is, the higher the heaps can be piled. This is necessary for space-saving considerations. In calculating the space needed it is not only necessary to take into account the proposed annual compost production but also the likely turnover. The agricultural industry has pronounced seasonal requirement peaks. Increased compost sales are to be expected outside the vegetative growth periods, that is in winter. The sale of compost in summer, on the other hand, is difficult because vehicular access to the cultivated areas is not possible. It is only to horticulture and forestry that a certain amount of sales can be expected during this period. The amount of area required for composting can be estimated roughly at 1 sq. m per cubic m compost per year. This includes access areas. A larger area may well be necessary because of the unfavourable annual sales pattern. However, since the completed compost can be piled high the amount of extra space required is relatively small.

5 The Composting Process

A schematic presentation of the composting process is illustrated in Fig. 5.1. Composting in heaps is basically the simplest technique but probably also the most time-consuming.

Various aerobic processes have established themselves on the market of late. This is not the place to describe the relative advantages and disadvantages of these particular techniques. There are two different groups of techniques; the first of which consists of the *dynamic* techniques and the second of the *static* techniques. The heap-composting technique is an example of the static type of technique in that the losse rotting material

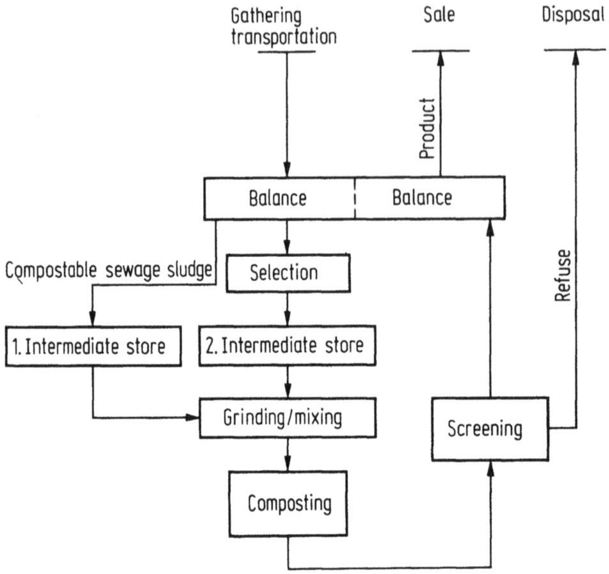

Fig. 5.1.

is not in continuous motion, but merely turned over from time to time. Various processes have also been tried out where the loose compost precursor is preshaped. The different processes involve the production of both regular and irregular-shaped bodies. The blocks can be the size of bricks. After shaping, the compost blocks are set out in layers to rot (Bricolare process). The intention of the shaping is to increase the relative surface area of the composting mass and to achieve faster rotting by improving the air supply. The rotting mass is kept in constant motion in dynamic techniques, the best known of which is probably the Danover process, in which the rotting mass passes through a rotating drum. This type of composting can take place under controlled conditions whereby oxygen and water can be supplied as required.

The RIKO company's biothermal process goes further and recovers heat produced in the aerobic rotting process [9]. This opens up new perspectives, since heat, which can be used for technical processes, is obtained in addition to the compost. The amounts of heat released during the exothermic rotting process can be considerable and amount to up to 50% of the heat of combustion. Combustion data for some organic materials are reported below:

1 ltr light fuel oil = 36.0 MJ = 8600 kcal
1 cbm biogas = 21.6 MJ = 5160 kcal
1 kg lignite = 30.5 MJ = 4900 kcal
1 kg firewood = 15.0 MJ = 3600 kcal
1 kg straw = 14.2 MJ = 3400 kcal
1 kg grain = 15.5 MJ = 3700 kcal

It can be calculated that the fuel oil equivalent of organic matter is about 3–4 kg in utilization of the heat of rotting, i.e. the same amount of heat can be produced on rotting 3–4 kg of easily decomposed organic matter as on burning 1 kg of fuel oil. This combined rotting process with simultaneous heat recovery is economical both to install and to use. The installation of such a process is particularly attractive for small throughputs of up to 5000 cbm/year.

Heat recovery is a more and more interesting aspect of organic waste treatment. Different ways of gaining the energy from the organic material are possible as shown in Fig. 5.2. Besides burning the aerobic rotting is the only way of gaining the energy directly.

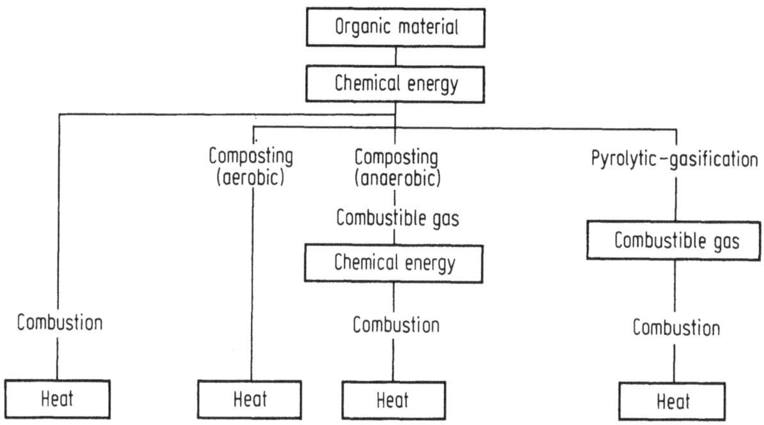

Fig. 5.2.

6 Process Hardware

As described in the previous section the process chosen determines the equipment required. Composition and surface structure are decisive for the composting time. Thus, as is well-known, the rotting of unaltered timber or bark wastes can last for years. For this reason in order to accelerate the rotting process it is necessary not only to ensure that sufficient nitrogen and moisture are available but also that the compost fibre is in a suitable physical form. The larger the surface area the faster the microbial attack. It is important to ensure that the body of the compost possesses a good capillary structure so that the proper regulation of its moisture content is possible. The preparation and composition required depends on the material. Too small a particle size leads to the formation of a sludge and, thus, brings about anaerobic conditions. Some organic wastes contain plant residues which possess a protective wax layer on their external surfaces. This wax layer, which occurs in various types of stalks, such as straw, can inhibit composting considerably. The wax-protected plant components must be exposed by suitable milling because this layer cannot be removed merely by mechanical chopping. It is advantageous to allow the preliminary rotting phase to reach high temperatures (ca. 65 °C) in order to destroy this layer. The type of machinery used for comminution is of great importance. Fine division consumes a great deal of power and moist plant residues require more than dry material. Moist plant tissue is much tougher than dry plant residues which are usually brittle. Drying during intermediate storage can save considerable amounts of energy which would otherwise have to be expended in disintegrating the material.

The power requirement for disintegration can only be discovered by experiment and is dependent upon the type of machine. Three types of machine are available for comminution:

- Choppers are usually equipped with a rotating knife and a fixed counter cutter. Such machines are easily damaged by stones or metallic impurities. The necessary hardening makes the knives particularly liable to break, so that when considering such machines it should be ensured that the knives are easily exchangeable.
- In *percussion mills* the working tools are blunt hammers which break the compost fibre against counter knives; their action is that of disintegration rather than cutting. Such comminution is advantageous with many materials as the area open to attack by the microbes is considerably increased by disintegration.
- *Ripping machines* have both the cutters and counter cutters rigidly fixed, usually in comb-like arrays. Very tough material can be broken down with such machines but the power consumption is very large.

The best type of machine can only be discovered by practical trials. It is necessary, however, that the material used in such trials corresponds to the actual raw material. Experience indicates that the moisture content, possible impurities and the extremes of particle size have a significant effect on the effeciency of preparation of the machines used. It is advantageous to store the raw material as dry as possible, to try to ensure that the particle size of the fibrous material is as uniform as possible and to feed the preparation machine with as uniform a stream of raw material as possible. Savings in energy requirements and reduced repair and maintenance costs, by virtue of the greater throughput, more than compensate for the expense of preparing the compost fibre. It is, of course, necessary to keep the whole preparation plant as free as possible from gross impurities, such as fragments of glass, metal or stone, since these inevitably cause wear and tear if not complete breakdown.

The finer the compost material is broken down the higher will be the power consumption at this stage. The comminution should, therefore, only be carried out as far as is necessary for the rotting process. Experience has shown that a particle size of 5 mm up to maximum of 40 mm is advantageous. This guarantees that the surface area of the material is sufficiently enlarged for optimal microbial action while the structure remains open enough to maintain adequate aeration and moisture content. An optimal particle size is recognizable by the rapid onset of rotting, combined with a rapid increase in temperature. Machinery for the removal of foreign bodies forms part of the necessary preparation plant. Stone, metal, glass and plastics are particularly troublesome impurities in compost. The raw material should be cleansed by sieving to remove coarse impurities of stone, metal or glass. Several types of plant are available. Bar graders have proved useful; these are devices where elastic metal bars arranged in tiers catch large particles as the material falls through. Ferrous metals are most easily removed by means of magnets above the material flow. Power screen machines are also suitable for sorting the input; in these, precomminuted material is projected from a fast moving conveyor against a plate standing at an oblique angle; this separates out large material and very light components like plastics. The removal of glass causes the greatest difficulties, since it breaks easily into small splinters. Unfortunately glass is also very undesirable in the finished compost, since it is optically very obvious and carries with it the danger of injury. The separate collection of the industrial wastes intended for composting without the admixture of general waste is therefore of the greatest importance. Not only is the quality of the compost produced strongly influenced by the presence of impurities but the necessary measures for the removal of unwanted bodies increase the costs of the composting process enormously.

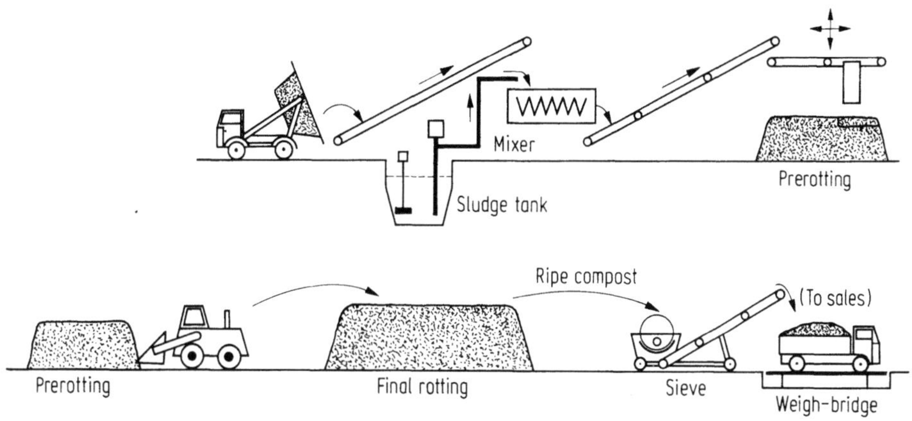

Fig. 6.1.

As illustrated in Fig. 6.1 mixing machines also belong to the preparation equipment; by this means pastes and sludges can be included in the composts. The amounts that can be added depend on their consistency and on the initial moisture content of the rest of the raw material. The moisture content of the compost can, thus, be adjusted by the addition of sludge. Double screw mixers have proved themselves in continuously operating plants for the preparation of compost on a large scale. Small amounts can be prepared by using screw pumps to dose the sludge before the comminution step so that mixing is achieved in the mill. Where the input material is very coarsely divided graters have proved of value; they also homogenize the raw material. Dump trucks can be used to heap the comminuted material. If the amounts to be heaped are large diagonal discharge conveyor belts are valuable. Fully automatic devices are available for turning over compost heaps. However grabs can also be used since the mixing effect obtained is usually adequate.

The addition of various additives such as fertilizers and composting starters can be done in several types of heaps. The space requirement for composting in individual heaps is usually too large. Therefore rows of suitable length of triangular or trapezoidal cross-section heaps are usually chosen. Various techniques have also proved suitable for turning the heaps over. In the wandering heap technique longitudinal turning over of the heap causes it to process. Various authors [10, 11, 12] have described the advantages and disadvantages of different types of heaps.

After rotting the compost should be size-graded by sieving. Two grades with diameters greater and less than 20 mm are adequate. Drum sieves are suitable for this process, since they are relatively immune to clogging even when processing moist compost. The necessary equipment capacity should be calculated from pilot scale experiments.

7 Additives in Composting

The growth of the decomposing micro-organisms can be assisted by a large number of additives or the quality of the raw material can be improved by certain additives.

The following come into consideration as additives:

- Agents for adjusting the nutrient content.
- Agents for adjusting the mineral content.
- Agents for adjusting the pH.

The raw compost can be improved by inoculation with special strains of micro-organisms [13]; the following types come into consideration:

- Aerobes which can utilize cellulose.
- Aerobes which can decompose organics at particularly high temperatures (thermophilic bacteria) [14].
- Aerobes which can decompose particularly persistent components of the raw compost.

The major purpose of additives for adjusting the fertilizer content is to provide nutrients for the micro-organisms. Nitrogen is, in the first instance, the most important nutrient for the micro-organisms. As has been previously mentioned the C:N ratio should be less than 30, this means that the optimal nitrogen concentration of a good compost is greater than 1.3%. Ammonium salts or nitrates are suitable, easily metabolized nitrogen sources. Such nitrogen compounds have proved to be particularly assimilable nutrients for micro-organisms. Urea, in contrast, has more of an inhibitory effect, at least at the beginning of the composting process [15]. The significantly higher costs rule out the use of organic nitrogen sources. However, it is worthwhile consulting the earlier section on "industries concerned" to see if suitable materials can be obtained as waste products. Organically bound fertilizers have the advantage that they are active over longer periods than inorganic nitrogen fertilizers. Indeed, the effects of such nitrogen fertilizers are also less initially, so that it may be necessary to supplement them with some quick-acting, inorganic nitrogen compounds. The addition of large amounts of nitrogen is advantageous when the compost is used since a further additional fertilizing effect is obtained.

Raw phosphate and superphosphate are particularly suitable for supplementing the phosphate content, since composting micro-organisms can utilize these forms without difficulty. Finely ground stone meal is suitable for supplementing the trace element or mineral content. The pH of acidic raw composts can be adjusted by the addition of chalk; carbonate chalk should be used in preference since other forms cause great losses of organic matter. Basic raw composts can be adjusted in pH by the addition of commercially available fertilizer grade ammonium sulphate. The optimal value for the pH lies in the neutral region, although most of the composting bacteria can tolerate a slightly acidic environment down to about pH 5.8. During rotting the pH is displaced towards the neutral or slightly alkaline region because of the formation of ammonia.

The rotting process should be initiated as quickly as possible in order to inhibit unwanted putrefaction processes. With organic materials having a high protein content there is the risk of unwanted fermentation with the production of unpleasant smells, even during short storage times, particularly in summer when outdoor temperatures are high. This is caused by the transformation of sulphur-containing compounds into mercaptans. The addition of a suitable preservative — e.g. naphthenic acids — allows the material to be stored until it is composted without the production of nauseous odours. Naphthenic acids have the property of only inhibiting microbial growth temporarily and having no effect on the subsequent composting process.

A variety of commercial products is now available for accelerating the rotting process. Products containing thermophilic cellulose-decomposing organisms have proved particularly effective. The resistance of these organisms to high temperatures, such as a regenerated in the preliminary rotting phase, is particularly advantageous in quickly sanitizing the compost and rapidly decomposing organic material. Various investigations have shown that hot fermentation at temperatures higher than 55 °C and lasting longer than 3 days is lethal to pathogenic organisms [16]. This is probably achieved by a combination of both the high temperature and the metabolites with antibiotic activity synthesized by the composting micro-organisms themselves. The micro-organisms dominating in the compost thus stop the development of other organisms or hinder their multiplication [17]. The addition of a combined composting starter, such as BIOROTT, that contains both specialized decomposing organisms and organic nutrients, makes possible both accelerated rotting and complete sanitization of the raw compost, as was confirmed by a 5-year comparative trial at the University of Hohenheim [18].

Interesting effects are obtained by the addition of sewage sludge. Here the iron compounds added to the sewage as flocculating agents act as complex exchange agent for trace heavy metals in the raw compost and thus have the positive effect of lowering the concentration of available heavy metals (see also the section on harmful materials in composts). At the same time the addition of sludge allows the adjustment of the water content of the raw compost to the required level. The ground itself possesses a considerable purifying ability because of the sorption capacity of the amorphous silicon-manganese-iron compounds it contains. The addition of a suitable quantity of organic material can significantly increase this buffer capacity. The redox potential plays a critical role here, the solubility, in particular, of the complexed metals being significantly dependent upon it. The redox potential can therefore be used as a measure of the ripeness and utility of the compost.

8 Compost Ripeness

The literature contains many attemps to produce schemes for evaluating utility and quality by the classification of the various stages of composting [19, 20]. However since composting, like the mineralization of soil components, is a dynamic process various stages are not amenable to exact definition. So that agreement has been reached on arbitrary definitions. Fresh compost is rotting material in a hygienically acceptable state. While ripe compost is defined as material in which the rotting process is virtually complete, the phytotoxic constituents have been decomposed and which contains no further sulphides or ammonia. So that testing for sulphide with lead acetate and for ammonia with test rods give a first indication of the usability of the compost. Tests of ripeness can be performed using cress germination tests.

The sowing of 10 g cress seed in pure compost should result in the formation of at least 30 g wet weight of cress in 5–6 days. These are only rough preliminary tests. Further plant tests and residue analyses should be used at regular intervals to check on the quality of the composted material. References concerning compost testing are given in Spohn [19].

9 Harmful Materials in Composts

9.1 Fertilizer and Plant Protection Agent Residues

Whenever plants are cultivated, plant protection agents and fertilizers are used to protect the crop or to increase the yield. This ubiquitous application of plant protection agents makes it possible that their residues are present when the plants are processed industrially. Fertilizer residues in plants do not play an important role of themselves. The inhibitory effects of fertilizers on the performance of compost micro-organisms seems minimal; fertilizers often have a stimulatory effect [15]. The nitrate problem only occurs during the intensive cultivation of vegetables. The plants themselves never contain so much nitrate as to cause problems during composting. Neither has the enrichment of heavy metals from certain fertilizers, cadmium from phosphates for example, any significance as far as composting is concerned [21]. The only real problem is the elution of easily soluble nutrients and humins during the preliminary rotting phase. During times of high precipitation leach water can escape from the compost heaps. Contamination with humins can be recognized by the dark coloration it imparts to the leach water. This can cause difficulties in drinking water wells in the immediate neighbourhood. In such cases the composting site should be chosen so that the subsoil is sandy and adequately aerated. The soil itself then takes over the function of purifying the leach water by filtering and oxidizing it [2]. This effect of the precipitation of humins in particular zones — localized zones — also occurs naturally.

The residues of plant protection materials are more difficult to judge. DDT, Endrin, Aldrin etc., which acquired notoriety in the past, are no longer used in the western world. The developing countries, however, for lack of better alternatives, often still use these substances, which have been replaced by more specific and sophisticated substances in the developed countries. Persistence is often an important factor in plant protection material. Persistence means that the substance is only slowly decomposed and thus remains and is active in the natural cycle for a long period of time. The metabolic processes of the soil micro-organisms are also affected by such substances. All the decomposition processes for organic substances depend, in the main, on microbial action. The metabolic processes of compost micro-organisms are, in principle, identical with those in the soil. Just as plant protection agents affect the metabolism of micro-organisms on the plants, they also affect growth of the compost micro-organisms. The only factors speaking for an accelerated decomposition are raised temperature and ample supply of organic material and hence nutrient material for the micro-organisms. This good supply of nutrients naturally results in high counts of decomposing micro-organisms. If the "correct" micro-organisms are present in the compost, something which can be achieved by the use of a suitable starter, then the rate of decomposition will be many times higher than it would be in the soil [22]. Lack of nutrients for the micro-organisms is not a limiting factor in compost since this contains large amounts of easily decomposed materials and nutrients, or they can be added to it. The C:N ratio, i.e. the nitrogen concentration, has already been mentioned in this connection as that value most regularly in need of adjustment. All organic materials can be decomposed. However, the process does not always go so far as reducing them to carbon dioxide, water and minerals. Sugars, for example, are only decomposed to the extent of about 80%, in spite of their easy metabolization by micro-organisms. As far as the time scale required for their decomposition is concerned it is advantageous if an agent is last applied some time before harvesting. This interval is usually laid down by law. Before industrial processing the materials are usually stored for some time since the whole harvest cannot be processed at once. Decomposition also occurs during this storage process. The main plant protection agents are herbicides. These are usually applied before the seed has germinated. Only a very small proportion of the agent is thus taken up by the cultivated plants. The amount of the agent in the plants is then drastically reduced during growth. For these reasons any acute danger from compost application is virtually eliminated. This applies also to plant growth agents. An investigation by Hentschel [23] indicated that even the intensive application of non-composted plant waste does not affect the growth of the plants that are cultivated.

Recently a group of american scientists found out published in Science Vol. 22, page 1434–36 that special microorganisms can reduce the content of pestizide residues drastically.

Under laboratory conditions a species called "phanerochaete chrysosporium" was able to metabolize more than 80% of an initial amount of DDT within 18 days. This species ist normally found on dead wood in the forest and turns over there resistant wood into fertile humus.

Under laboratory conditions the mushrooms produce a secretion that destroys also persistent substances as lignins and chlorinated combines. It has to be tested how this species works under outdoor conditions also in compost heaps.

Insecticides are rather insoluble in water and, for this reason, difficult to decompose. Such agents can have a great effect on the decomposition and metabolic processes occurring in composts and soils. Investigations by Bauchhenß [24] have demonstrated that the numbers of micro-organisms can be drastically reduced by treatment with insecticides. Therefore it should be checked to see if the plant residues to be processed contain measurable insecticide residues. Vegetable waste, in particular, can be considerably contaminated because, during industrial processing, it is often the outer parts of the plants that are removed and discarded as waste. Extraction processes can also enrich the proportion of plant protection agents in the extraction residues or slops. Although there are numerous investigations into the decomposition of plant protection agents in the soil there are scarcely any reliable data concerning the accelerated decomposition of plant protection agents in composts.

10 Heavy Metals

Until now no attention has been paid during the development of new chemicals to any negative effects that manufacturing wastes may have on the agricultural industry.

However the agricultural industry is predestined to accept organic residues. It is not just the maintenance of the nitrogen cycle but also the prospect of economic utilization that makes the employment of organic wastes in agriculture preferable to dumping them. The agricultural industry is, however, becoming more and more affected by environmental pollution. It cultivates, utilizes and tends a large proportion of the countryside, which has been subject to increasing pollution for decades.

Besides its own wastes the agricultural industry also utilizes a range of public authority wastes, such as activated sludge. A critical view of the cumulative effects of all foreign materials on plant growth makes it necessary to take great care in the introduction of further wastes onto the land. The heavy metals constitute one of the many pollutant groups. These are the metals with densities of more than 4 g/cm^3. The heavy metals are subclassified into the essential metals, which are also often called the trace elements, and the nonessential. The very small amounts of the trace elements are vitally necessary. Inadequate supplies can cause deficiency symptoms; however even slight excesses quickly damage the organism. Analogously to fertilization diagrams used in agriculture the regions of growth promotion can be discerned, as well as regions of highly toxic effect (see Fig. 10.1).

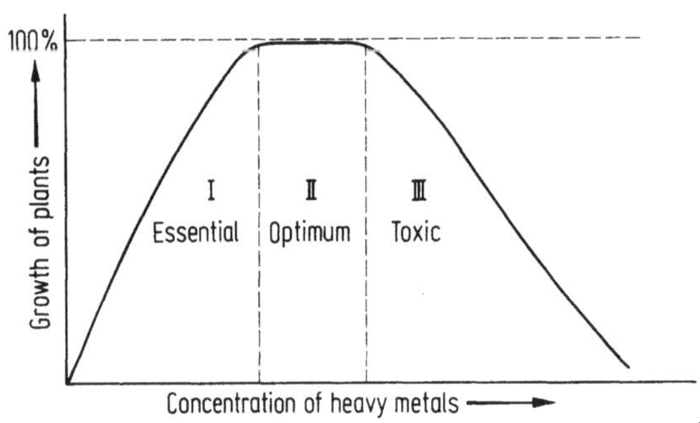

Fig. 10.1. Effect of the dosage of heavy metals on the growth of plants

The chemical form of the heavy metal is of importance; for instance, organically bound mercury is significantly more toxic than the metallic element. The question of the maximum permissible concentration is thus almost impossible to answer since it depends on so many factors. On the one hand, there is the actual concentration of the particular heavy metal in the compost raw material. Then during composting bonding processes can occur that considerably reduce the availability of the heavy metal. Then there is the amount of compost applied per unit area of land and finally the type of soil on which the compost is spread plays an important role. All these factors make it impossible to calculate an exact concentration range where agricultural application may lead to deleterious effects. Nevertheless, the table 10.1 gives some guidelines for heavy metal concentration limits, for the classification of composting raw materials according to their applicability.

The values given in the above table allow an estimation to be made concerning the permissible duration of agricultural composting (Kloke 1976).

$$n = \frac{(B_{tol} - B_G) \times 4200000}{K_G \times K_M}$$

n = number of allowable composting applications
B_{tol} = tolerable soil concentration
B_G = amount in soil
K_G = amount in soil
K_M = annual mass of compost
4200000 = weight of soil per hectare to a depth of 30 cm in kg

The formula applies for a mean pH of 6.5. Sandy soils accept less material and clay soils more. Most soils have a defined capacity for retaining heavy metals. This sorption capacity can, however, be destroyed by the application of large amounts of organic waste. For instance, the presence of the polyphenols, which are plentiful in organic waste, is known to greatly increase the solubility of boron [26].

Thus from the very beginning the concentrations of heavy metals should be minimized, as far as possible, by the application of suitable measures [27]. This begins with keeping the contact between compost raw materials and metallic containers to a minimum, since this contact can cause considerable enrichment with heavy metals. Alcohol distillation, for instance, can raise the copper content by considerable amounts, which can only be as a result of the dissolution of the distillation vessel. A considerable proportion of lead and cadmium contamination can also result from contact with pipelines and reaction vessels. Suitable precipitation reactions with H_2S or iron compounds are available for use during the production stage to reduce heavy metal concentrations [28]. There are also well-tried techniques for separating complexed heavy metals from organic wastes (Product leaflet Giulini Chemie, Ludwigshafen 1983).

Table 10.1. Guidelines for tolerable limits (Kloke/EPA) [25])

Element mg/kg		limit mg/kg	usual mg/kg
As	arsenic	20	0.1–20
B	boron	25	5–20
Be	beryllium	10	0.1–5
Br	bromine	10	1–10
Cd	cadmium	3	0.01–1
Co	cobalt	50	1–10
Cr	chromium	100	2–50
Cu	copper	100	1–20
F	fluorine	200	50–200
Ga	gallium	10	0.1 10
Hg	mercury	2	0.01–1
Mo	molybdenum	5	0.2–5
Ni	nickel	50	2–50
Pb	lead	100	0.1–20
Sb	antimony	5	0.01–0.5
Se	selenium	10	0.01–5
Sn	tin	50	1–20
Tl	thallium	1	0.01–0.5
Ti	titanium	5000	10–5000
U	uranium	5	0.01–1
V	vanadium	50	10–100
Zn	zinc	300	3–50
Zr	zirconium	300	1–300

Thus the problems of heavy metals in industrial composts can, in general, be regarded as having been solved [29] — in contrast to those in local authority activated sludges.

11 Toxin Formation in Composts

The ability of composts to decompose organic materials depends on the metabolic activity of micro-organisms. This metabolism can cause the formation of by-products of special toxicological significance. The toxins are of particular importance. Toxins, in particular mycotoxins, are substances synthesized by fungi growing on non-mineralized organic matter. It is often observed that organic wastes become covered by a lawn of fungus, which inhibits aerobic decomposition. Moulds which can excrete highly toxic metabolites are particularly common on oilseed residues. Care is therefore necessary in the utilization of mouldy animal feed, since intoxication phenomena can easily occur. Here too aerobic composting can have a considerable purification effect. Nilson [30] was able to demonstrate that, even after only 8 days' intensive hot fermentation, rejected peanuts which had contained large quantities of aflatoxins no longer contained measurable toxin concentrations. This indicates that the "cleansing" effect of aerobic composting goes much further than merely killing off weed seeds and cleaving long-chain organic compounds. The requirements of hygiene can be fulfilled with certainty by properly carried out composting. The soil itself is usually improved by the addition of compost because of the large amount of organic matter that is added. The "priming effect" [31], whereby the application of organic material to the soil results in an increase in microbiological activity, is well-known. At the same time the amounts of nutrients released are increased and air and water circulations are improved. The soil's buffering capacity, that is its ability to retain dissolved substances, is also increased by the addition of organic material. Because of their composition clay minerals mostly possess a negative surface charge, so that it is cations that are absorbed in the main. The buffering capacity thus causes the fixation of metals such as calcium or magnesium; cationic plant protection agents, such as paraquat, are also fixed. Organic substances mainly fix cations but they also possess some capacity for the binding of anions so that arsenate, chromate and inorganic pesticides are also bound. The positive properties of soil-improving composts should be reinforced by as high a content of valuable components as possible. The concentrations of undesirable constituents should be kept to a minimum right from the start. This is significantly easier to do with industrial wastes than with communal wastes by means of waste stream separation and collection. The addition of sufficient fertilizer and suitable composting additives allows the production of compost of such a quality that the receipts from its sale justify the expense involved.

12 Economic Aspects

From a long-term point of view the decision for or against composting should not be made solely by reference to the difference in costs between composting and dumping. It is quite true that, apart from the costs of machinery and equipment, the costs of composting are difficult to quantify [32]. Running costs consist overwhelmingly of personnel costs. Since time required for composting is very dependent upon climatic conditions, the costs and the proceeds obtainable cannot be estimated in advance. So the decision depends on whether the situation both concerning compost production and compost sales has been estimated correctly and properly planned. The costs of dumping vary tremendously from locality to locality and are also greatly influenced by transport costs. Before the decision is made for or against composting it is necessary to have a market available for the product.

This is agriculture in its widest sense, including special cultivation, public bodies involved in the maintenance of open spaces as well as horticulture and forestry. Agricultural customers will only be willing to accept compost insofar as there is a long-term guarantee of the necessary quality. It is for this reason that many authors have discussed the classification of compost [27, 33] not just in order to ensure a certain safety margin in the utilization of the compost but also to indicate to the producer how he can dispose of his product at a price which covers his costs.

These criteria of quality are based on external features, such as suitable appearance, the absence of residues and noxious substances, the absence of glass and plastic residues and also on hygenic aspects. The compost ought also to have a positive fertilizing effect. Naturally these criteria cannot be so uniformly summarized as is the case for normal industrial products. These criteria were formulated for composts from communal wastes which are of their nature contaminated with plastics and metals. Herein lies a great opportunity for composts from industrial wastes. The selection of wastes, which is often impossible and always expensive for communal wastes, is much more simply achieved with industrial wastes. It is, therefore, very necessary to use only uniform composting raw materials. This does not mean, however, that only single materials should be composted, but that composting cannot be regarded as a universal method of waste disposal. The requirement that composting raw materials be collected in separate containers at the site of their production may seem complicated and expensive, but the end product will more than repay this expenditure. Compare for example compost from local authority wastes, where the amounts of noncompostable materials or contaminating materials separated from the compost are often greater than the actual compost produced. The separation of contaminating ballast either during or after the composting process is very expensive and increases the disposal costs. If the compost raw materials available are not in a suitable physical form, whether they are too wet or not fibrous enough, then this can be corrected by mixing with wastes from other industries. The lists at the beginning of this chapter provide information concerning where suitable wastes may possibly be found. Thus very fine milling wastes, for example, can be composted together with coarse waste bark from the timber processing industry [34]. As far as receipts are concerned it can be very profitable to carry out the more elaborate separate collection of composting raw materials, since the amounts of waste can be considerably reduced thereby. Even small amounts of impurities can cause problems in composting under certain conditions and very much reduce the quality of the product obtained.

In the long term composting is favoured by the costs of disposal which are bound to increase greatly when nearby sites are filled and also by possible increasing freight charges. By greatly reducing the volume of waste for disposal composting can play its part in reducing industrial costs. If sufficient land is under agricultural cultivation, particularly as horticultural land, so that potential customers are available, then composting has a chance of being cheaper than waste disposal. The ecological aspects, such as the maintenance of natural cycles and the production of valuable humus should also not be forgotten.

13 Areas of Application for Composts

As previously mentioned agriculture, in the broadest sense, is a possible customer. The most important criterion for the range of applicability of compost is the starting materials. If they do not contain any impurities that are harmful to plants, then they are suitable for soil improvement, i.e. for humus enrichment. For this purpose at least 30% of the dry weight should be organic matter. The question of impurities has been discussed previously, whereby the most important criterion is the absence of sulphide and ammonia. The limits for heavy metals and other contaminants must naturally also be taken into account. Because of the transformations taking place in them composts containing less than 1% of the major nutrients do not usually have a fertilization effect and cannot therefore be sold as such. If the nitrogen content is more than 1.5% then a certain fertilizing effect can be assumed. It should, however, be remembered that the fertilizer content can constitute a limit to the amounts that can be applied. This applies, in particular, to phosphate. The limits set by fertilizer content are easily calculated using fertilization tables [35]. Care is recommended in the application of composts to young plants or seedbeds, since the transformation reactions, which are sometimes extensive, can damage young plants.

14 Summary

The utilization of vegetable waste by composting is an ecologically and economically sound method of disposing of industrially produced organic wastes. The outlook for economically worthwhile composting of industrial wastes is much better than for communal wastes, since standardized raw materials can be employed. Separate collection of wastes and, in particular, the exclusion of glass and plastics is hence particularly necessary. It can be economical to combine wastes from different industrial plants in order to achieve a suitable starting material and also in order to guarantee sufficient quantities of raw materials. Suitable industries are listed.

The following requirements must be fulfilled for the production of uniformly good compost:

Hygiene

Pathogens must be inactivated. The material must be free from agents that could cause infection.

Plant Cultivation

The compost must be rich in useful components; the organic matter content must be at least 30% and contain plant nutrients and fertilizers. It must contain very little glass, plastic or stone. The legally laid down limits for heavy metals and chemical residues must be met. The water content should, if possible, be less than 40%. The quality should be high (crumb quality, water content, nutrients). The compost should be tested by plant growth tests and for nutrient content.

If the raw materials are carefully chosen and the above conditions are satisfied, industrial composts have a good chance of finding customers in the agricultural industry. In view of the increasing costs of waste disposal and the growing scarcity of disposal sites, the possibility of utilizing vegetable wastes should be investigated in terms of the opportunity for sales to local agriculture. In this way industry could make a useful contribution to the maintenance of natural cycles and to the promotion of soil fertility.

References

1. Braun, R. (1967): Begriffsbestimmungen, Fachausdrücke, Begriffserklärungen, Kompostierung. Handbuch für Müll- und Abfallbeseitigung/Kumpf, Maas, Straub. 1224.
2. Quasim, S. R.; Burchinal, M. (1970): Leaching from simulated landfills. Journal Water Pollution Control Federatin, vol. 42, nr. 3.
3. Pfeiffer, E.: Biodynamics, The Art and Science of Composting, Biodynamic Farming and Gardening Ass. Dover Plains N.Y.
4. Stickelberger, D. (1966): Kompostierung von Industrieabfällen, insbesondere aus der pharmazeutischen Industrie. Chemiker-Zeitung — Chemische Apparatur 90 (18).
5. Fuller, W.; Bosma, S. (1965): The nitrogen requirement of some municipal composts, Compost Sience vol. 6, nr. 2.
6. Schuchard, F.; Baader, W. (1978): Untersuchungen über die Möglichkeiten der Steuerung der Selbsterhitzung beim mikrobiellen Abbau organischer Reststoffe aus der landwirtschaftlichen Produktion, mit dem Ziel maximaler Energieausbeute. FAL Braunschweig.
7. Brauss, F. W. (1966): Die hygienische Bedeutung der Müllkompostierung unter besonderer Berücksichtigung der Schnellkompostierung. Arch. Hyg. Bakteriol 150 (5).
8. Spillmann, P. (1981): Das Kaminzugverfahren, Forum Städtehygiene, 32 (15–24).
9. RIKO (1984): Firmenschrift D — 6800 Mannheim, Biowärme.
10. Pierau, H.; Müller, G. (1970): Die Bedeutung der Rottedeponie für eine hygienische einwandfreie Beseitigung von Klärschlamm zusammen mit festen häuslichen Abfallstoffen. Städtehygiene 21, 4.
11. Rodale, J. L.: The complete Book of Composting, Rodale Books, Pennsy.
12. Pfeiffer, E. (1957): Anleitung zur Kompostfabrikation aus städtischen und industriellen Abfällen, Gustav Fischer Verlag.
13. Dunlap, C. E. et al. (1976): Treatment Processes to increase Cellulose Microbial Digestibility. AICNE-Symposium 158.
14. Müller, G. (1964): Die praktische Bedeutung der thermophilen Mikroorganismen. Biologische Rundschau 1 (4).
15. Franz, H. (1968): Der Einfluß von Düngemitteln auf die Bodenlebewelt, Handbuch der Pflanzenernährung und Düngung II.
16. Strauch, D. (1966): Zur Hygiene der Beseitung und Verwertung fester Siedlungsabfälle. Deutsche Tierärztliche Wochenschrift 73 (12, 14).
17. Knoll, K. H. (1963): The influence of various composting processes on non-sporeforming pathogenic bacteria. International Research Group on Refuse Disposal. Information Bulletin, Nr. 20.
18. Gaul, D.; Strauch, W. (1974): Zeitschrift Umwelthygiene 10 (227–236).
19. Spohn, E. (1975): Selber kompostieren für Garten und Feld, Schnitzer-Verlag.
20. Könemann, E. (1981): Neuzeitliche Kompostbereitung, Bionomica-Verlag.
21. Sauerbeck, D. (1982): Zur Bedeutung des Cadmiums in Phosphatdüngemitteln, Landbauforschung 32, (192–197).

22. Leikanova, L. (1967): Utilization of microorganisms in biological treatment of industrial wastes. Folia Microbiol. 12 (4).
23. Hentschel, H. (1983): Rückstände von Pflanzenbehandlungsmitteln in Stroh und Stroh kultiviertem Gemüse. Gartenbauwissenschaften 48, (71–74).
24. Bauchhenß, J. (1983): Die Bedeutung der Bodentiere für die Bodenfruchtbarkeit und die Auswirkung landwirtschaftlicher Maßnahmen auf die Bodenfauna. Kali-Briefe (Büntehof) 16 (9).
25. Kloke, Umweltbundesamt (1976): Orientierungsdaten über die tolerierbaren Gehalte einiger Elemente in Kulturböden.
26. Rhode, G. (1972): Sind bedenkliche Anreicherungen von Schwermetallen in Böden oder Pflanze möglich, Wasser und Abwasser 11 (1–5).
27. Kampe, W. (1981): Praxis und Problematik der Anwendung von Klärschlämmen und Müllkomposten in der Landwirtschaft. LUFA Speyer.
28. Hartinger, L. (1968): Beseitigung von Schwermetallen aus Abwässern. Zeitschrift für Wasser und Abwasserforschung, Nr. 1.
29. Eidgen. Forschungsanstalt für Umwelthygiene (1977): Informationsberichte 3 „Klärschlammverwertung in der Landwirtschaft".
30. Nilson, G.; Kalju, V. (1974): Kompostering ab bark och aflatoxinhaltiga jordnöts expellers, Uppsalla.
31. Trolldeinier, G. (1971): Bodenbiologie, Kosmosverlag.
32. Davis, A. G. (1964): The economics of municipal composting. Mother Earth 13 (3).
33. Chrometzka, P. (1969): Bestimmung des Sauerstoffverbrauchs heranreifender Komposte. Müll, Abfall, Abwasser, Nr. 8.
34. Jonsson, E.; Kalju, V. (1971): Rapporter fran adelningen för växtnäringslära, Uppsalla, Lantbrukshögskolan.
35. Fink, A. (1979): Dünger und Düngung, Verlag Chemie, Weinheim.

Recycling

K. R. Müller

Recycling techniques opening new options for recycling chemical wastes have been extensively described in the recent past (1–4). Here we choose not to examine any details of this research and the currently employed techniques, but much more would like to point out some rarely discussed general aspects of the inherent limitations and obstacles to recycling.

The following six points shall be considered:

1. The *transportation* of residues is a severe problem. Certain process residues could easily be used elsewhere if exceedingly high transportation costs were not to render such an attempt uneconomic; the cleaning of transportation vessels would further add to these high costs. All this holds true especially for solvent mixtures and metal-containing sludges.
2. The *availability* of a residue must be guaranteed on a continuous or regular scale. The recycling of side products from batch processes require large storage capacities to accommodate such substances until needed.
3. The *quality* of residues is more and more becoming a limiting factor for their recycling. Whereas in the past the demands on the quality of chemicals varied throughout the world, today similar production techniques require similar quality standards. Thus, goods of lower quality cannot be and are not being processed as before. This means a setback to recycling and an increase of waste.
4. *Loss of know-how to competitors*, or at least the fear thereof, may lead a company to withhold recyclable byproducts from the open market. In many cases it is possible to determine the competitor's production volume and technique by evaluating the residues.
5. *Increased pollution* may result from the use of residues instead of refined raw materials. In the past many residues have been transported to regions where their processing is legal. Today environmental legislation is tightening up in all regions of the world.
6. *Legislation* is setting further limitations to the use of recyclable materials to protect the health of the population. For instance, the admissible content of heavy metal ions in sewage sludge to be used as fertilizer for agriculture has recently been lowered dramatically in various countries and, thus, the use of sewage sludge in agriculture has nearly ceased.

Furthermore, the member countries of the European Community have agreed to test in detail any new chemical substance for possible adverse effects. This, of course, further increases the cost of these products.

Contrary to common belief higher energy prices do not favor recycling. Quite the opposite is true, as high fuel costs may lead a company to combust spent solvents instead of distilling or fractionating them. Another example is the workup of spent sulfuric acid. Here the increasing fuel prices caused recycled sulfuric acid to cost eight times more than product made directly from sulfur.

To close this survey an explanation shall be given why certain companies can fare better with recycling than others. Let us consider, e.g., a large chemical company producing a wide range of products from relatively few starting materials (Fig. 1).

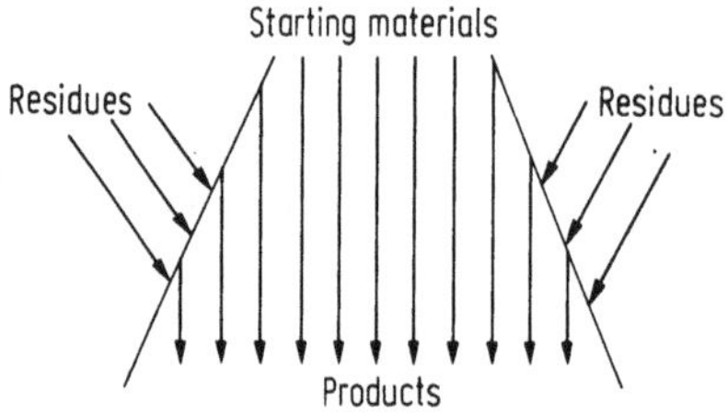

Fig. 1

This company might be able to accept residues from others to incorporate them into their stream of production. On the other side, a company that produces a restricted line of products, like cars and very specialized consumer goods, from a host of starting materials, will for evident reasons produce more residues than the company with a wide range of products (Fig. 2).

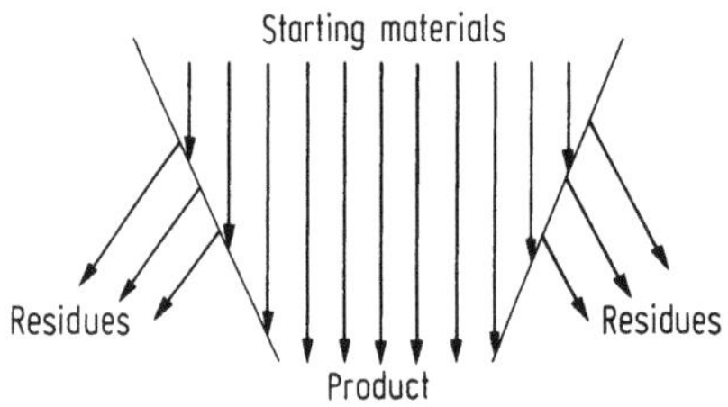

Fig. 2

A car manufacturer can not be expected to produce kettles from the scraps from car production; neither can a pharmaceutical company accommodate reaction residues from numerous processes leading to only a few refined end products.

We have witnessed an increased use of recycling processes in all industries, especially the chemical industry for obvious reasons. This has been achieved by ingenuity, new processes, and, somtimes, compromise. But the still widespread belief that wastes could be avoided by "low waste processes" and recycling of all residues is now and will be in future an unrealistic dream.

References

1. Marshall Sittig: Resource Recovery and Recycling Handbook of Industrial Wastes. Noyes Data Corporation. Park Ridge, N.J. — London, 1975
2. Ullmanns Encyklopädie der technischen Chemie. 24 Volumes, 4th Edition. Verlag Chemie, Weinheim–Deerfield Beach, Florida–Basel (1972–1983)
3. Kirk-Othmer Encyclopedia of Chemical Technology. 24 Volumes, 3rd Edition. John Wiley & Sons, New York–Chichester–Brisbane–Toronto–Singapore (1978–1984).
4. Three of Four Conferences on "Recycling International". Berlin 1979, 1982, 1983, 1984 Karl J. Thomé-Kozmiensky (ed.). E. Freitag-Verlag für Umwelttechnik, Berlin.

D. Hazards

Problems and Accidents

H. K. Hatayama

This chapter describes many of the common problems encountered in each major step of chemical waste handling from the point of generation to ultimate disposal. It reports on the kinds of accidents which can result from mismanagement of chemical wastes and attempts to show the interrelationships between the various problem areas. Mismanagement of such wastes has resulted in severe contamination of air, water supplies, homes and the environment in general. The major problem areas in the storage phase include; (1) Types of containers used, (2) Segregation of containers, (3) Labeling of containers, and (4) Mixing of wastes. In the collection and treatment step, the problem areas are: (1) Manifesting, (2) Mixing of wastes, (3) "Midnight dumping," and (4) Small generators. The problem of chemical waste treatment are related primarily to: (1) Characteristics of the waste and (2) Management of the residues and emissions from the treatment processes. Land disposal of chemical wastes has caused numereous problems related to: (1) Rupture of containers, (2) Mixing of chemically incompatible substances, (3) Leaking of toxic substances into groundwater, and (4) Emission of volatile toxic substances upon or after disposal.

Contents

1 Introduction

1.1 Purpose of Chapter

The primary purpose of this chapter on problems and accidents in Chemical Waste handling and treatment is to provide a frame of reference for the remainder of this volume. It is a brief statement of the problems of chemical waste handling from the point of generation to ultimate disposal. The chapter describes many of the common problems encountered in each step and attempts to show the interrelationships between these problems. It represents a critical review of the issue based on a review of selected literature and the author's experience in chemical waste management.

This chapter is organized according to the sequence of steps in the process of chemical waste handling. There are consequently sections on typical problems and accidents in chemical waste: 1) Storage, 2) Collection and transport, 3) Treatment, and 4) Disposal.

2 Typical Problems and Accidents in Chemical Waste Storage

For this chapter, storage refers to the accumulation of chemical wastes into such containers as drums and tanks, where the containers or waste are eventually removed for treatment and/or disposal. The subject of storage in ponds, surface impoundments or on land is not included here. It will, however, be discussed in the section on disposal since the problems with this activity are the same as with disposal to land.

2.1 Containers Used for Storage

It is common practice in many industries to store chemical wastes in drums and other similar containers which were used for transport of raw materials to the facility. These containers usually range in size from 23 liters to 210 liters, however, in certain cases smaller containers (1 to 4 liters) are used where the waste is generated in a batchwise fashion and the volumes are very small. The types of containers in terms of construction varies according to the types used in transporting raw or finished products. They are usually constructed of steel, glass, polyethylene, or high strength paper products.

The other common type of container used for storage of chemical wastes are large volume steel tanks which can range up to 40,000 liters or more. A facility may have one or more of these large tanks on-site to store a large volume of wastes or to store different types of wastes. These tanks are usually constructed of steel with the appropriate appurtenances for transfer of liquid chemical wastes to and from the tank, and for entering the tank.

2.2 Container Integrity and Location of Storage Area

Container integrity is generally the most significant problem in the storage of chemical wastes. Small containers made of steel or paper are most prone to deterioration while glass containers break easily upon handling. Larger containers like tanks are also prone to deterioration due to their steel construction.

Deterioration is a function of both the location of storage as well as the fact that most containers used for storage are recycled containers that are not re-conditioned. Because the stored material is a waste rather than a valuable raw material or product, storage areas are usually outdoors with minimal protection from the weather. The problem is compounded by the use of containers which are already deteriorated or are of questionable structural integrity. Leakage from such weathered waste containers is very common [1].

Failure of containers has also occurred from stacking of deteriorated waste drums, and from internal pressure generated by the waste. The latter can result from chemical interaction of the waste material with air or moisture leaking into the container, or from volatilization of solvents and other low boiling waste constituents by radiant heating [2].

Large tanks are also subject to leakage from faulty valves and connections which result from poor maintenance and use of recycled equipment. A factor which compounds the problem of leakage in these types of containers is that leakage from the bottom of the tanks is rarely detected until the underlying soil and groundwater become heavily contaminated. Until recently, these large tanks were placed on highly permeable foundations and no leak detection devices were installed. Even tanks containing raw material or products were rarely equipped with leak detection devices.

2.3 Labeling of Containers

Another major problem in storage of chemical wastes is inadequate labeling of containers, particularly small containers. Labeling practices have improved considerably within the last 5 years. Prior to this period, however, labeling of waste containers consisted largely of such generic descriptors as "wastes", "waste organics", "waste acid", "brine", "bottoms", "bad batch", "mud and water" and other even less specific terms or codes. Often the only description on the container is that of the original material stored by it. In reality, the waste may bear some resemblance to the original material such as paint wastes in a drum labeled "Paint thinner". However, this is an uncommon occurrence and it is prudent practice to disregard product labels on containers unless there is some other indication that the labels are sufficiently descriptive.

The consequences of unlabeled, mislabeled or otherwise inacequately labeled waste containers are serious indeed. Drums of chemically incompatible wastes may be stored, transported, and disposed of together. If failure of these containers should occur at any stage, violent reactions or reactions generating toxic or flammable gases could occur causing considerable harm to workers and the general public. Waste containing incompatible constituents may be carbined in bilk during storage, transport, and disposal [2].

In addition to potential problems related to waste compatibility, inadequate labeling results in other safety hazards for people handling the containers. Upon opening or accidentally punctering a container, a worker may be exposed to extremely toxic or corrosive chemicals which were not specifically or generically referred to on the container. The container may be under pressure due to the presence of volatile compounds or other gaseous reaction products without the knowledge of the worker. Public safety officials would not be adequately informed in case of a highway accident involving a transporter of such inadequately labeled, potentially harmful waste materials.

Inadequate labeling can also cause very serious problems during any treatment of the waste. Certain mixtures of solvents are known to co-distill and if the presence of these solvents is not identified, the products of solvent recovery processes can be significantly altered. Incinerator emissions can be adversely affected by the presence of other halogenated compounds. The presence of organics in metal wastes can severely affect the integrity of any chemically fixed residues. Other treatment processes which depend on chemical interaction rather than physical destruction as the mode of action are generally quite sensitive to influent concentrations and constituents (3).

The mode and location of disposal is highly dependent on the waste constituents and inadequate labeling can cause severe problems. As briefly mentioned previously, disposal to surface impoundments or landfill areas of wastes which are not chemically compatible with that which is already in the waste management area can result in extreme heat generation, toxic gas generation, or other uncontrolled releases of toxic substances. The presence of extremely toxic volatile compounds in the waste constitutes a severe hazard to workers in the area of disposal if the waste is exposed to the atmosphere during disposal.

Adequate labeling is thus of primary importance and should be accomplished at the very beginning of the waste management process, the point of generation and storage. The level of specificity depends on the ultimate disposition of the waste, but it should at least be sufficient to warn of the particular hazards which the waste may present. Adequate labeling can avoid many serious problems, save precious resources, and above all provide for the safe handling of chemical wastes throughout the waste management process.

2.4 Segregation of Containers

This problem relates primarily to the possible contact between wastes that may be chemically incompatible. Small waste containers are typically stored in one general area at a facility and are not segregated according to the type of waste. If they are segregated, they are usually not placed in separate spill containment areas. These practices allow contact of potentially incompatible wastes upon accidental spillage, rupture or leakage of deteriorated containers. Furthermore, if flammable wastes are stored alongside wastes that could generate toxic fumes upon heating or burning, the problems of a fire could be compounded severely.

Segregation of large containers is also of concern for similar reasons. Leakages and spills can come in contact if separate spill containment ateas are not provided.

2.5 Consolidation of Wastes

It is common practice and makes good economic sense to attempt to consolidate wastes as much as possible during storage. Consolidation of different wastestreams in ponds, tanks, or other containers is common for large as well as small generators. There are, however, several potential problems related to consolidation which require careful consideration.

One of the most serious problems with consolidation is the potential for combining chemically incompatible wastes. This may result in the generation of toxic fumes, violent reactions involving heat and other gases, or even explosions [2]. The potential for such accidents is usually lower during storage than during transport because the generator knows more about the waste than the hauler. However, whenever a variety of wastes are generated, consolidation can pose serious safety hazards if the wastes are not carefully evaluated for chemical compatibility.

Another problem with consolidation of wastes is that effective treatment of the wastes can be severely hindered [3]. The combining of halogenated solvents with other non-halogenated solvents often results in a waste that requires incineration in specially designed facilities capable of capturing acid halide fumes. Whereas, if the two were not combined, only a portion of the waste would need special treatment. This combination also happens to be a mixture which is extremely difficult to separate in normal solvents recovery treatment operations involving distillation processes.

The consolidation of organic wastes with aqueous heavy metal containing wastes results in a mixture which is very difficult to fix or solidity by currently available methods. The presence of organics tends to interfere with the podzolonic reactions typical of these treatment methods and results in limited metal fixation and a sludge or gel type material rather than a monolithic structure [4].

Thcsc arc but a fcw of thc complications which can be expected in the treatment of consolidatcd wastcstrcams [5]. Thcsc problems should be taken into consideration when planning for the storage of hazardous wastes. Depending on the ultimate disposition of the waste, whether direct disposal or treatment and disposal, segregation may be the optimal method of storage.

3 Waste Collection and Transport

This section deals primarily with the problems related to collection and transport of wastes by trucks because it is the prevalent method for handling hazardous chemical wastes [6]. Other less common methods of collection and transport includes railcar and waste transport.

3.1 Manifesting of Wastes

Early manifests or bills-of-lading which accompanied loads of hazardous wastes consisted of very limited information such as: (1) Generator name and location, (2) Transporter name, (3) Amount of material, (4) Date, and (5) Charges. These documents were primarily used as records of economic transactions with little information on waste constituents and potential health and safety hazards. This is one of the most significant problems related to manifesting of wastes. Although much more information on waste composition is now required, unless the information is accurate and verifiable, the consequences can be grave indeed.

In the event of a highway accident, inadequate or inaccurate information on waste manifests results in unknown hazards to those involved in the accident as well as by-standers, and public safety personnel. If the manifest has inadequate waste composition data, the situation must be treated as if the waste cintains extremely hazardous compounds. Its true contents may require much less stringent and costly safety measures.

If the waste composition information is inaccurate, this usually cannot be known at the scene of the accident. People involved in the accident, by-standers, and public safety officials may be exposed to toxic compounds which are significantly different from those on the manifest. Severe health and environmental effects can occur because public safety officials were misinformed by the manifest about the types of wastes carried by the truck involved in the accident.

3.2 Mixing of Wastes

As with storage of wastes, there are many economic reasons for transporters to mix different wastes during collection such that a full load can be carried to the transfer, treatment, or disposal center. The practice of "topping off" tank trucks is common and can be done in a safe and effective manner. However, there are many potential safety and treatment related problems associated with "topping off" or otherwise mixing hazardous chemical wastes during collection and transport.

The primary safety problem is that of mixing chemically incompatible wastes [2]. As discussed in the previous section, this can result in extreme heat and pressure build-up in the tank truck and eventual release of highly hazardous substances through the safety release valve or by rupture of the tank. The probability of occurrence of such a release is increased by the fact that most wastes are collected and transported by contract haulers who know much less about a given wastestream than the generator.

Furthermore, the potential for damage, both public health and environmental, is greatly increased. The release may occur at virtually any place along the route from collection to disposition. If it should occur on a crowded street or freeway, or near a residential or office complex, or any other highly populated area, the resulting impact would be extremely serious. This would include the impact on public health as well as property. If the release should occur in an ecologically sensitive area, the impacts could also be very serious.

As mentioned in the previous section, mixing of wastes can severely affect its treatability [3]. Unless previous arrangements are made between the generator, transporter, and treater of the wastes, there is little disincentive for the hauler to "top off" loads. For a waste treater or processor who operates on a continuous feed basis rather than batchwise, the quality control of incoming waste is critical to the plant operation. Mixing of wastes during transport is thus a significant factor in maintaining smooth plant operations and effluent quality.

"Topping off" can also severely limit disposal options. Where disposal to land of specific waste types is restricted, mixing of wastes during collection and transport can disqualify a waste for land disposal. Where deep well injection is practiced, the addition of materials to a waste which accelerates the formation of pore clogging precipitates can quickly cause the deterioration of the well. Where restrictions on volatile emissions from surface impoundments exist, topping off a load of aqueous waste with volatile solvents can eliminate this option.

3.3 Midnight Dumping

The term "Midnight dumping" in this chapter is used to refer to the illegal disposal of wastes at unpermitted or otherwise undesignated areas such as roadsides, and empty lots. Such clandestine activities usually occur under the cover of darkness.

These occurences are generally decreasing in frequency due to increased enforcement of waste disposal laws. However, the temptation to make a fast profit from transportation of hazardous chemical wastes is ever present. The hazardous nature of the cargo and the increased liability of such wastes result in high fees charged by all handlers of hazardous chemical wastes. These fees will continue to increase as regulatory controls increase and land disposal capacity decreases.

A typical practice is for a transporter to contract with a generator to carry its wastes to a licensed transfer, treatment or disposal facility. While en route, the transporter would take a detour and deposit the wastes in an empty lot or convenient sewer or stream. Waste manifest systems are usually designated to ensure that the generator as well as the local regulatory agency receives copies of the manifest for each load such that appropriate disposal can be confirmed. Due to the volume of paper generated by this system and the fact that the paper tracks wastes rather than something with positive economic value to the generator, the controls on the system are not adequate to identify potential violators.

3.4 Small Generators

In this chapter, the term "small generators" is used to refer to laboratories, small businesses, farms and households that usually generate small quantities of hazardous chemical wastes that range from several 1 liter containers to several drums of waste solvents per year. Laboratories typically generate small quantities of used or retrograde reagents and residual samples from analysis. These are then packed in larger containers with adsorbant material for eventual collection. Small businesses such as automobile repair shops, print shops, and custom plastics shops generate small quantities of waste solvent and other organic wastes which are accumulated in drums. Farmers generate empty pesticide containers and other residuals from chemicals used. Households generate such items as 1 liter to 4 liter containers of retrograde pesticides, solvents, cleansing agents, and other consumer chemical products, as well as asbestos wastes from home insulation.

The primary problem with small generators is the cost involved in appropriate collection, transport and disposal of such wastes. The transporter will generally charge the same price or more for a drum of laboratory wastes as for a drum of waste solvent from a paint manufacturer. These costs may be prohibitive for small businesses and even more prohibitive for households. The result is often disposal in the local sewer, on-site disposal or indefinite storage. All of which may pose significant threats to workers, members of the household, or the environment depending on the toxicity of the material. Although waste transfer stations can effectively alleviate some of these problems, the cost and risk of operating such facilities have deterred many potential private interests.

4 Chemical Waste Treatment

4.1 Waste Characteristics

Chemical wastes present some very challenging problems for treatment and detoxification. These problems are primarily related to the high variability in chemical, physical and hazard characteristics [3, 5, 7].

Chemically, these wastes may contain only one or two hazardous components in relatively dilute concentrations such as boiler blowdown wastewaters which may contain low concentrations (ppm) of a few metals. It is more likely that chemical wastes are very complex mixtures of inorganic as well as organic components which are present in relatively high (%) concentrations. Metal plating wastes are typically strongly acidic or basic solutions of a variety of metals. Equipment cleaning wastes contain cleaning solvents, such as trichloroethylene as well as various metals and other organics. Petroleum refining wastes can contain both organics and inorganics depending on the sources within the refinery. Ethyl chloride synthesis results in a mixture of chlorinated hydrocarbon solvents, aluminium chloride, and organic tars in the wastes [8].

Physically, chemical wastes range in characteristics from solid rock-like materials to highly viscous sludges to multiphased liquids. The primary refining of metal ores often results in a molten mass of tailings which solidifies into a rock-like material [9]. This waste often contains high levels of other metals which are not recovered and which tend to be more leachable under environmental conditions than they were in the ore.

The refining of petroleum produces a number of different wastes that are sludges. These include still bottoms from sulfuric acid treatment, alkaline sludges from the alkylation process, catalyst sludges, and others (8). These sludges as well as other liquid chemical wastes are often multiphased systems which generally include solids, aqueous and organic fractions.

The hazardous characteristics of chemical wastes are numerous and range considerably in degree. A particular waste can pose a number of hazards and the degree of each hazard characteristic varies with the concentration of the waste components, the physical state of the waste, and the manner in which it is handled. These hazard characteristics include toxicity, flammability, corrosivity, reactivity, explosivity, and irritability.

Because most chemical wastestreams pose multiple hazards, the treatment of such wastes to render them non-hazardous often requires several unit processes. A typical example is alkaline cyanide metal plating wastes. This waste is toxic due to the presence of cyanide as well as metals, and is corrosive due to its high pH. In order to render this waste non-hazardous, a combination of oxidation, neutralization, and precipitation/sedimentation is necessary.

Still bettoms from ethyl chloride production is another typical example where the residual chlorinated organics are very toxic while the aluminimum chloride can react with water to generate heat and HCl fumes. Treatment of this wastestream is further complicated by its high solids content (30–40%). The reaction of aluminium chloride with water can be done under controlled conditions to reduce its hazards. However, because of its high organic content and the versatility of incineration techniques, incineration is usually the method of choice for treating these types of wastes.

4.2 Residues and Emissions

One of the primary problems related to treatment of chemical wastes is the management of residues and emissions from the processes. Many of the processes which are available for treatment of chemical wastes are indeed successful in achieving volume reduction, degradation, immobilization, and/or some level of detoxification of the wastes. However, the level of treatment is often inadequate to render the residues, solid or liquid, non-hazardous under all probable environmental circumstances. Furthermore, the emissions from the treatment process, particularly air emissions can also pose significant hazards to workers, people, and the surrounding environment.

Separation processes usually result in a concentrated solid phase and a dilute liquid phase. These processes include flocculation/sedimentation, filtration, evaporation, and adsorption techniques [7]. The management of the liquid phase depends not only on the quality of the effluent but also on the quality and limitations of the area to which the effluent is discharged. The level of dissolved metals in effluent which is discharge to a sewer is usually not appropriate for discharge to surface waters.

Solid residues from separation processes usually require further treatment or special disposal in order to provide adequate protection. The hazardos constituents are still in their original form, but more concentrated and still relatively available for transport. PCBs that are adsorbed to the separated solids are not degraded by the separation process. Such sludge are usually candidates for thermal treatment or specialized land disposal. Toxic metals in metal sludges or filter cakes may have been transformed into various metal complexes by flocculating agents, but are still relatively available for leaching. These residues also require further treatment, perhaps solidification, or special disposal.

Chemical treatment techniques such as wet oxidation, and solidification can also produce residues which require continued special handling. Wet oxidation processes such as those applied to cyanide containing liquid wastes produces an effluent which is considerably less toxic to humans because of the oxidation of cyanide to various cyanates and isocyanales. However, the effluent may still be toxic to fish and wildlife if discharged to the environment because of the other waste constituents, particularly toxic metals, which are not significantly affected by the oxidation process.

Furthermore, chemical treatment techniques can result in various by-products which are still relatively toxic or of unknown toxicity and care must be exercised in management of these by-products.

Solidification techniques seek to immobilize toxic materials in a solid inorganic or organic matrix. These techniques employ such reagents such as lime, fly ash, cement, soluble silicates, and organic polymers, which are combined with inorganic sludges and result in products that range in physical properties from moderate strength solids to granular soil-like materials [4]. The degree to which immobilization is achieved depends on the type and quality of reagent used, the reaction of specific waste components with the reagents, and the presence of other materials such as organics in the waste which can interfere with the solidification process. These solid residues thus may still require special handling depending on their structural integrity and the degree of immobilization of toxic metals.

Physical and chemical destruction techniques such as molten salt combustion, and other incineration processes are applied to wastes which are primarily organic and contain particularly recalcitrant compounds such as PCBs and other halogenated organics [10]. The residues from these processes are highly oxidized solids which may pose problems of high pH and leachable metals, and gases which are acidic and contain significant levels of particularly stable halogenated organics such as hexaclorobenzene from the treatment of "Hex" wastes (11). These gases may also contain particularly toxic combustion by-products such as poly-chloro dibenzofurans and poly-chloro dibenzodioxins. The existence and control of these emissions are the primary issues in the large scale implementation of thermal destruction methods for the treatment of hazardous chemical wastes.

5 Typical Problems and Accidents in Disposal of Chemical Wastes

5.1 Disposal of Containers with Liquids

Disposal of containers with liquid chemical wastes at landfills has caused some very serious problems in the past. These problems relate primarily to spillage of highly toxic, flammable or reactive wastes during handling and disposal of the containers.

At some disposal sites, it was common practice to dispose of wastes by emptying the drum into the waste disposal area. Drivers responsible for disposing of the waste were commonly inadequately protected and such practice resulted in extremely hazardous situations for them.

Another common practice was to manually roll the containers off the back of a flat-bed truck. This resulted in rupture of many containers and subsequent release of wastes in the immediate area of workers and other users of the landfill.

The potential hazard is compounded at landfills which co-mingle municipal and chemical wastes by the presence of other people disposing of municipal refuse in the same area. Co-mingling also involves the use of heavy equipment to move and compact the refuse which causes further rupture of containers and release of wastes.

Metal containers in particular, usually decompose after being buried for some length of time. Leakage of wastes which then occurs can result in leachate of undesirable quality, emission into the atmosphere of toxic volatile compounds, or reaction of the wastes with the surrounding material (soil, moisture, minicipal wastes) that in turn produces hazardous consequences (toxic gases) [2].

5.2 Incompatible Reactions

Many chemical wastes are still highly reactive upon disposal. Such wastes include waste pickling liquors, or etching solutions which are usually very acidic and can react with cyanide waste already in the landfill to produce toxic HCN gas. Waste magnesium dross can react with moisture in the air to produce H_2 and fires. Waste titanium tetrachloride can react with moisture to form corrosive HCl fumes. Waste nitric acid can react with combustible material like municipal wastes to generate nitrogen oxide fumes.

These reactions can occur slowly as in the case of buried drums which decompose or they may occur very quickly when wastes are disposed in bulk directly from a tank truck into a pond, pit, or disposal well [2]. The latter results in the most serious hazard for workers and others in the area. These types of reactions can be expected when wastes containing strong acids, bases, oxidizers, reducers, and water reactive compounds are disposed of in concentrated form. Dilution or other treatment under controlled conditions at the generator or at the disposal site can greatly reduce the risk of such hazardous reactions occurring.

5.3 Leaching of Toxic Materials

The disposal of liquid and solid chemical wastes has caused numerous problems related to leaching of toxic chemicals off-site into ground and surface waters. Such migration has severely contaminated groundwater supplies in Lathrop, California. It has resulted in contamination of homes and subsequent evacuation of these homes at Love Canal, New York. It has resulted in contamination of water in supply lines in Lekkerkerk, The Netherlands. These and similar cases [12] of severe contamination from leachate from uncontrolled disposal sites have only surfaced within the past five years and is the focus of much activity in the United States and Europe [13, 14]. The leaching of toxic materials from wastes occurs as a result of disposal of of chemical wastes in areas where the wastes are subjected to contact with rainfall, runoff or groundwater. Upon contact with the wastes, many waste constituents dissolve in the water and are carried away as leachate. The problem arises when the leachate generated is allowed to migrate towards and contaminate groundwater, surface waters and surrounding soils. Where the soils are porous, migration can occur very quickly. Where the underlying strata is less porous, migration can occur around the relatively impermeable lenses.

A standard practice for many years was to dispose of liquid chemical wastes like pesticide rinse waters to ponds expressly designed to promote evaporation and percolation of liquids, and to take advantage of any natural attenuation capacity of soils. These practices have resulted in very severe groundwater problems at industrial facilities [1].

Old landfills, used for both municipal and industrial waste disposal have also resulted in severe groundwater contamination. Although operated as disposal sites, these sites were not subject to the stringent criteria that govern currently permitted sites. Moreover, little was known about the toxicity of the waste components, and the potential for migration at the time of disposal.

The extent of groundwater and surface water contamination from past practices is evident from the U.S. Environmental Protection Agency's *Hazardous Waste Sites: National Priorities List* [15]. This list shows that approximately 410 of the 546 sites have groundwater contamination. Surface water releases has been observed at 300 sites.

5.4 Air Emissions

The same U.S. EPA National Priority List indicates that approximately 100 sites were observed to have released contaminants into the air. These contaminants include such toxic compounds as volatile halogenated and aromatic solvents (TCE, benzene) inorganic gases (SO_2), and particulates (asbestos).

Many of the air emission problems have occurred at old or abandoned landfill sites where generation of landfill gas (CO_2, CH_4) promotes the emissions of toxic compounds deposited in the fill. This is a major problem of long term control and monitoring because the active life (microbiologically) of a landfill can be as long as 10 to 20 years. The problems can be exacerbated tremendously if the flammable CH_4 is ignited and formation of other toxic combustion products like polychlorinated dibenzofurans and dioxins occurs.

Another disposal practice which has resulted in air emission problems is the ponding or disposal to land of petroleum still bottoms and other similar petrochemical waste containing significant residues of sulfur compounds. These wastes are not only odorous but may also contain sulfurous acid which can be emitted as SO_2.

Several sites on the U.S. EPA National Priority List have problems related to emissions of asbestos into the air. These are typically mine tailings or piles of wastes from asbestos products manufacturing operations. The weathering of these wastes deposited near the surface of the land has resulted in a friable waste which is easily suspended and carried by the winds.

6 Summary

In general, the typical problems and accidents in chemical waste storage are related to the area of: (1) Containers used, (2) Labeling of containers, (3) Segregation of containers, and (4) The mixing of wastes. The types of problems that have been identified include the deterioration of containers and the subsequent release of hazardous constituents into the environment. Where a variety of wastes are stored there is a potential for incompatible wastes to interact upon contact and result in adverse consequences such as fires, explosins, and toxic gas generation. Past labeling practices have not been adequate to identify the components nor the hazards related to the waste in the container.

Collection and transport problems of chemical wastes are generally related to: (1) Manifesting, (2) Mixing of wastes, (3) "Midnight dumping", and (4) Small generators. Past manifesting practices have not been adequate to identify waste components, hazard characteristics, or destination of the wastes.

Consolidation of wastes by contract haulers have resulted in unexpected releases of toxic substances en route to the disposal site due to incompatible reactions of waste components. Disposal of wastes under cover of darkness at convenient but undesignated areas is still occurring and has resulted in significant damage to the environment. Small generators such as households and small businesses often find it overly costly to properly dispose of small quantities of hatardous wastes due to the inavailability of services such as transfer stations.

The problems in chemical waste treatment are related primarily to characteristics of the waste and management of the residues and emissions from treatment. Chemical wastes typically are complex mixtures of a wide variety of components with different chemical, physical and hazard characteristics. Treatment of such wastes is extremely difficult and costly. The residues and emissions from the treatment processes in use today are of concern because of the residual hazards they may pose.

The problems and accidents of past disposal practices are generally related to: (1) Rupture of containers, (2) The occurence of incompatible reactions, (3) Leaching of toxic substances, and (4) Emissions of volatile toxic substances upon or after disposal. The sites which are being addressed for the U.S. Environmental Protection Agency's "Superfund" program and those being addressed in other countries are prime examples of the problems and accidents which have been and can be caused by disposal of chemical waste to land.

References

1. Management of Uncontrolled Sites. Hazardous Materials Control Research Institute: Washington, D. C. 1980, 1981, 1982, and 1983.
2. Hatayama, H. K., J. J. Chen, E. R. de Vera, R. D. Stephens, D. L. Storm, A Method for Determining the Compatibility of Hazardous Wastes. Municipal Environmental Research Laboratory, Office of Research and Development, U.S. Environmental Protection Agency: Cincinnati, OH, 1980. EPA 600/2-80-076.
3. Metry, A. A., The Handbook of Hazardous Waste Management. Technomic Publishing Company Inc.: Wesport, CT, 1980.
4. U.S. Army Waterways Experiment Station. Physical Properties and Leach Testing of Solidified/Stabilized Industrial Waste. Municipal Environmental Research Laboratory, Office of Research and Development, U.S. Environmental Protection Agency: Cincinnati, OH, 1983. EPA-6001 52-82-099.
5. Management of Hazardous Waste: Policy Guidelines and Code of Practice. Edited by M. S. Suess and J. W. Huismans. World Health Organization, Regional Office for Europe: Copenhagen, 1983. WHO Regional Publications European Series No. 14.
6. Arthur, D., Little Inc. Characterization of Hazardous Waste Transportation and Economic Impact Assessment of Hazardous Waste Transportation Regulations. U.S. Environmental Protection Agency: Washington, D.C., 1979. SW-170C.
7. Hackman, E. E., Toxic Organic Chemicals: Destruction and Waste Treatment. Noyer Data Corp.: Park Ridge, New Jersey, 1978.
8. Storm, D. L., Handbook of Industrial Waste Compositions in California. California Department of Health Services: Berkeley, California, 1978.
9. Leonard, R. P., R. C. Ziegler, W. R. Brown, J. V. Yang, H. C. Reif. Assessment of Industrial Hazardous Waste Practices in the Metal Smelting and Refining Industry. U.S. Environmental Protection Agency: Washington, D.C. 1977. EPA/530/SW-145.

10. Reed, J. C. and B. L. Moore, "Ultimate Hazardous Waste Disposal By Incineration". In Toxic and Hazardous Waste Disposal Volume 4, Edited by R. B. Pojasek. Ann Arbor Science Publishers, Inc.: Ann Arbor, Michigan, 1980.
11. Duvall, D. S., W. A. Rubey, J. A. Mescher. "High Temperature Decomposition of Organic Hazardous Waste". In Treatment of Hazardous Waste: Proceedings of the Sixth Annual Research Symposium. Municipal Environmental Research Lanoratory, Office of Research and Development, U.S. Environmental Protection Agency, 1980. EPA-600/9-80-011.
12. Damages and Threats Caused by Hazardous Material Sites. U.S. Environmental Protection Agency: Washington, D. C., 1980 EPA/430/9-80/004.
13. Hazardous Waste "Problem" Sites: Report of An Expert Seminar. Organization for Economic Cooperation and Development: Paris, 1981. ENV/WMP/81.10.
14. Pilot Study on Contaminated Lands. M. A. Smith, Study Director. North Atlantic Treaty Organization, Committee on Challenges for Modern Society: Brussels, 1984.
15. Hazardous Waste Sites: National Priorities List. U.S. Environmental Protection Agency: Washington, D. C., 1983. HW-7.1.

Epilogue

Industrial waste treatment has evolved into a science rooted in extensive research efforts. Certainly the scope of this handbook would not enable a comprehensive treatment of the topic. The contributions to this volume intend to give a survey of currently important aspects of chemical wastes and their handling.

At this point we would like to give a brief account of the specific concept — i.e., selection of topics, contributions, and authors — of this book:

Denmark pioneered the centralized treatment of industrial wastes by installing a large-scale waste treatment facility at Nyborg. From Denmark, M. Palmark with his extensive experience and expertise has contributed a survey of the different types of chemical waste. His scheme hopefully will be considered for legislature in the future.

All presently important methods for the treatment of industrial wastes have been critically evaluated by D. Martinetz. His review points out which types of waste can be treated by the individual methods. This survey, making reference to most advantages, disadvantages, and inherent limits to the use of each procedure, should stimulate factually effective discussions. Cost/benefit evaluations have been excluded as energy, labor, realty, and construction costs vary considerably around the world.

All other contributions were selected by the editors in view of the current importance of certain techniques, their wide-spread use, and the expertise of the authors. The chosen techniques thus may serve as representitive examples:

The VRC procedure, as described by A. Robin, used for treating chlorinated hydrocarbons certainly is not the only possible method, but it is the most widely used.

The successful sanitation of a former waste disposal site is described by H. Schirmer. This may serve as an example; however, individual waste disposal sites have specific features which must be taken in account for a specific treatment plan. Therefore, generalization would be worthless.

Industrial waste can be biologically degraded, processed, and utilized, as H. Propfe illustrates in his survey. This is of special current interest as the separation into different categories of waste can reduce costs and render waste processing less hazardous; all this is accomplished by selective composting.

Analysis of chemical wastes is of crucial importance. E. Thomanetz and O. Tabasaran have developed a system which has repeatedly proved efficient in practice.

Industrial waste sludges (M. Chambon, A. Navarro) are of special concern, as most industrial wastes belong to this category. However, it must be mentioned that most waste sludges are still not satisfactorily processed. In France, the cooperation of the industry with the University of Lyon and the federal waste processing company ANRED has led to the development of a highly advanced sludge treatment procedure by dehydration, detoxification, and solidification. Further development along these lines can be expected, especially for on-site treatment.

The highly controversial issue of waste disposal at sea will be treated in a following volume.

The pyrolysis of wastes has not been separately dealt with, firstly as the extensive research in this field has not yielded the expected results and, secondly as new improved methods have not matured to the point allowing separate consideration here.

The US has gained considerable insights into the securing and safe-guarding of waste disposal sites, in part due to dreadful experiences. H. K. Hatayama gives a survey of current opinions on this matter.

The suveillance and control of storage and transport of wastes is an equally important topic. In Great Britain, as reported by J. Bromley, these aspects have been perfectionized to enable a high degree of reliability and efficiency. New trends along these lines are presently evolving.

We have reached a crossroad in waste processing: Industry and the science community have reacted to the justified claims and demands of environmental protection by providing a considerable "arsenal of weapons" to combat environmental deterioration. The suspicious and often insufficiently informed public demands extreme security and by means of their representatives exert pressure on the legislature to develop a reliable network to avoid environmental damage. Communication problems still exist. Maybe this book will contribute new insights.

K. R. Müller

Subject Index

MIX
Papier aus verantwortungsvollen Quellen
Paper from responsible sources
FSC® C105338

If you have any concerns about our products,
you can contact us on
ProductSafety@springernature.com

In case Publisher is established outside the EU,
the EU authorized representative is:
Springer Nature Customer Service Center GmbH
Europaplatz 3, 69115 Heidelberg, Germany

Printed by Libri Plureos GmbH
in Hamburg, Germany